# Future Manufacturing Systems

edited by

**Dr. Tauseef Aized**

**Future Manufacturing Systems**

Edited by Dr. Tauseef Aized

**Republished by InTech**

Janeza Trdine 9, 51000 Rijeka, Croatia



**Publishing Process Manager** Jelena Marusic

**Technical Editor** Goran Bajac

**Cover Designer** Martina Sirotic

Additional hard copies can be obtained from orders@intechopen.com

Future Manufacturing Systems, Edited by Dr. Tauseef Aized
p. cm.
ISBN 978-953-307-128-2

For EU product safety concerns:
IN TECH d.o.o., Prolaz Marije Krucifikse Kozulić 3, 51000 Rijeka, Croatia,
info@intechopen.com or visit our website at intechopen.com.

# Meet the editor

Dr. Tauseef Aized received his Ph.D. from Tokyo Institute of Technology. He held two research fellowships; first Endeavour Research Fellowship at Monash University and the second Commonwealth Fellowship at the University of Cambridge. He has a number of papers in reputed journals and conferences. His teaching and research interest include energy technology and policy, computer integrated manufacturing and operations management. Currently, he is a professor and chair of department of mechanical engineering, UET-KSK campus, Lahore-Pakistan.

# Contents

# Preface

Manufacturing is a vital activity for a society from a strategic point of view. It has a long history in human civilizations and gives a society a definite edge over its competitors. A manufacturing system can be viewed as an arrangement of tasks and processes, properly put together, to transform a selected group of raw materials and semi-finished products to a set of finished products. Historically, manufacturing activities have grown over centuries and their evolution can be divided into three stages. These are craft, mass and lean production methods. The development of electrical devices has led to better control of machines and resulted in machines with greater flexibility. Recent developments in information technology have made it feasible to achieve the purpose of rapid product prototyping, concurrent engineering, flexible and agile automation and computer integrated manufacturing. Many manufacturing system paradigms have been developed throughout the history of manufacturing, such as mass production, just in time manufacturing, lean manufacturing, flexible manufacturing, mass customization, agile manufacturing and others. All these systems are working efficiently under particular conditions attached to them. With the overall evolution of human society, product demand patterns are changing which force manufacturers to adjust their system paradigms according to changes. Thus, a change of product demand patterns always demands to remodel and improve manufacturing system designs, layouts, facilities and provisions which lead to an ongoing search of development of new ways and means of designing modern manufacturing systems. This book is a collection of articles aimed at finding new ways of manufacturing systems developments. The articles included in this volume comprise of current and new directions of manufacturing systems which I believe can lead to the development of more comprehensive and efficient future manufacturing systems. People from diverse background like academia, industry, research and others can take advantage of this volume and can shape future directions of manufacturing systems.

Editor

**Dr. Tauseef Aized**
*Professor, Mechanical Engineering,*
*University of Engineering and Technology (UET)- KSK campus,*
*Lahore, Pakistan*
*and*
*Research Fellow, Institute for Manufacturing (IFM),*
*University of Cambridge, UK*

August 29, 2010.

1

# Flexible manufacturing system: hardware components

Dr. Tauseef Aized
*Professor, Department of Mechanical , Mechatronics and Manufacturing Engineering, KSK Campus, University of Engineering and Technology, Lahore, Pakistan*

A flexible manufacturing system is a highly automated system consisting of a group of workstations interconnected by an automated material handling and storage system and controlled by a distributed computer system. It is capable of processing a variety of different part styles simultaneously at various workstations and the mix of part styles and quantities of production can be adjusted in response to changing demand patterns. A flexible manufacturing system comprises of processing stations, material handling and storage systems and requires hardware and software provisions. The hardware components typically required for a FMS are;

- Machine tools, for example, machining centers, turning centers, etc.
- Load/unload stations
- Guided vehicles
- Robots
- Inspection facilities like coordinate measuring machines
- Programmable Logic Controllers (PLC).

This chapter describes the hardware provisions required for a flexible manufacturing system.

## Introduction

Flexible manufacturing systems consist of hardware and software components. The hardware components typically comprise of processing stations, material handling systems and automated material storage and retrieval systems. The processing stations are the workstations that perform different operations on part families. These workstations are CNC machine tools, inspection equipments, assembly stations and material loading/ unloading areas. Material handling systems include automated guided vehicle systems, roller conveyors, tow line, shuttle cars etc whereas automated storage and retrieval systems are used to store and retrieve work parts automatically. Various types of storage and retrieval systems are pallets, carousels etc which help in convenient access of different types of parts from stores and increase machine utilization of flexible manufacturing systems. The processing and assembly equipments used in a flexible manufacturing system depend upon the type of work being accomplished by the system. In a system designed for machining operations, the principal types of processing stations are CNC machines like CNC machining and turning centers. However, the FMS concept is applicable to various other processes like automated assembly lines, sheet metal fabrication etc.

### Machining Stations

One of the most common applications of flexible manufacturing system is in the machining operations. The workstations designed in these systems, therefore, predominantly consist of CNC machines tools. The most common CNC machines tools used include CNC machining center, in particular, horizontal machining turning centers. These CNC machine tools possess the features that make them compatible with the FMS. These features include automatic tool changing and storage, use of palletized work parts, etc.

### CNC Machining Center

A CNC machining center is a highly automated machine tool capable of performing multiple machining operations under CNC control in one setup with minimal human attention. Machining centers generally include automated pallet changers to load the work part to the machine and to unload the finished part that can be readily interfaced with the FMS part handling system. A CNC machining center is a sophisticated CNC machine that can perform milling, drilling, tapping, and boring operations at the same location with a variety of tools.

Fig. 1. CNC Horizontal Machining Center

There are several special features that make a machining center more productive machine are as follows:

**Automatic tool-changing**

As there is a variety of machining operations to be performed by the machines on different part styles in a FMS environment, so cutting tools must be changed to switch from one machining operation to another. This is done on a machining center under NC program control by an automated tool-changer designed to exchange cutters between the machine tool spindle and a tool storage drum. The capacities of these drums commonly range from 16 to 80 cutting tools.

Fig. 2. Tool Storage

**Pallet shuttles**

Some machining centers in FMS are equipped with two or more pallet shuttles, which can automatically transfer the work part to the spindle of the machining center to perform the machining operation on it. With two shuttles, the operator may unload the finished part and load the next raw part on load/unload station while the machine tool is engaged in machining the current part. This reduces nonproductive time on the machine.

**Automatic work part positioning**

To enhance the productivity of a machine tool and to reduce the manufacturing lead time, machine tools in FMS are equipped with automatic work part positioning system that exactly position the work part before the machining operation starts. Many machining centers have more than three axes. One of the additional axes is often designed as a rotary table to position the part at some specified angle relative to the spindle. The rotary table permits the cutter to perform machining on four sides of the part in a single setup.

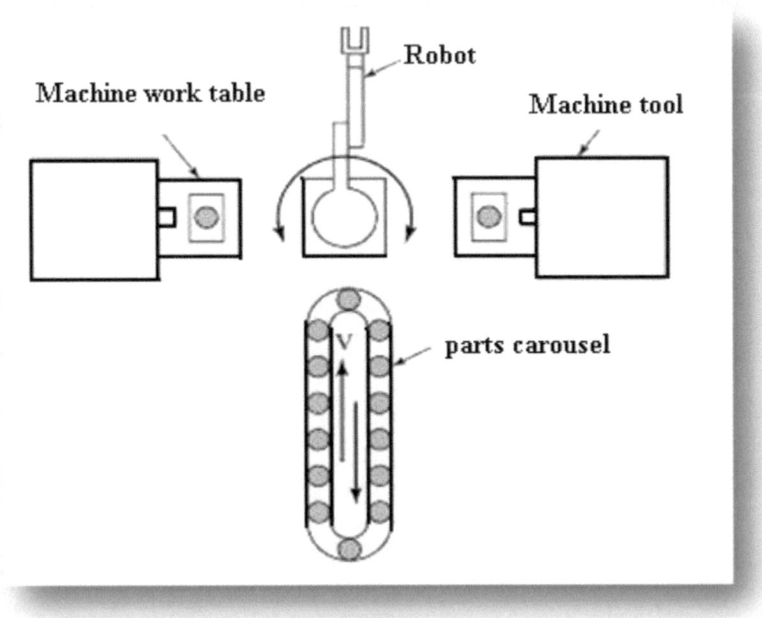


Fig. 3. Automated manufacturing cell with two CNC machine tools and robot.

**CNC Turning Centers**

A modern CNC turning center is capable of performing various turning and related operations, contour turning, and automatic tool indexing, all under computer control. A program is fed to the CNC turning center for a particular class of work parts and this program repeat itself on every new part. In addition, the most sophisticated turning centers can accomplish work part gauging (checking key dimensions after machining), tool monitoring (sensors to indicate when tools are worn), automatic tool changing, automatic work part changing at the completion of the work cycle. A recent development in the CNC machine tool technology is the *CNC mill-turn center.* This machine has the general configuration of a turning center; in addition, it can position a cylindrical work part at a specified angle so that a rotating cutter can machine features into the outside surface of the work part.

Fig. 4. CNC Turning Center

Fig. 5. CNC mill-turn center

**Load/Unload Stations**

Load/unload station is the physical interface between an FMS and the rest of the factory. It is the place where raw work parts enter the system and finished parts exit the system. Loading and unloading can be accomplished either manually (the most common method) or by automatic handling systems. The load/unload stations should be ergonomically designed to permit convenient and safe movement of work parts. Mechanized cranes and other handling devices are installed to assist the operator with the parts that are too heavy to lift by hand. A certain level of cleanliness must be maintained at the workplace, and air houses and other washing facilities are often used to flush away chips and ensure clean mounting and locating points. The station is often raised slightly above the floor level using as open-grid platform to permit chips and cutting fluid to drop through the openings for subsequent recycling or disposal.

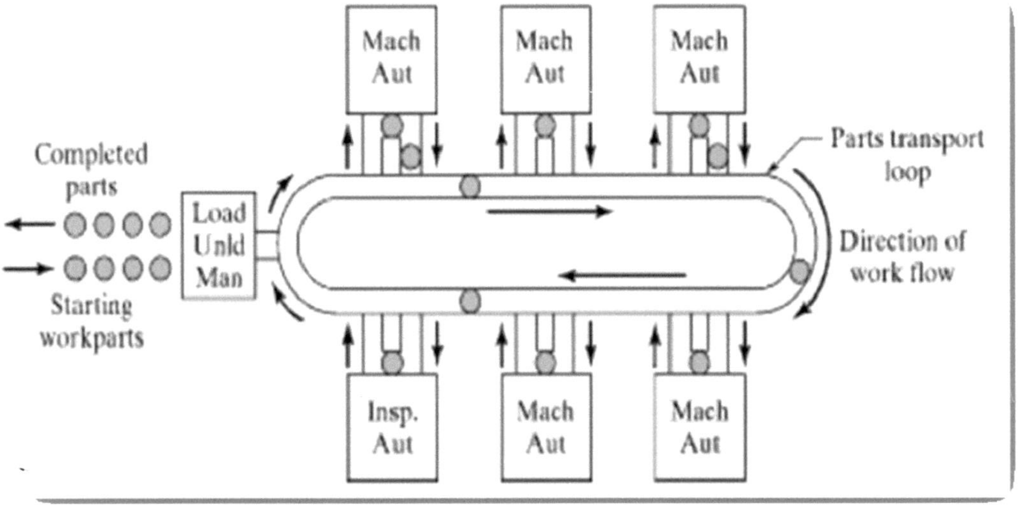


Fig. 6. Load/ unload stations in relation to overall system shown.

The load/unload station includes a data entry unit and a monitor for communication between the operator and the computer system. Through this system, the operator receives the instructions regarding which part to load on the next pallet in order to adhere to production schedule. When different pallets are required for different parts, the correct pallet must be supplied to the station. When modular fixing is used, the correct fixture must be specified and the required components and tools must be available at the workstation to build it. When the part loading procedure is completed, the handling system must launch the pallet into the system, but not until then; the handling system must be prevented from moving the pallet while the operator is still working. All of these conditions require communication between the computer system and the operator at the load/unload station.

### Robots
An industrial robot is a general-purpose, programmable machine possessing certain anthropomorphic (human like) characteristics. The most obvious anthropomorphic characteristic of an industrial robot is its mechanical arm, which is used to perform various industrial tasks. Other human-like characteristics are the robot's capabilities to respond to sensory inputs, communicate with other machines and make decisions. These capabilities permit robots to perform a variety of useful tasks in industrial environments. One of the most common applications of robots in FMS may be loading the raw work part and unloading the finished part at the loading/unloading stations. Robots can be found in manufacturing industry, military, space exploration, transportation, and medical applications.

### Types of Robots
Typical industrial robots do jobs that are difficult, dangerous or dull. They lift heavy objects, paint, handle chemicals, and perform assembly work. They perform the same job hours after hours, days after days with precision. They don't get tired and they don't make errors associated with fatigue and so are ideally suited to performing repetitive tasks. The major categories of industrial robots by mechanical structure are:

### Cartesian robot /Gantry robot
These robots are used for pick and place work, assembly operations and handling machine tools and arc welding. It's a robot whose arm has three prismatic joints, whose axes are coincident with a cartesian coordinator.

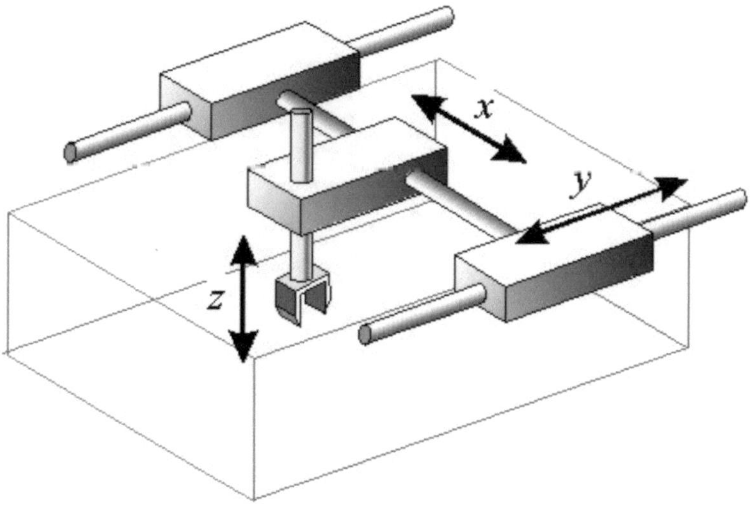


Fig. 7. A Cartesian robot

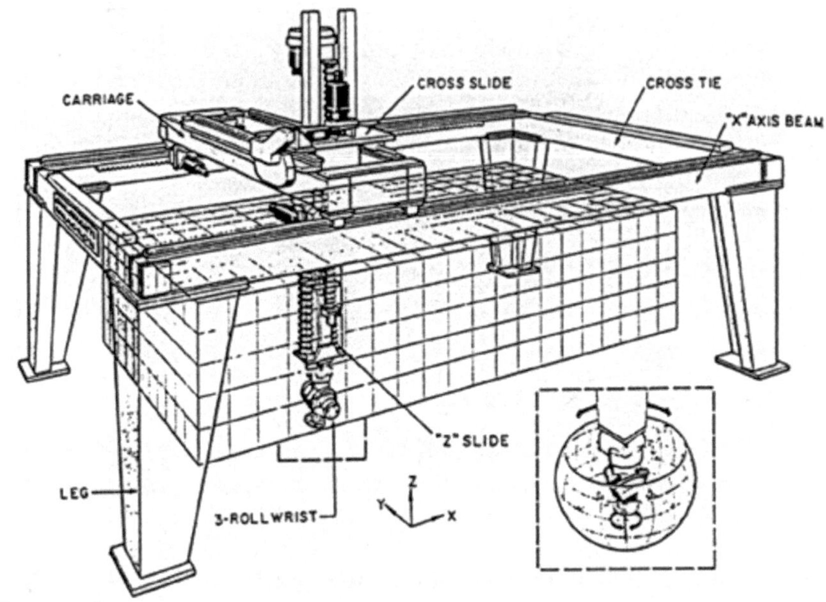


Fig. 8. A Gantry robot

- **Cylindrical robot**

These robots are used for assembly operations, spot welding, and handling at die-casting machines. It's a robot whose axes form a cylindrical coordinate system.

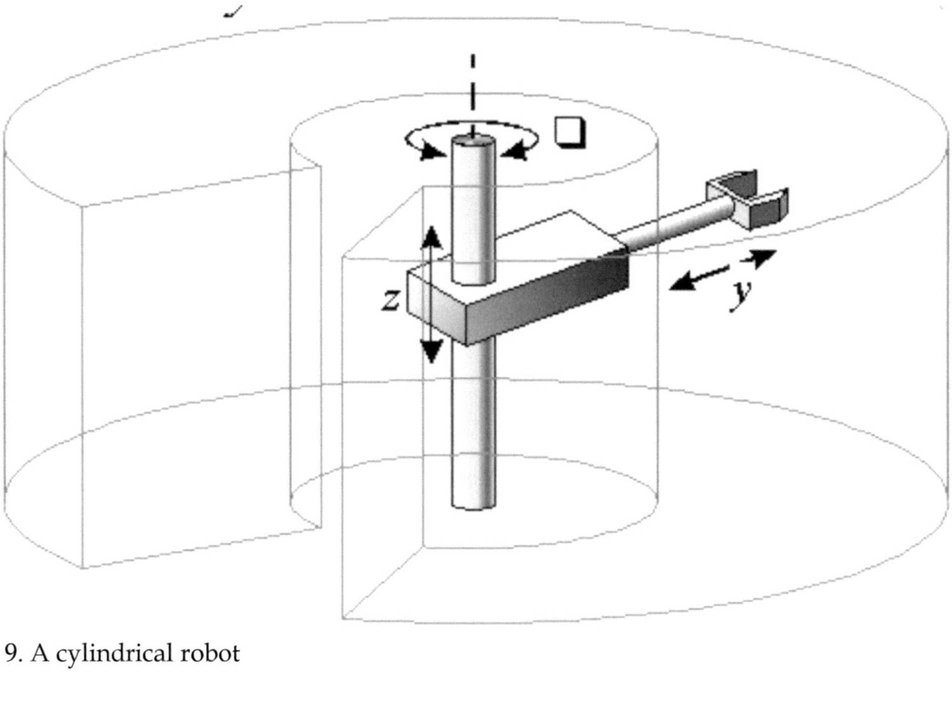


Fig. 9. A cylindrical robot

- **Spherical/Polar robot**

The spherical robots are used for handling work parts at machine tools, spot welding, die-casting, fettling machines, gas welding and arc welding. It's a robot whose axes form a polar coordinate system.

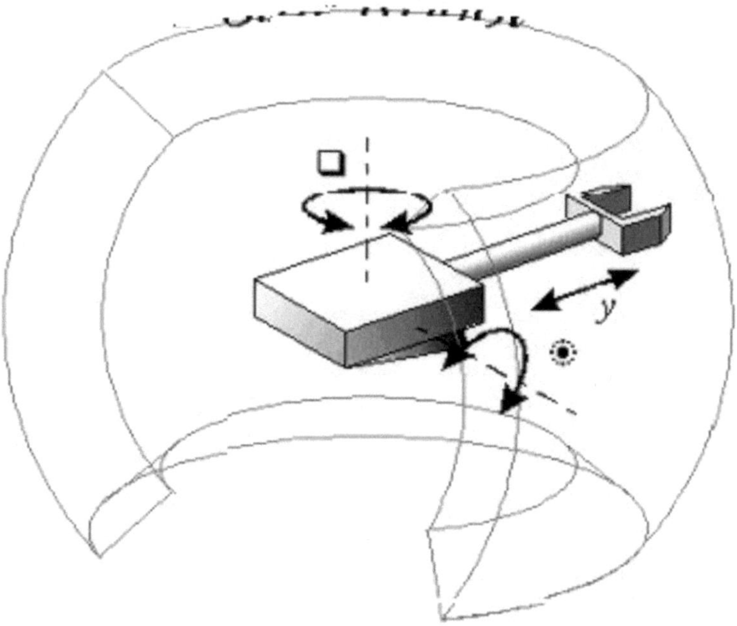


Fig. 10. A spherical robot configuration.

- **SCARA robot**

The SCARA robots are used for pick and place work, assembly operations and handling machine tools. It's a robot which has two parallel rotary joints to provide compliance in a plane.

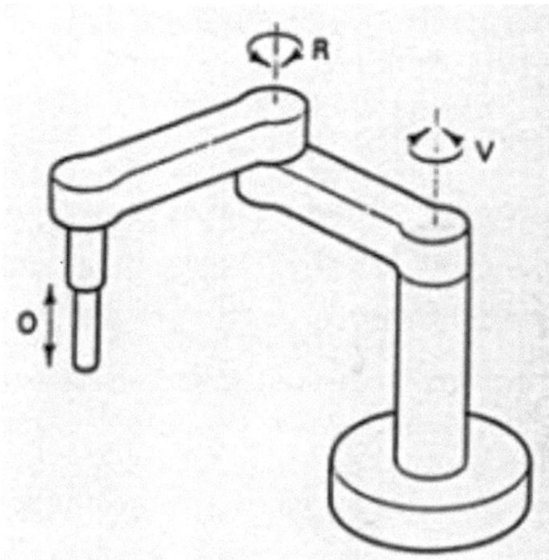


Fig. 11. SCARA robot configuration.

- **Articulated robot**

An articulated robot is used for assembly operations, die-casting, fettling machines, gas welding, arc welding and spray painting. It's a robot whose arm has at least three rotary joints.

Fig. 12. Articulated robot configuration

**Robot Applications**

Due to the diverse nature of robots and their flexibility in motion, there are various forms of applications in the flexible manufacturing system.

**1. Pick and Drop Operations**

The most common application of robot within FMS is pick and drop operations, where point to point control devices are sufficient. These applications include tool changing, loading/unloading un-fixtured parts into work tables. The following figure shows a pick and drop robot arm.

**2. Contouring Operations**

A second major application area for robot is in contouring type operations. These include welding, limited machining, deburring, assembly/disassembly and inspection. In case of welding, robots have been proven reliable, effective and efficient. However in other areas such as machining, deburring and inspection robots' limited accuracy and repeatability limited their applications. In addition, whenever a tool change is required, such as in deburring, the cost of robotic change is almost as expensive as that of three-axis machining center. It is possible that simple operations which require multiple tools are most efficiently performed in the machining centers.

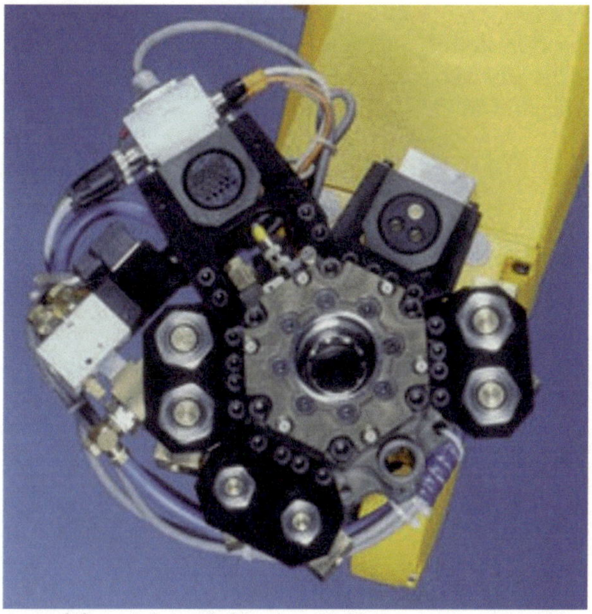

Fig. 13. A robotic arm used for pick and drop operation

### 3. Assembly/Disassembly

The use of robot in FMS is wide spread in assembly and disassembly. Robot have been proven effective for assembly of small parts and printed circuit board (PCB's). The following figure shows a PCB assembled by robots.

Fig. 14. Printed Circuit Boards (PCB) assembled by robots

### Inspection Equipments

Since an FMS is a closed system (feedback control system), it is necessary to provide some means to monitor the quality of operations being performed. This monitoring can take place in many different places and by different components.

Fig. 15. Multi Function Gantry CMM

Fig. 16. Coordinate measuring machine

**1. Coordinate Measuring Machine**

The most obvious type of inspection equipments is coordinate measuring machine (CMM). This machine can be programmed to probe a piece part and identify depth of holes, flatness of surfaces and perpendicularity.

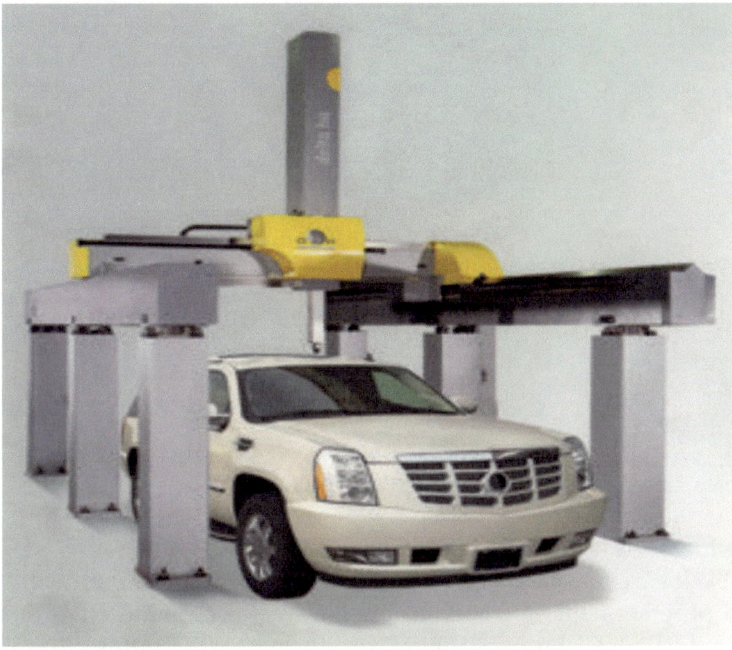

Fig. 17. A Large Scale CMM

Special requirements usually include constant temperature congruity environment and piece part. Also, because of the slow movements necessary to precisely measure surfaces, the inspection time is usually long compared to machining time.

**2. Probing Machining Centers**

Probe marching centers are also used as for inspection purposes in addition to CMM station. These machines inspect equipment in work centers by inserting a probe into the gripper or spindle and then moving the probe contacting the work piece or fixture.

Fig. 18. Example of on-machine checking and inspection

Programmable Logic Controllers (PLC's):

A programmable logic controller (PLC) is a microcomputer-based controller that uses stored instructions in programmable memory to implement logic, sequencing, timing, counting, and arithmetic functions through digital or analog input/output (I/O) module, for controlling machines and processes. PLC is universally called 'Work Horse' of industrial automations. Various systems like material handling system, material storage system, load/unloading stations, etc. are programmed through PLC in order to streamline the operations in a flexible manufacturing system.:

- PLCs consist of input modules or points, a central processing unit (CPU), and output modules or points. The basic components of PLC are the followings:
- Processor
- Memory unit
- Power supply
- I/O module
- Programming device

These components are housed in a suitable cabinet for the industrial environment. The processor is the central processing unit of the programmable controller. It execute various logic and sequencing functions by operating on the PLC input to determine the appropriate output signal. Connected to the CPU is the PLC memory unit, which contains the programs of logic, sequencing, and I/O operation. It also holds data files associated with these programs including I/O status bits, counter and timer constants, and other type variable and parameter values. This memory unit is referred to as the user or applicant memory because its contents are entered by the user. A power supply is typically used to drive a PLC. The I/O module provides the connections to the industrial equipments or process that is to be controlled. Inputs to the controller are signals from limits switches, push buttons, sensors, and other on/off devices. Outputs from the controller are on/off signals to operate motors, valves and other devices required to actuate the process. The PLC is programmed by means of a programming device. The programming device is usually detachable from the PLC cabinet so that it can be shared among different controllers. The following figure shows a PLC input/ output module.

Fig. 19. PLC Input /Output Module

## References


1. Mikell P. Groover, "Automation, Production Systems and Computer Integrated Manufacturing", 3rd Edition, Pearson Education, Inc., 2008.
2. Implementing Flexible Manufacturing Systems by Greenwood, Nigel. Published by M Macmillan Education. 1988
3. Flexible manufacturing system (FMS): the investigative phase By David L. Setter, Published by Technical Communications, Kansas City Division, Allied-Signal Aerospace, 1993
4. Flexible Manufacturing Systems: Decision Support for Design and Operation" H. Tempelmeier and H. Kuhn John Wiley and Sons 1993
5. Production and Operations Management, By Chary
6. Rapid prototyping: theory and practice, By Ali K. Kamrani, Emad Abouel Nasr
7. http://www.robots.com

# 2

# Discrete event models for flexible manufacturing cells

Constantin Filote and Calin Ciufudean
*Stefan cel Mare University of Suceava*
*Romania*

## 1. Introduction

A manufacturing system includes a set of machines performing different operations, linked by a material handling system. A major consideration in designing a manufacturing system is its availability. When a machine or any other hardware component of the system fails, the system reconfiguration is often less than perfect. It is shown that, if these imperfections constitute even a very small percent of all possible system faults, the availability of the system may be considerably reduced. The system availability is computed as the sum of probabilities of the system operational states. A state is operational when its performance is better than a threshold value. In order to calculate the availability of a manufacturing system, its states (each corresponding to an acceptable system level) are determined. A system level is acceptable when its production capacity is satisfied. To analyze the system with failure/repair process, Markov models are often used. As a manufacturing system includes a large number of components with failure/repair processes, the system-level Markov model becomes computationally intractable. In this paper, a decomposition approach for the analysis of manufacturing systems is decomposed in manufacturing cells. A Markov chain is constructed and solved for each cell $i$ to determine the probability of at least $N_i$ operational machines at time $t$. $N_i$ satisfies the production capacity requirement of machine cell $i$.

The probability is determined so that the material handling carriers provide the service required between $N_i$ operational machines in machine cell $i$, and $N_{i+1}$ operational machines in machine cell $i$+1.

The number $i$=1,…,n at time t, where n is the number of machine cells in the decomposed system.

Production lines are sets of machines arranged so as to produce a finished product or a component of a product. Machines are typically unreliable and experience random breakdowns, which lead to unscheduled downtime and production losses. Breakdown of a machine affects all other machines in the system, causing blockage of those upstream and starvation of those downstream. To minimize such perturbations, finite buffers separate the machines. The empty space of buffers protects against blockage and the full space against starvation. Thus, production lines may be modeled as sets of machines and buffers connected according to a certain topology. From a system theoretic perspective, production lines are discrete event systems. Two basic models of machine reliability are mentioned in the literature: Bernoulli (Jacobs & Meerkov, 1995) and Markov (Gershwin, 1994), (Lim et al., 1990). The first model assumes that the process of Bernoulli trials determines the status of a machine in each cycle (i.e. the time necessary to process a part). In Markov model the state of a machine in a cycle is determined by a conditional probability, with the condition being the state of the machine in the previous cycle. Both model Bernoulli and Markov reflect practical situations: Bernoulli reliability model is more appropriate when downtime is small and comparable with the cycle time. This is often the case in assembly operations where the downtime is due to quality problems. Markov models reflect operations where the downtime is due to mechanical failures, which could be much longer than the cycle time. In this paper we address the Markov model. Intuitively, bottleneck (BN) of a production line is understood as a machine that impedes the system performance in the strongest manner. Some authors define the BN as the machine with the smallest isolation production rate (i.e. the production rate of the machine when no starvation and blockages are present). Others call the BN the machine with the largest inventory accumulated in front of it. Any may identify the machine that affects the bottom line, i.e. the system production rate, because the above definitions are local in nature and do not take into account the total system properties, such as the order of the machines in the production line, capacity of the buffers, etc. Identification of BNs and their optimal capacity for avoiding the machine downtime is considered as one of the most important problems in manufacturing systems. An illustrative example will emphasize our approach.

## 2. The System Model of Production Lines

The following model of a production line is considered:
1) The system consists of N machines arranged serially and N+1 buffers separating each consecutive pair of machines.
2) Each buffer $B_i$ is characterized by its capacity $C_i < \infty$, $2 \leq i \leq N$, the first and the last buffer are considered to be of an infinite capacity.
3) Each machine has two states: up and down. When up, the machine produces with the rate of 1 part per unit of time (cycle); when the machine is down, no production takes place.
4) The uptime and the downtime of each machine $M_i$ are random variables distributed exponentially with parameters $\lambda_i$ and $\mu_i$ respectively.
5) Machine $M_i$ is starved at time t if buffer $B_{i-1}$ is empty at time t, machine $M_1$ is never starved.
6) Machine $M_i$ is blocked at time t if buffer $B_{i-1}$ is full at time t, machine $M_N$ is never blocked.
The isolation production rate of each machine (i.e. the average number of parts produced per unit time if no starvation or blockage takes place) is:

$$\eta_i = \frac{Tup_i}{Tup_i + Tdown_i} = \frac{1}{1 + \frac{\mu_i}{\lambda_i}} \tag{1}$$

Machine $M_i$ is the uptime bottleneck if:

$$\frac{\partial \eta}{\partial Tup_i} > \frac{\partial \eta}{\partial Tup_j}, j \neq i \tag{2}$$

and is the downtime bottleneck if:

$$\left|\frac{\partial \eta}{\partial Tdown_i}\right| > \left|\frac{\partial \eta}{\partial Tdown_j}\right|, j \neq \tag{3}$$

Machine $M_i$ is the bottleneck (BN) if it is both uptime bottleneck and downtime bottleneck. Let $M_i$ be the bottleneck machine. Then it is referred to as the uptime preventive maintenance bottleneck if:

$$\frac{\partial \eta}{\partial Tup_i} > \left|\frac{\partial \eta}{\partial Tdown_i}\right| \tag{4}$$

If the inequality is reversed, the bottleneck is referred to as the downtime preventive maintenance bottleneck.

*Notice:* a) The absolute values of $\frac{\partial \eta}{\partial Tdown_i}$ are used because otherwise this number is negative: increase in *Tdown* leads to a decrease of $\eta$.

b) In some instances, the downtime of a machine is due to lapses in the performance of manual operators, rather than machine breakdown, thus the identification of downtime bottlenecks provides guidance for the development of production automation.

c) Preventive maintenance, as part of the total production maintenance, leads to both an increased uptime and a decrease of automated machine downtime. Some of the preventive maintenance measures affect the uptime and others the downtime. We refer to them as uptime preventive maintenance and downtime preventive maintenance. Thus, the classification of the bottleneck in either uptime bottleneck or downtime bottleneck has an impact on planning actions that lead to the most efficient system improvement.

### 2.1 Bottleneck indicators

We are seeking bottlenecks identification tools that are based either on the data available on the factory floor by means of real time measurements (such as average up - and down - time, starvation and blockage time, etc.), or on the data that can be constructively using the machines and buffers parameters $(\lambda_i, \mu_i, N_i)$. We refer to these tools as bottleneck indicators.

#### 2.1.1 A single machine case

A single machine defined by the assumptions made in the second paragraph is uptime bottleneck if $Tup < Tdown$ and it is downtime bottleneck if $Tdown < Tup$.

We can easily show that this assumption is true from (1) since:

$$\left|\frac{\partial \eta}{\partial Tdown}\right| = \frac{Tup}{(Tup + Tdown)^2} \tag{5}$$

and

$$\frac{\partial \eta}{\partial Tup} = \frac{Tdown}{(Tup + Tdown)^2} \tag{6}$$

We may say that the smallest average uptime or down-time of a machine defines its nature as bottleneck. The primary focus of the preventive maintenance and automation should be placed on the downtime further decrease, if $Tdown < Tup$. If $Tup < Tdown$, the attention should be concentrated on the increase of the uptime. In most practical situations $Tdown < Tup$, therefore the above indicator states that the reduction of the downtime is more efficient than a comparable increase of the uptime (Proth. & Xie, 1994).

### 2.1.2 Two machine cases

It is well known that, given a constant ratio between $Tup_i$ and $Tdown_i$, the machine with the longer up - and down - time is more detrimental to the system's production rate than with a shorter up - and down - time.

In view of this property, one might think that the bottleneck is the machine with the longer up - and down - time. This is not true. The reason is that an improvement of the machine with a shorter up - and down - time leads to a better utilization of the disturbance attenuation capabilities of the buffer than a comparable improvement of the machine with a longer up - and down - time. Therefore, an improvement of the "better" machine is the best for the system as a whole (Chiang et al., 2000).

In a production line with two machines of equal efficiency (i.e., $\frac{Tup_1}{Tdown_1} = \frac{Tup_2}{Tdown_2}$), the machine with the smaller downtime is the bottleneck (Narahari & Viswandham, 1994). If the downtime of this machine is smaller than its uptime, preventive maintenance and automation should be directed toward the downtime decrease. If the downtime is sufficiently longer than the uptime, preventive maintenance and automation should be directed toward the increase of the uptime.

In the most practical situations, the isolation production rate of the machines (i.e., the faction $Tup/(Tup+Tdown)$) is greater than 0,5. Therefore, the most usual bottleneck is the downtime bottleneck. To identify the downtime bottleneck in the case of machine with unequal efficiency (i.e. $\frac{Tup_1}{Tdown_1} \neq \frac{Tup_2}{Tdown_2}$) in (Laftit et al., 1992) the following bottleneck indicator is given:

If $mb_1 Tup_1 Tdown_1 < ms_2 Tup_2 Tdown_2$, machine $M_1$ is the downtime bottleneck.

If $mb_1 Tup_1 Tdown_1 > ms_2 Tup_2 Tdown_2$, machine $M_2$ is the downtime bottleneck.

The probability of manufacturing blockage $mb_i$ is defined as:

$$mb_i = Prob(\{M_i \text{ is up at time } t\} \cap \{B_i \text{ is full at time } t\} \cap \{M_{i+1} \text{ fails to take parts from } B_i \text{ at time } t\}).$$

The probability of manufacturing starvation $ms_i$ is defined as:

$$ms_i = Prob(\{M_{i-1} \text{ fails to put parts into } B_{i-1} \text{ at time } t\} \cap \{B_{i-1} \text{ is empty at time } t\} \cap \{M_i \text{ is up at time } t\}).$$

### 2.2 Extreme status for buffers

In the sequel we will try to determine the bottleneck behavior of the machines as a function of their efficiency correlated with buffer size. We will also try to anticipate the events like buffers full or empty, which determine the bottlenecks. We consider a segment consisting of two machines $M_i$ and $M_{i+1}$ with intermediate storage $B_i$ at any time between successive events. Let TA be the apparent time of an event occurrence at $B_i$. This event may occur or not if, in the mean time, another cancelling event takes place.
Let $P_i$ be the number of parts which are scheduled in process by $M_i$ until the occurrence of the event. We examine two different situations, which result in a buffer event.
We define the following:

| | |
|---|---|
| $pr_i$ | The nominal production rate of machine $M_i$, $i = 1,...,N$ |
| $BL(j,t)$ | Level of buffer $B_j$, $j = 2,..., N-1$ |
| $T_{1j}(t)$ | Delay time until the next arrival to $B_j$ |
| $T_{2j}(t)$ | Delay time until the next departure from $B_j$ |
| $BC_j$ | The capacity of buffer $B_j$, $j = 2,..., N$ |

#### 2.2.1 Buffer-full event

Although the buffer $B_i$ has enough space to accept the parts produced by $M_i$ during the transient time $T_{2i}$, since $M_i$ produces at a faster rate than $M_{i+1}$ ( or the delay time $T_{2i}$ is too long), buffer $B_i$ will be full (see Fig. 1).

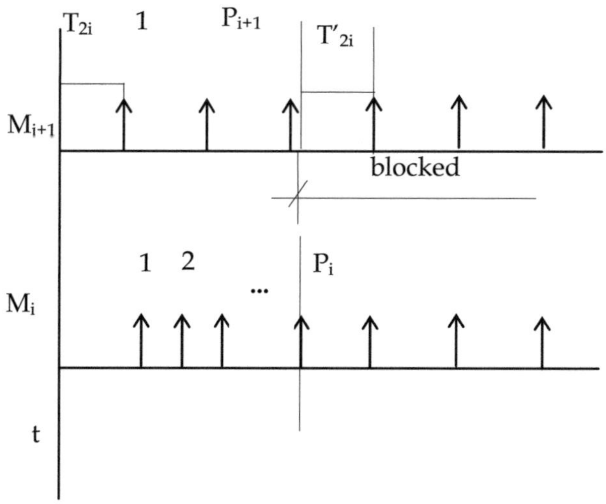


Fig. 1. Buffer - full event

In Fig. 1. the continuous line represents a machine operation on a work-part and the arrows represent arrivals to the succeeding buffer. Blank intervals indicate idle periods due to blockage or starvation of machines. The function depicted in Fig.1. is encountered when:

$$(pr_i > pr_{i+1}) \cap [pr_i (T_{2i} - T_{1i}) > BC_i - BL(i)] \tag{7}$$

The buffer - full event will occur when the $P_i$-th part leaves from $M_i$. The number of parts produced by $M_i$ after $t + T_{1i}$ is $P_i$ -1. From Fig. 1. the sequel relations hold:

$$P_i - P_{i+1} = BC_i - BL(i) \tag{8}$$

$$P_i = 1 + (TA - t - T_{1i}).\ pr_i \tag{9}$$

$$P_{i+1} = (TA - t - T_{2i} + T'_{2i}).\ pr_{i+1},\ i = 1,...,N \tag{10}$$

Time interval between departure and the processing end of the first blocked part of $M_i$, lies in an inter-departure interval of $M_{i+1}$:

$$\frac{1}{pr_i} \leq T'_{2i} \leq \frac{1}{pr_{i+1}} \tag{11}$$

From (8) - (11), we obtain:

$$(T_{1i} - T_{2i}) + \frac{BC_i - BL(i)}{pr_{i+1}} < P_i \left(\frac{1}{pr_{i+1}} - \frac{1}{pr_i}\right) \leq (T_{1i} - T_{2i}) + \frac{BC_i - BL(i)}{pr_{i+1}} + \frac{1}{pr_{i+1}} - \frac{1}{pr_i} \tag{12}$$

which yields:

$$P_i = 1 + int\{[BC_i - BL(i) + pr_{i+1}(T_{1i} - T_{2i})].\frac{pr_i}{pr_i - pr_{i+1}}\} \tag{13}$$

### 2.2.2 Buffer-empty event

This event is dual to the blockage and analogous results will be derived. The buffer-empty event is encountered when buffer $B_i$ is exhausted and its succeeding machine $M_{i+1}$ has just transmitted a work-part downstream.

Although the buffer $B_i$ has enough parts for the transient period $T_{1i}$, because machine $M_{i+1}$ produces faster than $M_i$ (see Fig. 2.), or the delay time $T_{1i}$ is too long, finally $B_i$ becomes empty.

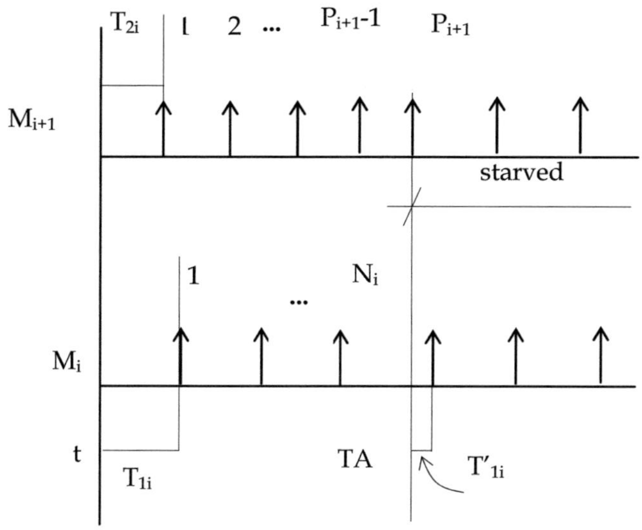


Fig. 2. Buffer - empty event

The situation depicted in Fig. 2. is encountered when:

$$(pr_i < pr_{i+1}) \cap [pr_{i+1} (T_{i1} - T_{2i}) > BL(i)] \tag{14}$$

The inter-departure interval of $M_{i+1}$ just before the occurrence of the empty buffer event satisfies the following:

$$0 \leq T_{i1}^{'} \leq \frac{1}{pr_i} - \frac{1}{pr_{i+1}} \tag{15}$$

$$P_{i+1} - N_i = BL(i) + 1 \tag{16}$$

$$P_{i+1} = 1 + (TA - t - T_{2i}).\ pr_{i+1} \tag{17}$$

$$P_i = (TA - t - T_{1i} + T_{1i}^{'}) \cdot pr_i \tag{18}$$

Analogous results with those of the previous section are obtained:

$$P_{i+1} = 1 + int\,\{[BL(i) + pr_i\,(T_{2i} - T_{1i})].\ \frac{pr_{i+1}}{pr_{i+1} - pr_i}\} \tag{19}$$

## 3. The System Model of Flexible Manufacturing Cells

In this paper, a flexible manufacturing system (FMS) is treated as a discrete event system and we consider that the system evolution constitutes a discrete state-space stochastic process. In particular, we focus on Markov chain models. Such a model could be generated directly or using higher level models such as stochastic Petri nets or discrete event simulation.

Markov models with absorbing states have a trivial steady-state, namely the chain ends up in some absorbing state, remaining there forever; therefore, transient analysis alone emphasizes the system performance.

We assume that a manufacturing system evolves in time as a homogenous continuous time Markov chain $\{x(t);\ t \leq 0\}$ with state space $S = \{0, 1, ...\}$ and infinitesimal generator $W$. Let $i, j \in S$ and:

$$p_{ij}(t) = P\{x(t) = j;\ x(0) = i\} \tag{20}$$

$$A(t) = [p_{ij}(t)] \tag{21}$$

The forward and backward differential equations that govern the behavior of this Markov chain are respectively given by (Gershwin, 1994):

$$\frac{d}{dt}[A(t)] = A(t).\ W \tag{22}$$

$$\frac{d}{dt}[A(t)]^{*} = W.\ A(t) \tag{23}$$

with initial conditions $A(0) = I$ in both cases. Note that these are first order, linear, ordinary differential equations. In terms of the individual matrix elements, the above equations become:

$$\frac{d}{dt}[p_{ij}(t)] = w_{ij}.\ p_{ij}(t) + \sum_{k \neq j} w_{kj} \cdot p_{ik}(t) \tag{24}$$

$$\frac{d}{dt}[p_{ij}(t)] = w_{ii}.\ p_{ij}(t) + \sum_{k \neq i} w_{ik} \cdot p_{kj}(t) \tag{25}$$

The forward and backward equations have the same unique solution given by

$$A(t) = e^{Wt} \tag{26}$$

where, $e^{Wt}$ is the exponential matrix defined by the Taylor series.

$$e^{Wt} = \sum_{k=0}^{\infty} \frac{(W \cdot t)^k}{k!} \tag{27}$$

To find out the state probabilities $Y(t) = [p_0(t), p_1(t),...]$ where $p_j(t) = P\{x(t) = j\}, j \in S$, we need to solve the differential equation:

$$\frac{d}{dt}[\,Y(t)] = Y(t).\,W \tag{28}$$

The solution is given by:

$$Y(t) = Y(0).\,e^{Wt} \tag{29}$$

### 3.1 Discrete-event model of a flexible manufacturing cell line

A flexible manufacturing production line is a series arrangement of machines and buffers, as shown in Fig. 3.

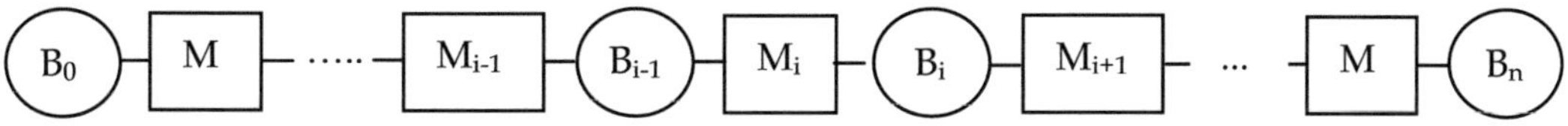


Fig. 3. The flexible manufacturing cell line

The parts enter the first machine and they are processed and transported to the succeeding components, until they finally leave the system. The machines produce at different rates, fail, and are repaired randomly, thus causing changes in the flow of parts. The changes propagate to neighboring machines and may render them starved or blocked. Buffers of finite capacity are inserted in order to reduce these effects. The operation of the production line is ruled by the following:

a) The line consists of $N+1$ buffers. There is one buffer $B_0$ at the beginning of the line, with finite capacity and another $B_n$ at the end, with unlimited capacity.

b) The uptimes and the downtimes of machines are assumed for convenience to be exponential random variables, although any type of distribution may be considered.

c) In each machine there is space for a single work-part. A machine $M_i$ is starved if it has no part to work on and the inventory of the upstream buffer $B_{i-1}$ has been exhausted. Moreover, $M_i$ is blocked if it is prevented from releasing a finished part downstream because $B_i$ is full.

d) Starved or blocked machines remain forced down until a work-part or a unit space is available. During these idle periods, machines do not deteriorate.

e) Transportation time of work-parts to and from buffers is negligible or is incorporated in the processing time.

As we have seen, absorbing states occur in manufacturing system models that capture phenomena such as deadlocks. Interesting for such systems is the time until an absorbing state is reached. Let $\{X(u);\ u \geq 0\}$ be the Markov chain under consideration. Let the state space be finite and given by $S = \{0, 1,..., m, m+1,..., m+n\}$, where $m \geq 0$, $n > 0$, the first $(m+1)$ states are transient states, and the rest of the states are absorbing states. Let 0 be the initial state and $T$, the time to reach any absorbing state. Define:

$$p_{ij}(t) = P\{X(t) = j / X(0) = i\} \tag{30}$$

Then, we have, for any $t > 0$

$$P\{T > t\} = P\{X(t) \notin \{m+1,..., m+n\}\} \tag{31}$$

So we have:

$$P\{T > t\} = 1 - \sum_{j=1}^{n} P_{0,m+j}(t) \tag{32}$$

The cumulative distribution function of T is given by

$$F_T(t) = \sum_{j=1}^{n} p_{0,m+j}(t) \tag{33}$$

The individual probabilities $p_{0,m+j}(t)$ have to be computed by solving the differential equations shown in relation (22) or (23).

### 3.2 The basic cell of the flexible manufacturing system

The basic cell of the proposed model for flexible manufacturing system analysis consists of a machine, for example $M_i$, its upstream buffer $B_{i-1}$ and its downstream buffer $B_i$. In Fig. 4 we have the Markov chain representation of the basic cell of our model for flexible manufacturing system analysis and in Fig. 5 we depicted our Makov chain model for flexible manufacturing cell system. The interpretation of basic cell is that the machine $M_i$ is in state 0 when there is no part being processed, but only transfers of the parts from the machine to its buffers. In state 1, when there is a part being processed and a part in state 2, there is a deadlock in the system. The arrival rate of parts is $\lambda_i$ and the service rate of each part is $\mu_i$.

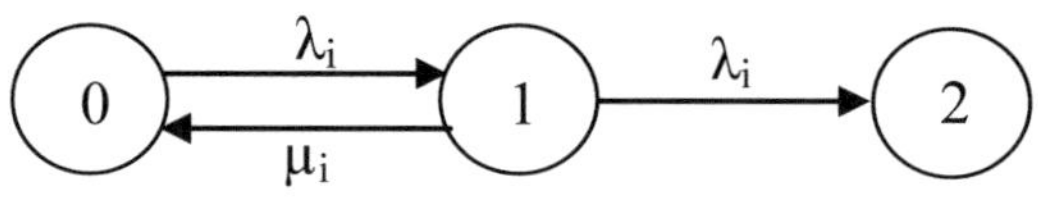


Fig. 4. Markov chain of the model basic cell for flexible manufacturing system analysis

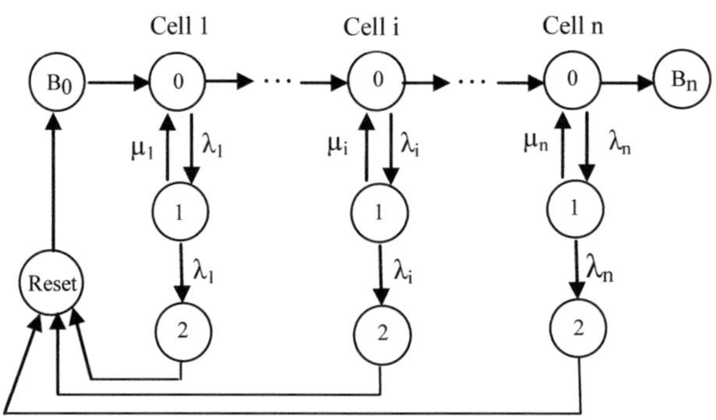


Fig. 5. Markov chain model for flexible manufacturing cells system analysis

Here, the time to absorption is the time elapsed before a deadlock is reached. We know in this case that $F_T(t) = p_{02}(t)$. To compute $p_{02}(t)$, we first write down the infinitesimal generator $W$ of this Markov chain:

$$W = \begin{bmatrix} -\lambda_i & \lambda_i & 0 \\ \mu_i & -(\lambda_i + \mu_i) & \lambda_i \\ 0 & 0 & 0 \end{bmatrix} \tag{34}$$

First, consider the backward equation (6) for $P_{02}(t)$:

$$\frac{d}{dt}(p_{02}(t)) = w_{00}.\ p_{02}(t) + w_{01}.\ p_{12}(t) + w_{02}.\ p_{22}(t) \tag{35}$$

Since $w_{02} = 0$, the above becomes:

$$\frac{d}{dt}(p_{02}(t)) = -\lambda_i p_{02}(t) + \lambda_i p_{12}(t) \tag{36}$$

The backward equation for $p_{12}(t)$ is given by:

$$\frac{d}{dt}(p_{12}(t)) = w_{10}.\ p_{02}(t) + w_{11}.\ p_{12}(t) + w_{12}.\ p_{22}(t) \tag{37}$$

Since $p_{22}(t) = 1$, the above becomes:

$$\frac{d}{dt}(p_{12}(t)) = -\mu_i p_{02}(t) - (\lambda_i + \mu_i) \cdot p_{12}(t) + \lambda_i \tag{38}$$

Let $p_{ij}(s)$ denote the Laplace transform of $p_{ij}(t)$. Taking the transform on either side of the equation above, we get:

$$sP_{02}(s) = -\lambda_i P_{02}(s) + \lambda_i P_{12}(s) \tag{39}$$

$$sP_{12}(s) = \mu_i P_{02}(s) - (\lambda_i + \mu_i) \cdot P_{12}(s) + \frac{\lambda_i}{s} \tag{40}$$

Simplifying using (20) and (21), we get:

$$P_{02}(s) = \frac{\lambda_i^2}{s\left[s^2 + s(2\lambda_i + \mu_i) + \lambda_i^2\right]} \tag{41}$$

Now, $p_{02}(t)$ can be obtained from equation (41) by inverse Laplace transformation:

$$p_{02}(t) = A + B.\ e^{-at} + C.\ e^{-bt} \tag{42}$$

where, the constants are given by:

$$a = \frac{2\lambda_i + \mu_i + \sqrt{\mu_i^2 + 4\lambda_i\mu_i}}{2}; \quad b = \frac{2\lambda_i + \mu_i - \sqrt{\mu_i^2 + 4\lambda_i\mu_i}}{2} \tag{43}$$

$$A = \frac{\lambda_i}{ab}; \quad B = \frac{\lambda_i(b-2a)}{ab(b-a)}; \quad C = \frac{\lambda_i}{b(b-a)} \tag{44}$$

The works (Viswanadham & Ram, 1994), (Lanzon et al., 1996), (Sethi et al., 1997), contain a similar discussion on computing the mean time to absorption.

### 3.3 The buffer events

In a production line, the next event of a component depends on its current state and on the state of adjacent components. An interesting quantity to study is the relation between the buffer size and the machine duration of service. This is because a failed machine, being coupled with a large buffer, may be delayed enough so that the blocking event is avoided, if in the mean time the machine is repaired. So, we may say that in a production line the parameter which determines the deadlock of a basic cell, as well as of the entire production line, is the buffer size. The phenomenon of blocked and starved states occurs frequently when a machine produces at a faster rate than its adjacent ones. In this case, machine $M_i$ is located between an empty and a full buffer. It is then forced to wait until a part arrives from the upstream cell and upon completion of part processing it is blocked until an empty space is available in the downstream buffer. We examine two possibilities: a blocked machine empties its upstream buffer (see case A) or a starved machine fills its downstream buffer (see case B). In both situations the event is conventionally encountered when a work-part is released from $M_i$ to the downstream buffer (see Fig. 6 and Fig. 7). A starved or blocked machine alternates between two states for some time until either a not full or a not empty event occurs. We consider the segment of machines $M_{i-1}$, $M_i$ and $M_{i+1}$, and buffers $B_{i-1}$ and $B_i$. Let $t$ be the time when the starved and blocked event is encountered and $TA$ the apparent time of the next event.

$$M(i,t) = \begin{cases} 1, \text{ if machine } M_i \text{ is up} \\ 0, \text{ if } M_i \text{ is under repair} \end{cases}$$

$$B(j,t) = \begin{cases} 0, \text{ if buffer } B_j \text{ is empty} \\ 2, \text{ if buffer } B_j \text{ is full} \\ 1, \text{ otherwise state} \end{cases}$$

$$BE_j(t) = \begin{cases} 1, \text{ if } B_j \text{ empties at time } t \\ 0, \text{ otherwise} \end{cases}$$

We define the following (Narahari & Viswandham, 1994), (Chiang et al., 2000):

$\mu_i$ = The nominal production rate (workparts/time-unit) of machine $M_i$, $i = 1,\ldots, n$

$T_{1j}(t)$ = Delay time until the next arrival to $B_j$

$T_{2j}(t)$ = Delay time until the next departure from $B_j$

We discuss the following situations: *Case A:* Machine $M_{i+1}$ is faster than $M_{i-1}$. This situation is depicted in Fig. 4 and the condition is:

$$(T_{21} > T_{1,i-1} + \frac{1}{\mu_i}) \cap (\mu_{i+1} > \mu_{i-1}) \tag{45}$$

We note that in Fig. 6 and in Fig. 7 the continuous line represents a machine operation on a work-part and the arrows represents arrivals to the succeeding buffer. Blank intervals indicate idle periods due to blockage or starvation of machines. The wavy lines denote a machine under repair. For Fig. 6, buffer $B_i$ is scheduled to switch from full to an intermediate state. The not-full event occurs upon the departure of the last blocked part from $M_i$. We notice that the end-of-processing time of the $(1 + N_i)^{th}$ work-part in $M_i$ is greater than the time when a single space for this part is available in $B_i$.

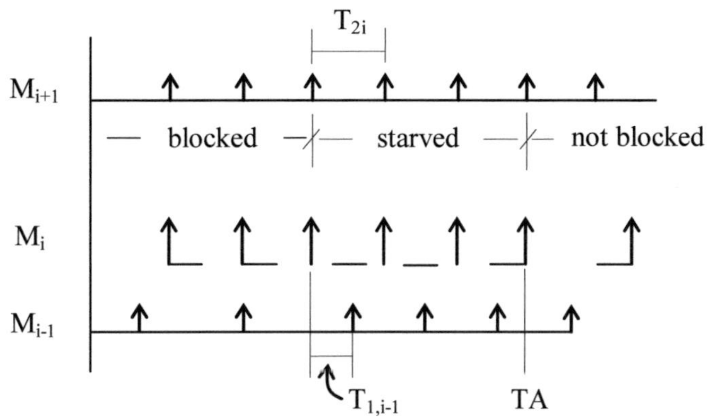


Fig. 6. Starved and blocked machine states when $M_{i+1}$ is faster than $M_{i-1}$

The opposite holds for the first $N_i$ work-parts. This observation leads to (Ciufudean et al., 2005):

$$t + T_{2i} + \frac{N_i}{\mu_{i+1}} < t + T_{1,i-1} + \frac{N_i}{\mu_{i-1}} + \frac{1}{\mu_i} \tag{46}$$

and

$$TA = t + T_{2i} + \frac{N_i - 1}{\mu_{i+1}} \geq t + T_{1,i-1} + \frac{N_i - 1}{\mu_{i-1}} + \frac{1}{\mu_i} \tag{47}$$

From relation (26) and (27) we compute the parts until next event, for $B_i$:

$$N_i = 1 + \text{Int}\left\{\frac{T_{2i} - T_{1,i-1} - \frac{1}{\mu_i}}{\frac{1}{\mu_{i-1}} - \frac{1}{\mu_{i+1}}}\right\} \tag{48}$$

*Case B*: Machine $M_{i-1}$ produces at a faster rate than $M_{i+1}$ (machine $M_{i-1}$ is faster than $M_{i+1}$) and the starved machine $M_i$ fills its upstream buffer $B_i$ (see Fig. 5). The condition is:

$$(T_{2i} > T_{1,i-1} + \frac{1}{\mu_i}) \cap (\mu_{i-1} > \mu_{i+1}) \tag{49}$$

This is dual to case A. After that, machine $M_{i-1}$ processes $N_{i-1}$ parts, a non-empty event will occur.

In Fig. 5 we see that the arrival time of $(1 + N_{i-1})^{th}$ work-part at buffer $B_{i-1}$ is less than the time machine $M_i$ is ready to receive it. The opposite holds for the first $N_{i-1}$ work-parts (Ciufudean, 2008).

This observation leads to:

$$t + T_{2i} + \frac{N_{i-1} - 1}{\mu_{i+1}} > t + T_{1,i-1} + \frac{N_{i-1}}{\mu_{i-1}} \tag{50}$$

and

$$TA = t + T_{2i} + \frac{N_{i-1} - 1}{\mu_{i+1}} \le t + T_{1,i-1} + \frac{N_{i-1} - 1}{\mu_{i-1}} \tag{51}$$

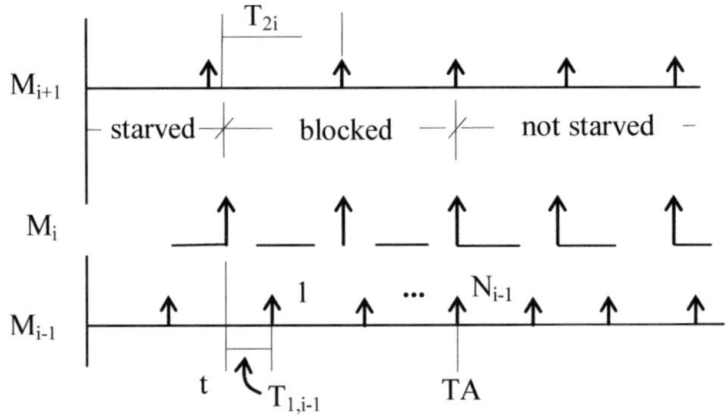


Fig. 7. Starved and blocked machine states when $M_{i-1}$ is faster than $M_{i+1}$

In a dual situation to case A, we get the parts until the next event, for $B_{i-1}$ (Ciufudean & Filote, 2010):

$$N_{i-1} = 1 + \text{Int}\left\{\frac{T_{1,i-1} - T_{2i} - \dfrac{1}{\mu_{i+1}}}{\dfrac{1}{\mu_{i+1}} - \dfrac{1}{\mu_{i-1}}}\right\} \tag{52}$$

From relation (48) and (52) we notice that the number of parts until the next event depends on the service rate of each part. We may say that the buffer dimensions can be calculated in such a manner as to avoid the failed state of machines, considering that the time to calculate the number of parts until next event is set to $T_{21} = P_{02}$ in relation (48) and, respectively, to $T_{1,i-1} = P_{02}$ in relation (52); where $P_{02}$ is given by relation (42).
As we discussed before, the failed state of machines can be avoided if the buffer size is bigger than the critical size (the size determined from relation (48) and respectively (52)). The condition to be accomplished is that the average time to repair a machine is less than the average time to fill the upstream buffer of that machine.

## 4. An Illustrative Example

The manufacturing system considered in this paper consists of two cells linked together by a material system composed of two buffers A and B and a conveyor. Each cell consists of a machine to handle within cell part movement. Pieces enter the system at the load/unload station, where they are released from those two buffers, A and B, and then are sorted in cells (pieces of type "a" in one cell, and pieces of type "b" in the other cell).
We notice that in the buffer A there are pieces of types "a", "b", and others, where the number of pieces "a" is greater than the number of pieces "b". In the buffer B there are pieces of types "a", "b", and others, where the number of pieces "b" is greater than the number of pieces "a". The conveyor moves pieces between the load/unload station of the various cells.
The sorted piece leaves the system, and an unsorted piece enters the system, respectively one of those two buffers A or B. The conveyor along with the central storage incorporates a sufficiently large buffer space, so that it can be thought of as possessing infinite storage capacity. Thus, if a piece routed to a particular cell finds that the cell is full, its entry is refused and it is routed back to the centralized storage area. If a piece routed by the conveyor is of a different type than the required types to be sorted, respectively "a" and "b", then that piece is rejected from the system.
We notice that once a piece is blocked from cell entry, the conveyor does not stop service; instead it proceeds with its operation to the other pieces waiting for transport.
At the system level, we assume that the cells are functionally equivalent, so that each cell can provide the necessary processing for a piece. Hence, one cell is sufficient to maintain production (at a reduced throughput). We say that the manufacturing system is available (or, operational) if the conveyor and at least one of the cells are available. A cell is available if its machine is available (Rodriguez et al., 2010).
Over a specified period of operation, due to the randomly occurring subsystem failures and subsequent repairs, the cellular automated manufacturing system will function in different

configurations and exhibit varying levels of performance over the random residence times in these configurations.
The logical model of our manufacturing system is showed in Fig.8.

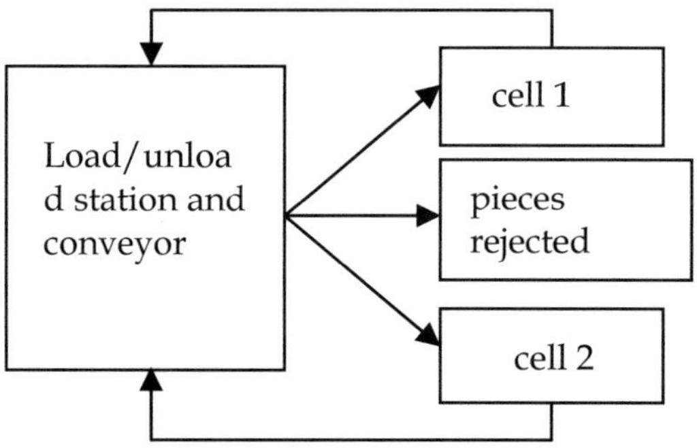


Fig. 8. Logical model for a manufacturing system

### 4.1 A Markov model for evaluating the system availability

For the flexible manufacturing system depicted in Fig. 8, we assume that the machines are failure-prone, while the load/unload station and the conveyor are extremely reliable. Assuming the failure times and the repair times to be exponentially distributed, we can formulate the state process as a continuous time Markov chain (CTMC). The state process is given by $\{X(u), u \geq 0\}$ with state space $S = \{(ij), i \in \{0,1,2\}, j \in \{0,1\}\}$, where $i$ denotes the number of working machines, and $j$ denotes the status of the material handling system (load station and conveyor): up "1", and down "0". We consider the state independent (or, time dependent) failure case and the operation dependent failure case separately.

#### 4.1.1 Time dependent failures

In this case, the component fails irrespective of whether the system is operational or not. All failure states are recoverable. Let $r_a$ and $r_m$ denote the repair rates of the material handling system, and a machine respectively. The state process is shown in Fig. 9, a.

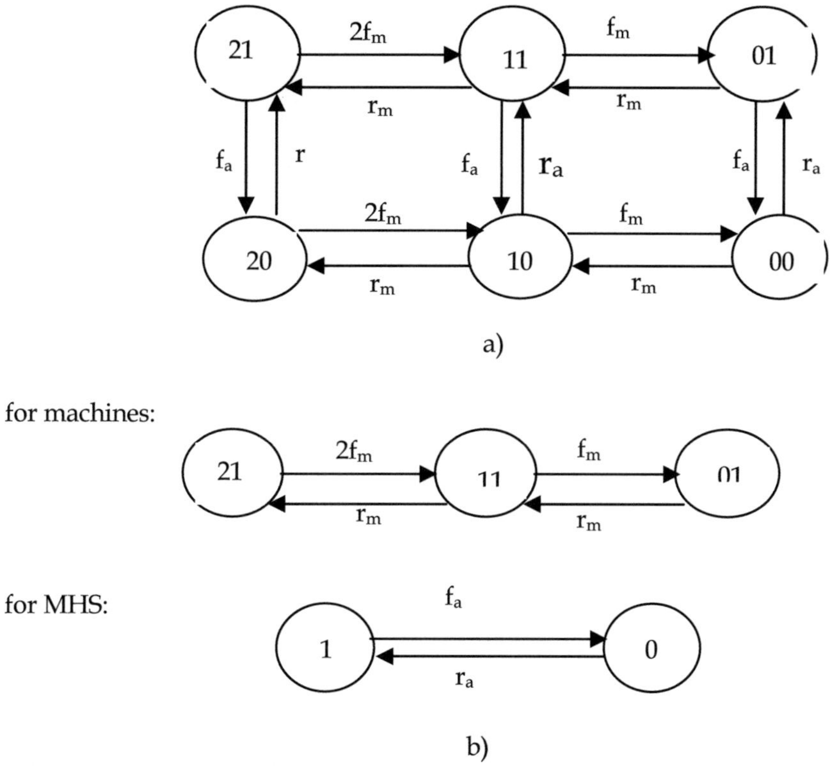


Fig. 9. State process of a FMC with time-dependent failures, (a) State process for a state-independent failure model, (b) Decomposed failure/repair process.

Because the failure/repair behavior of the system components are independent, the state process can be decomposed into two CTMCs as shown in Fig. 9, b. Analytically, the state process is expressed by relations: $S_0 = \{(21), (11)\}$ and $S_F = \{(20),(10), (00)\}$. For each state in $S_F$ no production is possible since the M0HS or both machines are down. In Fig. 2, b the failure/repair behavior of each resource type (machines or MHS) is described by a unique Markov chain. Thus, the transient state probabilities, $p_{ij}(t)$ can be obtained from relation:

$$p_{ij}(t) = p_i(t) \cdot p_j(t) \tag{53}$$

where $p_i(t)$ is the probability that $i$ machines are working at time $t$ for $i = 0,1,2$. The probability pi(t) is obtained by solving (separately) the failure/repair model of the machines. $P_j(t)$ is the probability that j MHS (load/unload station and conveyor) are working at instant $t$, for $j = 0,1$. Let fa and fm denote the failure rates of MHS and of a machine respectively.

#### 4.1.2 Operation dependent failures

We assume that when the system is functional, the resources are all fully utilized. Since failures occur only when the system is operational, the state space is: S = {(21), (11), (20), (10), (01)}, with $S_0$ = {(21), (11)}, $S_F$ = {(20), (10), (01)}. The Markov chain model is shown in Fig. 10. Transitions representing failures will be allowed only when the resource is busy. Transitions rates can however be computed as the product of the failure rates and percentage utilization of the resource. If $T_k^{ij}$ represents the average utilization of the $k^{th}$ resource in the state (i j), the transition rates are given in Fig. 10.

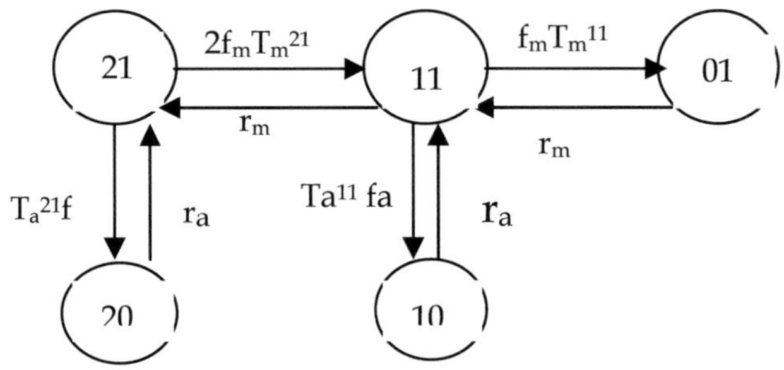


Fig. 10. State process of a FMC with state-dependent failures

#### 4.1.3 A numerical example

For the FMC presented in this paper in Table 1, the failure/repair data of the system components are given. We notice that $T_k^{ij}$ (the system average utilization of the $k^{th}$ resource in state (ij), $T_k^{ij}$ = 1 since the utilization in each operational state is 100% for all i, j, k, i = {0,1,2}, j = {0,1}, k = 4.

The other notations used in Table 1 are: f, the exponential failure rate of resources; r, the exponential repair rate of resources; $N_p$, the required minimum number of operational machines in cell p; p = {1,2} and $n_p$, the total number of machines in cell p.

| | R | F | $N_p$ | $n_p$ | $T_k^{ij}$ |
|---|---|---|---|---|---|
| Machines | 1 | 0,05 | 1 | 2 | 1 |
| MHS | 0,2 | 0,001 | 1 | 1 | 1 |

Table 1. Data for the numerical study

From Fig. 9 and Fig. 10 we calculate the corresponding infinitesimal generators, and after that, the probability vector of CTMC. With relation (1) we calculate the availability of FMC given in this article.

The computational results are summarized in Table 2 for the state process given in Fig. 9 (FMC with time-dependent failures), and respectively in Table 3 for the state process given in Fig. 10 (FMC with state-dependent failures). We consider the system operation over an interval of 24 hours (three consecutive shifts).

| Time [hour] | Machines | MHS | System Availability |
|---|---|---|---|
| 0 | 1.0000 | 1.0000 | 1.0000 |
| 1 | 0.9800 | 0.9548 | 0.9217 |
| 4 | 0.9470 | 0.8645 | 0.7789 |
| 8 | 0.9335 | 0.8061 | 0.7025 |
| 12 | 0.9330 | 0.7810 | 0.6758 |
| 16 | 0.9331 | 0.7701 | 0.6655 |
| 20 | 0.9330 | 0.7654 | 0.6623 |
| 24 | 0.9328 | 0.7648 | 0.6617 |

Table 2. Computational results for the FMC in Fig. 9

| Time hour | Machines | MHS | System Availability |
|---|---|---|---|
| 0 | 1.0000 | 1.0000 | 1.0000 |
| 1 | 0.9580 | 0.9228 | 0.9001 |
| 4 | 0.9350 | 0.8228 | 0.7362 |
| 8 | 0.9315 | 0.8039 | 0.7008 |
| 12 | 0.9310 | 0.7798 | 0.6739 |
| 16 | 0.9320 | 0.7688 | 0.6632 |
| 20 | 0.9318 | 0.7639 | 0.6598 |
| 24 | 0.9320 | 0.7636 | 0.6583 |

Table 3. Computational results for the FMC in Fig. 10

The results of the availability analysis of the flexible manufacturing system are illustrated in Fig.11, which depicts the availability of the system as a time function. The numbers x = 2, 3 indicate the system in Fig. 9, respectively Fig. 10. One can see from Fig.11 that the layout with FMC with time-dependent failures is superior to that with FMC with state-dependent failure.

An analytical technique for the availability evaluation of the flexible manufacturing systems was presented. The novelty of the approach is that the construction of large Markov chains is not required. Using a structural decomposition, the manufacturing system is divided into cells.

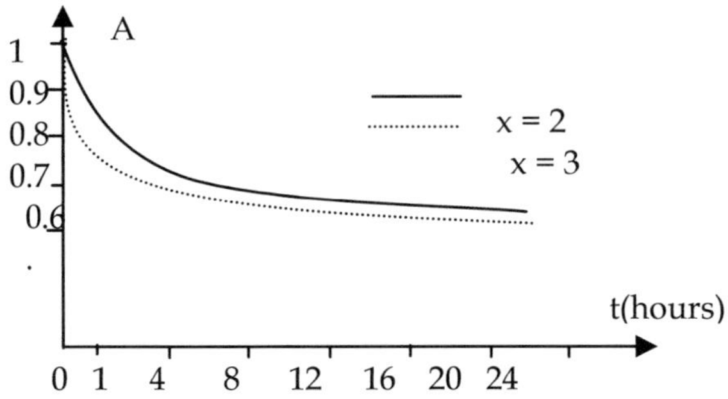


Fig. 11. Availability analysis of the flexible manufacturing system

For each cell, a Markov model was derived and the probability was determined of at least $N_i$ working machines in cell *i*, for *i* = 1,2,..,n and *j* working material handling system at time *t*, where $N_i$ and *j* satisfy the system production capacity requirements. The model presented in this paper can be extended to include other components, e.g., tools, control systems. The results reported here can form the basis of several enhancements, such as conducting throughput studies of cellular flexible manufacturing types with multiple part types.

## 5. Conclusion

A model for flexible manufacturing cellular systems analysis has been introduced in this paper. Such a model could be generated directly or using higher level models such as stochastic Petri nets or discrete event simulation. A discrete-event system formulation and state partition into basic cells and fast and slow varying section, lead to a reduced computation cost. Further research in this area should focus on systems modeled with Markov chains which exhibit a cut-off phenomenon, as the existence of a cut-off phenomenon is a good indicator to whether a transient or a steady-state analysis is appropriate in a given setting. For example, if the cut-off time is known and the duration of observation is less than the cut-off time, then transient analysis is more meaningful than steady-state analysis.

Identification and measurement of the bottleneck times in production lines has implications for both natures concerning the preventive maintenance and the production automation. In this paper we address the Markov model of production lines with bottlenecks. In lines where machines have identical efficiency, the machine with the smaller downtime is the bottleneck. In two-machine lines, the downtime bottleneck is the machine with the smallest value of $p{\cdot}Tup{\cdot}Tdown$, where p is the probability of blockage for the first machine and the probability of starvation for the second. Anticipation of events like full buffer or empty buffer, which determine bottlenecks, has also implications for the preventive maintenance of the manufacturing system. Future work in this area should focus on extensions of the results obtained in manufacturing systems with high failure rates.

## 6. References


Chiang, S. I.; Kuo, C. T. & Meerkov, S. M. (2000) DT-Bottlenecks in serial production lines: theory and applications, *IEEE Trans. Autom. Contr.*, vol 16, no 5, pp. 567-580

Ciufudean, C.; Petrescu, C. & Graur, A. (2005) Determining the Performances of Cellular Manufacturing Systems, *International Conference on Advances in the Internet, Processing, Systems, and Interdisciplinary Research, IPSI-2005,* Spain, Barcelona, April 29-May 1, pp. 44-48, ISBN: 86-7466-117-3

Ciufudean, C.; Filote, C. & Popescu, D. (2005) Determining the Availability of Flexible Manufacturing Cells, *International Conference on Advances in the Internet, Processing, Systems, and Interdisciplinary Research, IPSI-2005,* France, Carcassone, April 23-26, pp. 36-40, ISBN: 86-7466-117-3

Ciufudean, C.; Petrescu, C. & Filote, C. (2005) Performance Evaluation of Distributed Systems, *International Conference on Control and Automation, ICCA 2005,* June 26-29, Budapest, Hungary, pp. 21-25, ISBN0-7803-9138-1, IEEE Catalog Number: 05EX1076C

Ciufudean, C.; Larionescu, A. & Popa, V. (2005) Virtual Diagnosis of Cellular Manufacturing Systems, *11th International Conference on Electrical Machines, Drives and Power Systems, ELMA 2005,* 15-16 September, Sofia, Bulgaria, pp. 195-200, ISBN 954-902-09-5-9

Ciufudean, C. (2008) Risk and Reliability Analysis of Flexible Construction Robotized Systems, In: *Robotics and Automation in Construction,* Editors: C. Balaguer & M. Abderrahim, pp. 139-158, IN-TEH, ISBN 978-953-7619-13-8, Vienna, Austria

Ciufudean, C. & Filote, C. (2010) Workflow Diagnosis Using Petri Nets Charts, In: *Petri Nets: Applications,* Editor: P. Pawlewski, pp. 445-468, IN-TEH, ISBN 978-953-307-047-6, Vukovar, Croatia

Gershwin, S. B. (1994) *Manufacturing Systems Engineering*, Englewood Cliffs, N. J.: Prentice Hall

Jacobs, D. & Meerkov, S. M. (1995) A system - theoretic property of serial production lines: Improvability, *Int. J. Syst. Sci.*, vol 26, no 4, pp. 755-785

Laftit, S.; Proth, J. M. & Xie, X. L. (1992) Optimization of invariant criteria for event graphs, *IEEE Trans. Autom. Contr.*, vol 37, no. 12, pp. 547-555

Lanzon, S. C.; Ma, A. K. L.; Mills, J. K. & Benhabib, B. (1996) Application of discrete-event-system theory to flexible manufacturing, *IEEE Trans. Contr. Syst.*, vol.21, pp.41-48.

Lim, J. T.; Meerkov, S. M. & Top, F. (1990) Homogeneous, asymptotically reliable serial production lines: Theory and a case study, *IEEE Trans. Autom. Contr.*, vol 35, pp. 524-532

Narahari, J. & Viswandham, N. (1994) Transient Analysis of Manufacturing System Performance, *IEEE Trans. Autom. Contr.*, vol. 10, no. 2, pp. 230-243

Proth, J. M. & Xie, X. L. (1994) Cycle time of stochastic event graphs: evaluation and marking optimization, *IEEE Trans. Autom. Contr.*, vol 39, no. 7, pp. 1482-1486

Rodriguez, L.; Garcia, E.; Morant, F.; Correcher, A. &Quilea, E. (2010) Fault diagnosis for complex systems using Coloured Petri Nets, In: *Petri Nets: Applications,* Editor: P. Pawlewski, pp. 445-468, IN-TEH, ISBN 978-953-307-047-6, Vukovar, Croatia

Sethi, S. P.; Zhang, Q. & Zhou, X. Y. (1997) Hierarchical production controls in a stochastic two-machine flow-shop with a finite internal buffer, *IEEE Trans. Robot. Autom.*, vol.13, pp. 1-12

Viswanadham, N. & Ram, R. (1994) Composite performance dependability of cellular manufacturing systems, *IEEE Trans. Robot. Autom.*, vol.10. pp. 245-258

# Process rescheduling: enabling performance by applying multiple metrics and efficient adaptations

Rodrigo da Rosa Righi
*Programa Interdisciplinar de Pós-Graduação em Computação Aplicada - Universidade do Vale do Rio dos Sinos*
*Brazil*

Laércio Pilla, Alexandre Carissimi and Philippe Navaux
*Programa de Pós-Graduação em Computação - Universidade Federal do Rio Grande do Sul*
*Brazil*

Hans-Ulrich Heiss
*Kommunikations- und Betriebssysteme - Technische Universität Berlin*
*Germany*

## 1. Introduction

As grid technologies gain popularity, separate clusters and computers are increasingly being interconnected to create computing architectures for the processing of scientific and commercial applications (Sánchez et al., 2010). These constituent parts may be different from each other as well as be located either within a single organization or across various geographical sites. Concerning this, the task of allocating processes to processors on such architecture often becomes a problem requiring considerable effort. In order to fully exploit this kind of environments, the programmer must know both the machine architecture and the application code properly. Moreover, each new application requires another analysis for process scheduling.

Since both resource management and scheduling are key services for grid environments, issues like load balancing represent a common concern for most developers. Thus, a possibility is to explore the automatic load balancing at middleware level, linking the balancer tool with the programming library. For instance, an allocation scheme where the processes with longer computing times are mapped to faster machines can be used. On the other hand, this approach is not the best one for irregular applications and dynamic distributed environments, where a good processes-resources assignment performed in the beginning of the application may not remain efficient with time (Low et al., 2007). At this moment, it is not possible to recognize either the amount of computation of each process or the communication patterns among them. Besides fluctuations in the processes' computation and communication actions, the processors' load may vary and networks may become congested while the application is running. An alternative is to perform process rescheduling by applying the migration of the processes to new resources, offering runtime load balancing (Chen et al., 2008).

In this context, we developed a model called MigBSP. MigBSP controls process rescheduling on dynamic and heterogeneous environments, like multi-cluster ones. It acts over BSP (**Bulk Synchronous Parallel**) applications (Valiant, 1990) and works with the concept of hierarchy in two levels using Sets (considering different networks) and Set Managers. MigBSP's adjusts the conclusion of both local computation and global communication phases of BSP processes to be faster taking benefit from data collected at runtime. This adjustment happens through the migration of those processes whose have long computation time, perform several communication actions with other processes that belong to a same Set and present low migration costs. The use of the Computation, Communication and Memory (migration costs) metrics aims to offer performance and better quality for scheduling decisions. Besides multiple metrics, other keyword of MigBSP is adaptivity. Contrary to existing approaches (Bonorden et al., 2005; Vadhiyar & Dongarra, 2005), MigBSP performs the rescheduling launching according to the system state in order to reduce its impact on application execution.

The present chapter describes MigBSP and its novel ideas for process rescheduling. We developed three different BSP-based scientific applications in order to verify the impact and the efficiency of MigBSP's algorithms. Besides the choice of the applications, we modeled a multi-cluster infrastructure and varied the number of processes. Summarizing the results, we achieved a mean performance gain of 19% when applying process migration over our applications. Moreover, a mean overhead lower than 7% was observed when migrations are not applied (if the model decides that migrations are not recommended during all application execution). Therefore, the use of multiple metrics and efficient adaptations configure MigBSP as a viable solution when treating migration of BSP processes. Besides BSP, MigBSP's algorithms can be used in several other situations where load balancing takes place, such as in Web servers, data centers and synchronous computations in general (Bonorden et al., 2005).

The remaining of this chapter is organized as follows. Section 2 shows related work and some opportunities of research. Section 3 describes the rationales of MigBSP, emphasizing its contribution and novel ideas. Section 4 presents our evaluation methodology. Section 5 discusses the results and points out the performance gain/loss when executing MigBSP. Finally, the concluding remarks of the chapter are displayed in Section 6.

## 2. Related Work

GridWay resource broker treats with time and cost optimization on scheduling and migration areas (Moreno-Vozmediano & Alonso-Conde, 2005). Both migration mechanisms consider only data from CPU, like speed and load. (Bhandarkar, Brunner & Kale, 2000) presented a support for adaptive load balancing in MPI applications. Periodically, MPI application transfers control to the load balancer using a special call MPI_Migrate(). This mechanism implies in modifications in the application code. Besides this last work, a fixed period for rescheduling launching is also demonstrated in the following approaches (Hernandez & Cole, 2007; Utrera et al., 2005). Adaptive MPI (AMPI) (Huang et al., 2006) uses Charm++ framework to offer load balancing. Charm++ uses workload data and objects communication pattern to redistribute the workload at each load balancing time.

A system for autonomic rescheduling of MPI (Message Passing Interface) programs is presented in (Du et al., 2004). This work presents an extensible rule-based mechanism for policy making. When a policy is satisfied, its actions are done. Besides the consideration of monitored data, this system also uses an application description in order to estimate the execution time. Vadhiyar and Dongarra presented a migration framework and self-adaptivity in GrADS system (Vadhiyar & Dongarra, 2005). However, they computed the migration costs as a fixed value. In addition, the gain with rescheduling is based on the remaining execution time prediction over a new specified resource. Thus, this framework must work with applications in which their parts and durations are known in advance.

(Heiss & Schmitz, 1995) developed a load balancer where the load of each task is represented by a particle. Such work considers the processors load, the communication among the tasks and the amount of data to be migrated. This work considers static information about the behavior of the tasks (number of instructions, interactions among the tasks and amount of memory). Furthermore, the migration of tasks is performed only to a neighbor node (direct connection in the processors graph). (Du, Sun & Wu, 2007) measured the migration costs at application runtime. For that, they described a model that considers the process, the memory, the I/O and the communication states. Nevertheless, these authors specify neither when to launch the process migration, nor which processes will be migrated actually. Kondo et al. (Kondo et al., 2002) described a client-server scheduling model for global computing. Their model measures the processor speed, the network bandwidth and the disk space to set the number of work units that can be sent to a client. However, these values are not combined and the minimum of them gives the number of work that the server will pass to a client.

Concerning the BSP scope, we can cite two works that present migration on BSP applications. The first one describes the PUBWCL library which aims to take profit of idle cycles from nodes around the Internet (Bonorden et al., 2005). PUBWCL can migrate a process during its computation phase as well as after the barrier. All proposed algorithms just use computation data about processes and the the nodes. Other work comprises an extension of PUB library to support migration (Bonorden, 2007). The author explains that a load balancer decides when to launch the process migration. Nevertheless, this issue is not addressed in (Bonorden, 2007). Bonorden proposed both a centralized and a distributed strategies for load balancing. In the first one, all nodes send data about their CPU power and load to a master node. The master verifies the least and the most loaded node and migrates one process between them. In the distributed approach, every node chooses $c$ other nodes randomly and asks them for their load. One process is migrated if the minimum load of $c$ analyzed nodes is smaller than the load of the node that is performing the test. The drawback of this strategy is that it can create a lot of messages among the nodes. Moreover, both strategies take into consideration neither the communication among the processes, nor the migration costs.

## 3. MigBSP: Process Rescheduling Model

A BSP application is divided in one or more supersteps, each one containing both computation and communication phases followed by a barrier synchronization. Since the barrier always wait for the slowest process, MigBSP's final objective is to adjust the processes' location in order to reduce the supersteps' times. Figure 1 (a) shows a superstep $k$ of an application in which the processes are not balanced among the resources. Figure 1 (b) depicts the expected result with processes redistribution at the end of superstep $k$, which will influence the execution of the following supersteps. MigBSP offers automatic load balancing at middleware level, requiring no changes in the application code nor previous knowledge about the system/application. All necessary data for its functioning can be captured directly in both communication and barrier functions as well as in other sources like the operating system. The final result of MigBSP is a formalism that answers the following issues regarding process migration: (i) "When" to launch the mechanism for process migration; (ii) "Which" processes are candidates for migration and; (iii) "Where" to put the elected processes. We are not interested in the keyword "How", that treats the mechanism employed to perform migrations.

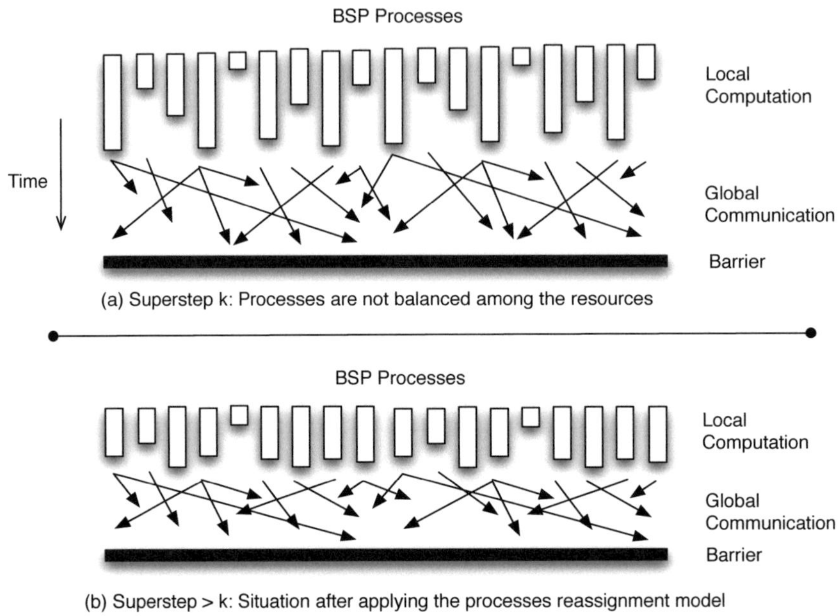

Fig. 1. BSP Supersteps in two different situations

MigBSP can be seen as a scheduling middleware. Concerning this area, (Casavant & Kuhl, 1988) proposed a scheduling taxonomy for general purpose distributed computing systems in order to formalize the classification of schedulers. MigBSP can be enclosed on the global and dynamic items. The dynamic feature considers that information for process scheduling are collected at application runtime. The role of scheduling is spread among several processes that cooperate among themselves in order to improve resource utilization (processors and network). Thus, the model performs a physically distributed and cooperative scheduling. The achieved scheduling is sub-optimal and employs heuristics. Finally, following the horizontal classification of Casavant and Kuhl, the idea is to present an adaptive scheduling that can change its execution depending on the environment feedback.

### 3.1 Model of Parallel Machine and Communication

MigBSP works over an heterogeneous and dynamic distributed environment. The heterogeneous issue considers the processors' capacities (all processors have the same architecture, *e.g.* i386), as well as the network bandwidth and level (Fast and Gigabit Ethernet and multi-clusters environments, for instance). The dynamic behavior deals with environment changes which were perceived at runtime (such as network congestion and fluctuations on processors' load). Moreover, the dynamic behavior can also occur at process level, since some processes may need more computational power or increase their network interaction with other processes during application runtime. Each process is mapped to a real processor which can execute more than one process. In order to turn the scheduling more flexible and efficient, MigBSP proposes a hierarchical scheduling. The nodes are gathered to create the abstraction of a Set. A Set could be a LAN network or a cluster. Each Set is composed by one or more nodes (each with one or more processors) and a Set Manager. The scheduling mechanism is located inside every process (additional code in barrier function) and inside each Set Manager. This last entity captures scheduling data from a Set and exchanges it among other managers. Our communication model affirms that the Sets are fully interconnected, meaning that there exists at least one communication path between any two nodes. The communication is asynchronous, where the sending is non blocking while the receiving is blocking. In addition, the underlying network protocol always guarantee reliability and the fact that the messages sent across the network are received in the order sent previously.

### 3.2 Question "When": Process Rescheduling Activation

The decision for process remapping is taken at the end of a superstep. We are employing the reactive migration approach (Milanés et al., 2008), where the process relocation is launched from outside the application (in this case, at middleware level). The migration point was chosen because in this moment it is possible to analyze data about the computation and communication from all processes. We applied two adaptations aiming to put as less intrusion in the application as possible. They provide an adaptable interval between migration calls.

#### 3.2.1 First Adaptation: Controlling the Migration Interval based on the Processes' Balance

We are using an index $\alpha$ ($\alpha \in N^*$) in order to turn viable the adaptivity on process rescheduling calling. $\alpha$ will inform the interval between supersteps to apply process migration. This index increases if the system tends to the stability in the conclusion time of each superstep and decreases on the contrary. The last case means that the frequency of calls increases to turn the system more stable quickly. In order to allow a sliding $\alpha$, it is necessary to verify if the distributed system is balanced or not. To treat this issue, the time of each BSP process is collected at the end of every superstep. Thus, the times of the slowest and the fastest processes are captured, and an arithmetic average of the times is computed. The distributed system is considered stable if both Inequalities 1 and 2 are true. In both inequalities, $D$ informs the percentage of how far the time of the slowest and the fastest processes can be from the average. The $D$ value is passed in model initialization. Figure 2 shows the algorithm that reveals how the $\alpha$ value is computed along the application execution. A variable called $\alpha'$ was employed to save the temporary value of $\alpha$. $\alpha'$ will show the next interval to trigger the load balancing. $\alpha'$ suffers a variation of one unity at each superstep depending on the state of the system.

$$time\ of\ the\ slowest\ process < average\ time\ .\ (1 + D) \tag{1}$$

$$time\ of\ the\ fastest\ process > average\ time\ .\ (1 - D) \tag{2}$$

In Figure 2, $t$ ($k \leq t \leq k + \alpha - 1$) is the index of a superstep and $k$ represents the superstep that comes after the last call for load rebalancing or it is 1 when the application is beginning ($k$ and $\alpha$ will have the same meaning in all algorithms). $\alpha'$ does not have an upper bound, but its lower value is the initial value of $\alpha$. In the best case, the system is always in equilibrium and $\alpha'$ always increases. For example, if the system is always stable and the initial value of $\alpha$ is 10, after 10 supersteps the new value of $\alpha$ will be 20. The idea of the model is to minimize its intrusion in application execution while the system stays stable, postponing the process rescheduling activation according to $\alpha$. In implementation view, BSP processes save their times in a vector and pass them to their Set Managers when rescheduling is activated. Following this, all Set Managers exchange their information. Taking into account the the times of each process, the Set Managers compute both Inequalities 1 and 2. Therefore, each manager knows the $\alpha'$ variation locally.

**for** $t$ from superstep $k$ to superstep $k + \alpha - 1$ **do**
  **if** Inequalities 1 and 2 are true **then**
    Increase $\alpha'$ by 1
  **else if** $\alpha' >$ initial $\alpha$
    Decrease $\alpha'$ by 1
  **end if**
**end for**
Call for BSP process rescheduling
$\alpha = \alpha'$

Fig. 2. Interval of supersteps $\alpha$ for the next call for BSP process rescheduling

### 3.2.2 Second Adaptation: Controlling the Rescheduling Interval based on the Number of Calls without Migrations

The other adaptation considers the management of $D$ (see Inequalities 1 and 2) based on the frequency of migrations. Figure 4 depicts the impact of $D$ when defining the situation of the processes. The idea is to increase $D$ if process rescheduling is activated for $\omega$ consecutive times and none migrations happen. The increase of $D$ enlarges the interval in which the system is considered stable, causing the increase of $\alpha'$ consequently. In contrast, $D$ can decrease down to a limit if each call produces the migration of at least one process. The algorithm depicted in Figure 3 presents how $D$ is controlled at each rescheduling call.

$\gamma \leftarrow$ Consecutive rescheduling calls without migrations
**if** $\gamma \geq \omega$ **then**
  **if** D + $\frac{D}{2} < 1$**then**
    D $\leftarrow$ D + $\frac{D}{2}$
  **end if**
**else if** D $>$ initial D and $\gamma = 0$ **then**
    D $\leftarrow$ D - $\frac{D}{2}$
**end if**

Fig. 3. Stability of the system according to $D$

The computation of $D$ is done by each Set Manager, which knows if migrations occurred during the migration call. This adaptation is important when the migration costs are high. Thus, although a process is selected for migration, its transferring will not take place and the system will remain with the same scheduling configuration. Consequently, it is pertinent to increase $D$ in order to minimize MigBSP impact on application execution in this situation.

### 3.3 Question "Which": Choosing the Candidate Processes for Migration

The answer for "Which" is solved through our decision function called **Potential of Migration** (PM). Each process $i$ computes $n$ functions $PM(i, j)$, where $n$ is the number of Sets and $j$ means a specific Set. The key idea consists in not performing all available processes-resources tests at the rescheduling moment. $PM(i, j)$ is found through the combination of Computation, Communication and Memory metrics. The first two work at the computation and communication phases of a superstep. The Memory metric acts as an idea of migration costs.

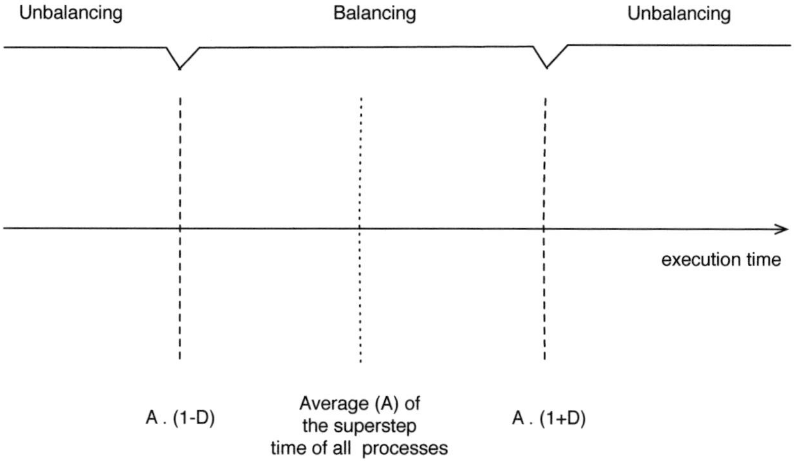


Fig. 4. Balancing situations which depend on the distance $D$ from the average $A$

### 3.3.1 Computation Metric

Each process $i$ computes $Comp(i,j)$ functions, where $Comp(i,j)$ informs the Computation metric for a process $i$ and a specific Set $j$. Set $j$ is used in $Comp(i,j)$ calculus to simulate the performance of process $i$ on different sites of the parallel architecture. The data used to calculate this metric start at superstep $k$ and finish at superstep $k+\alpha-1$. For every superstep $t$ ($k \leq t \leq k+\alpha-1$), the number of processor's instructions ($I_t$) and the conclusion time of the computation phase ($CT_t$) are stored. The value of $I_t$ is used to evaluate the process stability (regularity), that is represented by the Computation Pattern called $P_{comp}(i)$. This pattern is a real number that belongs to the [0,1] interval. A $P_{comp}(i)$ close to 1 means that the process $i$ is regular in the number of instructions that executes at each superstep. On the other side, this pattern will be close to 0 if the process suffers large variations in the amount of executed instructions. Its initial value is 1 for all processes because it is made an assumption that all processes are stable. Logically, this value goes down if this is not proven.
$P_{comp}(i)$ of process $i$ increases or decreases depending on the prediction of the amount of performed instructions at each superstep. $PI_t(i)$ represents this prediction for superstep $t$ and process $i$. It is based on the Aging concept (Tanenbaum, 2003). For instance, $PI_t(i)$ at superstep $k+3$ needs data from supersteps $k+3$, $k+2$, $k+1$ and $k$. The Aging concept uses the idea that the prediction value is more strongly influenced by recent supersteps. The generic formula to compute the prediction $PI_t(i)$ for process $i$ and superstep $t$ is shown below.

$$PI_t(i) = \begin{cases} I_t(i) & if\ t = k \\ \frac{1}{2}PI_{t-1}(i) + \frac{1}{2}I_t(i) & if\ k < t \leq k+\alpha-1 \end{cases}$$

The advantage of this prediction scheme is that only data between two process reassignment activations (among the supersteps $k$ and $k+\alpha-1$) is used. This scheme saves memory and contributes to decrease the prediction calculation time. On the other hand, the value of $P_{comp}(i)$ persists during the BSP application execution independently of the amount of calls for reassignment. $P_{comp}(i)$ is updated following the algorithm described in Figure 5. We consider the system stable if the forecast is within a $\delta$ margin of fluctuation from the amount of instructions performed. For instance, if $\delta$ is equal to 0.1 and the number of instructions is 50, the prediction must be between 45 and 55 to increase the $P_{comp}(i)$ value.

1. **for** $t$ from superstep $k$ to superstep $k+\alpha-1$ **do**
2. **if** $PI_t(i) \geq I_t(i).(1-\delta)$ and $PI_t(i) \leq I_t(i).(1+\delta)$ **then**
3. Increases $P_{comp}(i)$ by $\frac{1}{\alpha}$ up to 1
4. **else**
5. Decreases $P_{comp}(i)$ by $\frac{1}{\alpha}$ down to 0
6. **endif**
7. **endfor**

Fig. 5. Computation Pattern $P_{comp}(i)$ of process $i$

The computation pattern $P_{comp}(i)$ is an element in the $Comp(i,j)$ function. Other element is a computation time prediction $CTP_{k+\alpha-1}(i)$ of the process $i$ at superstep $k+\alpha-1$ (last superstep executed before process rescheduling). Analogous to $PI$ prediction, $CTP$ also works with the Aging concept. Supposing that $CT_t(i)$ is the computation time of the process $i$ during superstep $t$, then the prediction $CTP_{k+\alpha-1}(i)$ is computed as follows.

$$CTP_t(i) = \begin{cases} CT_t(i) & if\ t = k \\ \frac{1}{2}CTP_{t-1}(i) + \frac{1}{2}CT_t(i) & if k < t \leq k+\alpha-1 \end{cases}$$

Finally, $Comp(i,j)$ presents an index $ISet_{k+\alpha-1}$(j). This index informs the average capacity of performance of the Set $j$ at the $k+\alpha-1^{th}$ superstep. For each processor in a Set, its load is multiplied by its theoretical capacity. Concerning this, the Set Managers compute a performance average of their Sets and exchange this value. Each manager calculates $ISet(j)$ for each Set normalizing their performance average by its own average. In the sequence, all Set Managers pass $ISet(j)$ index to the BSP processes under their jurisdiction.

$$Comp(i,j) = P_{comp}(i)\ .\ CTP_{k+\alpha-1}(i)\ .\ ISet_{k+\alpha-1}(j) \tag{3}$$

Equation 3 shows the function to calculate the Computation metric for process $i$ to Set $j$. The value of the equation is high if the BSP process presents stability on its executed instructions, has a large computation time and an efficient Set is involved. However, $Comp(i,j)$ is close to 0 if the process is unstable and/or it finishes its computation phase quickly. The model aims to migrate a delayed BSP process that presents a good behavior (amount of instructions that performs is regular) on the resource which belongs currently, because it can follow this actuation in another resource. In addition, we are considering the target Set in order to evaluate its capacity to receive a process.

### 3.3.2 Communication Metric

Communication metric is expressed through $Comm(i,j)$, where $i$ denotes a BSP process and $j$ means the target Set. This metric treats the communication (just receiving actions) involving the process $i$ and all processes that belong to Set $j$. In order to compute $Comm(i,j)$, data collected at superstep $k$ up to $k+\alpha-1$ is used. Besides this, each process maintains a communication time for a specified Set at each superstep and a pattern of communication called $P_{comm}(i,j)$. This pattern is a real number within the [0,1] interval. Its alteration depends on the prediction $PB_t(i,j)$, which deals with the number of bytes involved during receptions performed by process $i$ from sendings executed by processes that belong to Set $j$ at superstep $t$. $PB_t(i,j)$ is based on the Aging concept and is organized as follows.

$$PB_t(i,j) = \begin{cases} B_t(i,j) & if\ t = k \\ \frac{1}{2}PB_{t-1}(i,j) + \frac{1}{2}B_t(i,j) & if\ k < t \leq k+\alpha-1 \end{cases}$$

In $PB(i,j)$ context, $B_t(i,j)$ is a notation used to assign the number of received bytes by process $i$ at superstep $t$ from sendings of processes that belong to Set $j$. Figure 6 presents the algorithm which uses this prediction to compute $P_{comm}(i,j)$. This algorithm uses a variable $\beta$ which informs the acceptable variation in communication prediction. Similarly to $\delta$, if $\beta$ is 0.1 and $B_t(i,j)$ is 100, we must have our prediction between 90 and 110 in order to configure superstep $t$ as regular. $P_{comm}(i,j)$ is the first element in function $Comm(i,j)$. The second one is communication time prediction $BTP_{k+\alpha-1}(i,j)$ involving the process $i$ and Set $j$ at superstep $k+\alpha-1$. In order to compute this prediction, the communication time of receivings $BT_t(i,j)$ from process $i$ of sendings from processes that belong to Set $j$ at superstep $t$ is used. Concerning this, $CommTP_{k+\alpha-1}(i,j)$ is achieved as follows.

$$BTP_t(i) = \begin{cases} BT_t(i) & if\ \ t = k \\ \frac{1}{2}BTP_{t-1}(i) + \frac{1}{2}BT_t(i) & if\ k < t \leq k+\alpha-1 \end{cases}$$

```
for t from superstep k to superstep k + α − 1 do
    if (1 − β).B_k(i,j) ≤ PB_k(i,j) and (1 + β).B_k(i,j) ≥ PB_k(i,j) then
        Increases P_comm(i,j) by 1/α up to 1
    else
        Decreases P_comm(i,j) by 1/α down to 0
    endif
endfor
```

Fig. 6. Communication Pattern $P_{comm}(i,j)$

$$Comm(i,j) = P_{comm}(i,j)\ .\ BTP_{k+\alpha-1} \tag{4}$$

The function that computes Communication metric is presented in Equation 4. The result of Equation 4 increases if the process $i$ has a regularity considering the received bytes from processes of Set $j$ and performs slower communication actions to this Set. The value of $Comm(i,j)$ is close to 0 if process $i$ presents large variations in the amount of received data from Set $j$ and/or few (or none) communications are performed with this Set.

### 3.3.3 Memory Metric

Function $Mem(i,j)$ represents the Memory metric and evaluates the migration cost of the image of process $i$ to a resource in Set $j$. This metric just uses data collected at the superstep in which the load rebalancing will be activated (where $\alpha$ is achieved). Firstly, the memory space in bytes of considered process is captured through $M(i)$. After that, the transfer time of 1 byte to the destination Set is calculated through $T(i,j)$ function. The communication involving process $i$ is established with the Set Manager of each considered Set. Finally, the time spent on migration operations of process $i$ to Set $j$ is calculated through $Mig(i,j)$ function. These operations are dependent of operating system, as well as the tool used to offer process migration. They can include, for example, connections reorganizations, memory serialization, checkpoint recovery, time spent to create another process in the target host, and so on. However, $Mig(i,j)$ does not depend on the load of Set $j$.

$$Mem(i,j) = M(i) \cdot T(i,j) + Mig(i,j) \tag{5}$$

Equation 5 shows the elements of $Mem(i,j)$. Analyzing Memory metric, each BSP process will compute $n$ times $Mem(i,j)$, where $n$ is the number of Sets in the environment. The lower the value of $Mem(i,j)$ the easier is the transferring of process $i$ to Set $j$. On the other hand, as $Mem(i,j)$ increases, the migration cost of the process $i$ to Set $j$ increases as well.

### 3.3.4 Potential of Migration Analysis

We used the notion of force from Physics to create the Potential of Migration ($PM$) of each process. In Physics, force is an influence that can make an object accelerate and is represented by a vector. A vector has a size (magnitude) and a direction. Analyzing the force idea, each studied metric can be seen as a vector that acts over an object. In our case, this object is the migration of a process. Vectors $\vec{Comp}$ and $\vec{Comm}$ represent the Computation and Communication metrics, respectively. Both have the same direction and stimulate the process migration. On the other hand, the Memory metric means the migration costs and is symbolized by vector $\vec{Mem}$. $\vec{Mem}$ works against the migration, since its direction is opposite to $\vec{Comp}$ and $\vec{Comm}$.

$$PM(i,j) = Comp(i,j) + Comm(i,j) - Mem(i,j) \tag{6}$$

$\vec{Comp}$, $\vec{Comm}$ and $\vec{Mem}$ vectors are combined to create the resultant vector called $\vec{PM}$ (Potential of Migration). Then, $\vec{PM}$ means the resultant force that will decide if a process is a candidate for migration or not. Considering MigBSP context, $\vec{PM}$ will be denoted by $PM(i,j)$ function where $i$ means a process while $j$ represents a specific Set (see Equation 6). Thus, Figure 7 shows the actuation of Computation, Communication and Memory metrics to compute $PM$. $Comp(i,j)$, $Comm(i,j)$ and $Mem(i,j)$ represent the Computation, Communication and Memory metrics, respectively. The greater the value of $PM(i,j)$, the more prone the process will be to migrate. A high $PM(i,j)$ means that process $i$ has high computation time, high communication with processes that belong to Set $j$ and presents low migration costs to $j$.

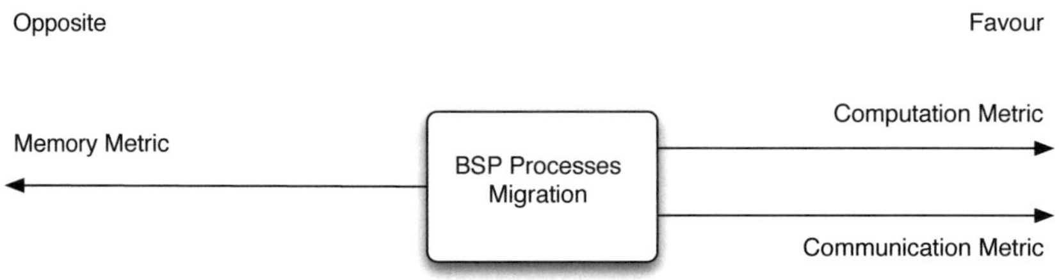


Fig. 7. Resultant force (Potential of Migration): (i) Computation and Communication metrics act in favor of migration; (ii) Memory works in the opposite direction as migration costs

Each process $i$ will compute $n$ times Equation 6, where $n$ is the amount of Sets in the environment. After that, process $i$ sends its highest Potential of Migration to its Set Manager. All

Set Managers exchange their $PM$ values. Concerning this, we applied list scheduling in order to select the candidates for migration. Each Set Manager creates a decreasing ordered list based on the highest $PM$ of each BSP process. MigBSP uses this list to apply one of two possible heuristics to select the candidates for migration. The first heuristic chooses processes that have $PM$ higher than a $MAX(PM).x$, where $MAX(PM)$ is the highest $PM$ and $x$ a percentage. The second heuristic takes one process, the first of the list, whose has the highest $PM$.

### 3.4 Analyzing Destination of Elected Processes

Process migration happens after the barrier synchronization of the superstep in which $\alpha$ is reached (see subsection 3.2). An elected process $i$ has a target Set $j$ informed on its Potential of Migration $PM(i,j)$. Thus, the pertinent question is to select which node/processor of this Set can be the destination of the process. Firstly, the Set Manager of process $i$ contacts the manager of the Set $j$ asking it for a processor to receive a process. This manager verifies the resources under its responsibility and elects the destination processor.

The manager of the destination Set calculates the time which each processor takes to compute the work assigned to it. This is performed through Equation 7. $time(p)$ captures the computation power of processor $p$ taking into account the external load (processes that do not belong to the BSP application). $load(p)$ represents the CPU load average on the last 15 minutes. This time interval was adopted based on work of (Moreno-Vozmediano & Alonso-Conde, 2005). Equation 7 also works with instruction summing of each BSP process assigned to processor $p$ in the last executed superstep. In this context, $S(i,p)$ is equal to 1 if a process $i$ is executing on processor $p$. The processor $p$ with the shortest $time(p)$ is chosen to be tested to receive a BSP process. After that, this Set Manager computes Equation 8 based on data from process $i$, as well as from its own Set.

$$time(p) = \frac{\sum\limits_{i,p:S(i,p)=1} I_{k+\alpha-1}(i)}{(1 - load(p)) \,.\, cpu(p)} \tag{7}$$

$$t1 = time(p) + B_{k+\alpha-1}(i,j) \,.\, T(i,j) \; + \; Mem(i,j) \tag{8}$$

$$t2 = time(p') + B_{k+\alpha-1}(i,j) \,.\, T(i,j) \tag{9}$$

The idea of Equation 8 is to simulate the execution of the considered process in the destination Set taking into account the migration costs. In this situation, $time(p)$ is the simulation of the execution of process $i$ on target processor $p$. In the same way, $T(i,j)$ refers to the transferring rate of 1 byte of process $i$ inside the Set $j$ (communication established with the Set Manager). $Mem(i,j)$ is the Memory Metric and is associated with the migration cost ($W_{mem}$ equal to 1) . Contrary to $time(p)$ and $T(i,j)$, $Mem(i,j)$ involves the current location of process $i$ and target Set $j$. The manager of Set $j$ sends to the manager of process $i$ the destination processor $p$ and $t1$ value. This last Set Manager computes Equation 9. This equation is used to analyze the execution of process $i$ considering its current execution. In this situation, $p'$ is the current processor of process $i$ and $T(i,j)$ means the transfer rate between the Set of process $i$ and Set $j$. On both Equations 8 and 9, $B_{k+\alpha-1(i,j)}$ is the amount of received bytes of process $i$ from sendings of processes that belong to Set $j$ at superstep $k+\alpha-1$. Process $i$ will migrate from $p'$ to $p$ if $t1 \; < \; t2$.

## 4. Evaluation Methodology

The main objective of this evaluation is to observe the changes on performance when MigBSP controls the process relocation in different scientific applications. Concerning this, we are testing MigBSP with three applications, which are listed below.

(i) Lattice Boltzmann application - It is widely used in the computational fluid dynamics area. Its algorithm may be easily adapted to a large serie of other computing areas.

(ii) Smith-Waterman application - This application is based on dynamic programming and it is characterized by the variation in the computation intensity along the matrix cells.

(iii) LU decomposition application - This application presents an algorithm where a matrix is written as the product of a lower triangular matrix and an upper triangular matrix. The decomposition is used to solve systems of linear equations.

While the first application is regular, the other two present an irregular behavior. The regularity issue treats the processes' activities at each superstep. The behaviors of the processes on the last two applications change along the execution due to fluctuations on the number of instructions and/or on the amount of communicated bytes that the processes perform at each superstep. The evaluation comprises the simulation of the applications on three scenarios.

- Scenario (*i*): Application execution simply;
- Scenario (*ii*): Application execution with scheduler without applying migrations;
- Scenario (*iii*): Application execution with scheduler allowing migrations.

Scenario *ii* consists in performing all scheduling calculus about which processes will migrate, but it does not comprise any migrations. Scenario *iii* enables migrations and adds the migrations costs on those processes that migrate from one processor to another. The difference between scenarios *ii* and *i* represents exactly the overhead imposed by MigBSP. We are using the **SimGrid Simulator** (Casanova et al., 2008) (MSG module), which makes possible application modeling and process migration. This simulator is deterministic, where a specific input always results in the same output. In addition, a time equal to $Mem(i,j)$ is paid for each migration of process $i$ to Set $j$ (see subsection 3.3). We assembled an infrastructure with five Sets, which is depicted in Figure 8. This infrastructure represents the clusters and the network connections that we have at UFRGS University, Brazil. Each node has a single processor. For the sake of simplicity, we hide the network of each cluster. Clusters Labtec, Corisco and Frontal have their nodes linked by Fast Ethernet, while ICE and Aquario use Gigabit connection. The migration costs are based on executions with AMPI (Huang et al., 2006) on our clusters.

Figure 8 also reveals the initial processes-recourses mappings. The basic idea is to fill one cluster and then to pass to another one. We map one process per node owing to each one has a single processor. If the amount of process is greater than processors, the mapping begins again from the first Set. The notation $\{(p,m)\}$ is used in the following sections and means that process $p$ is running over machine $m$ (or $p$ will migrate to $m$). Finally, the tests were executed using $\alpha$ equal to 2, 4, 8 and 16. Furthermore, we employed $\omega$ equal to 3 and initial $D$ equal to 0.5. The first application used the heuristic two to select the process for migration, while the other two employ the heuristic one to choose the candidates with $x$ equal to 80%.

## 5. Results Remarks and Discussions

This section is divided in three subsections, which explain in details the results of each tested application separately. The overall analysis of the results will be presented in the last section.

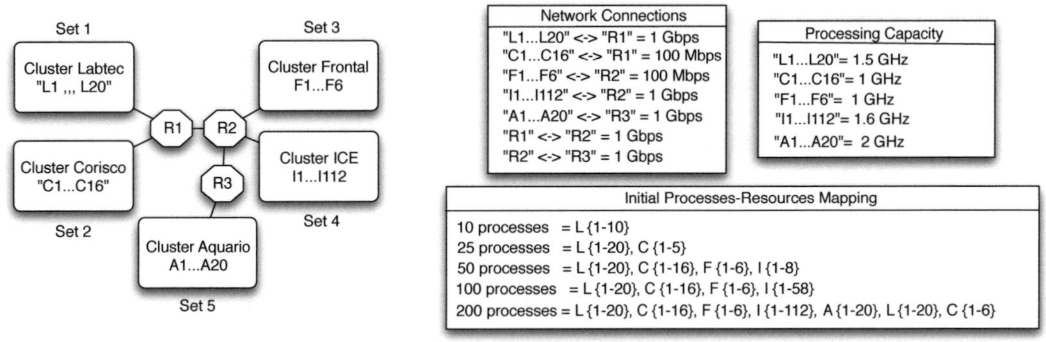


Fig. 8. Testbed infrastructure and the initial processes-resources mappings

### 5.1 Lattice Boltzmann Method

This method is a powerful technique for the computational modeling of a wide variety of complex fluid flow problems (Schepke, 2007). It models the fluid consisting of particles whose perform consecutive propagation and collision processes over a discrete lattice mesh.

#### 5.1.1 Modeling de Problem

We modeled a BSP implementation of a 2D-based Lattice Boltzmann Method to SimGrid using vertical domain decomposition. The data volume is divided into spatially contiguous blocks along one axis. Multiple copies of the same program run simultaneously, each operating on its own block of data. At the end of each iteration, data that lie on the boundaries between blocks are passed between the appropriate processes and the superstep is completed. An abstract view of the problem is illustrated in Figure 9.

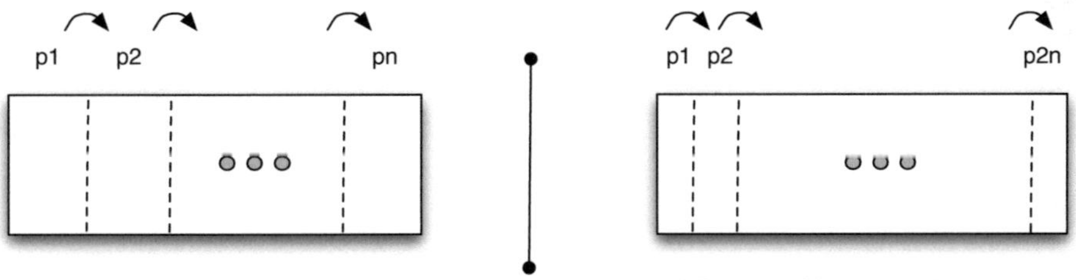


(a) Decomposition among n processes (b) Decomposition among 2n processes

Fig. 9. Different matrix partition organizations when varying the number of processes

Besides Lattice Boltzmann, the developed scheme encompasses a broad spectrum of scientific computations, from mesh based solvers, signal processing to image processing algorithms. The considered matrix requires the computation of $10^{10}$ instructions and occupies 10 Megabytes in memory. As we can observe in Figure 9, matrix partition will influence the number of instructions to be executed per process and, consequently, the size of each process in memory. Nevertheless, the quantity of communication remains the same independent of the used partition scheme. It is important to emphasize that our modeling may be characterized as regular, where each superstep presents the same number of instructions to be computed by processes as well as the same communication behavior. When using 10 processes, each one is responsible for a sub-lattice computation of $10^9$ instructions, occupies 1.5 Megabyte

(500 Kbytes is fixed to other process' data) and passes 100 Kilobytes of boundary data to its right neighbor. In the same way, when 25 processes are employed, each one computes $4.10^8$ instructions and occupies 900 Kbytes in memory.

### 5.1.2 Results and Discussions

Table 1 presents the times when testing 10 processes. Firstly, we can observe that MigBSP's intrusivity on application execution is short when comparing both scenarios *i* and *ii* (overhead lower than 5%). The processes are balanced among themselves with this configuration, causing the increasing of $\alpha$ at each call for process rescheduling. This explain the low impact when comparing scenarios *i* and *ii*. Besides this, MigBSP decides that migrations are inviable for any moment, independing on the amount of executed supersteps. In this case, our model causes a loss of performance in application execution. We obtained negative values of $PM$ when the rescheduling was tested. This fact resulted in an empty list of migration candidates.

| Super-step | Scenario *i* | $\alpha = 4$ | | $\alpha = 8$ | | $\alpha = 16$ | |
|---|---|---|---|---|---|---|---|
| | | Scen. ii | Scen. iii | Scen. ii | Scen. iii | Scen. ii | Scen. iii |
| 10 | 6.70 | 7.05 | 7.05 | 7.05 | 7.05 | 6.70 | 6.70 |
| 50 | 33.60 | 34.59 | 34.59 | 34.26 | 34.26 | 34.04 | 34.04 |
| 100 | 67.20 | 68.53 | 68.53 | 68.20 | 68.20 | 67.87 | 67.87 |
| 500 | 336.02 | 338.02 | 338.02 | 337.69 | 337.69 | 337.32 | 337.32 |
| 1000 | 672.04 | 674.39 | 674.39 | 674.06 | 674.06 | 673.73 | 673.73 |
| 2000 | 1344.09 | 1347.88 | 1347.88 | 1346.67 | 1346.67 | 1344.91 | 1344.91 |

Table 1. Evaluating 10 processes on three considered scenarios (time in seconds)

The results of the execution of 25 processes are presented in Table 2. In this context, the system remains stable and $\alpha$ grows at each rescheduling call. One migration occurred {(p21,a1)} when testing 10 supersteps and using $\alpha$ equal to 4. Our notation informs that process p21 was reassigned to run on node a1. A second and a third migrations happened when considering 50 supersteps: {(p22,a2), (p23,a3)}. They happened in the next two calls for process rescheduling (at supersteps 12 and 28). When evaluating 2000 supersteps and maintaining this value of $\alpha$, eight migrations take place: {(p21,a1), (p22,a2), (p23,a3), (p24,a4), (p25,a5), (p18,a6), (p19,a7), (p20,a8)}. We analyzed that all migrations occurred to the fastest cluster (Aquario). The first five migrations moved processes from cluster Corisco to Aquario. After that, three processes from Labtec were chosen for migration. Concluding, we obtained a profit of 14% after executing 2000 supersteps when $\alpha$ equal to 4 is used.

| Super-steps | Scenario *i* | $\alpha = 4$ | | $\alpha = 8$ | | $\alpha = 16$ | |
|---|---|---|---|---|---|---|---|
| | | Scen. ii | Scen. iii | Scen. ii | Scen. iii | Scen. ii | Scen. iii |
| 10 | 3.49 | 4.18 | 4.42 | 4.42 | 4.44 | 3.49 | 3.49 |
| 50 | 17.35 | 19.32 | 20.45 | 18.66 | 19.44 | 18.66 | 19.42 |
| 100 | 34.70 | 37.33 | 38.91 | 36.67 | 37.90 | 36.01 | 36.88 |
| 500 | 173.53 | 177.46 | 154.87 | 176.80 | 161.48 | 176.80 | 179.24 |
| 1000 | 347.06 | 351.64 | 297.13 | 350.97 | 303.72 | 350.31 | 317.96 |
| 2000 | 694.12 | 699.47 | 592.26 | 698.68 | 599,14 | 697.43 | 613.88 |

Table 2. Evaluating 25 processes on three considered scenarios (time in seconds)

Analyzing scenario *iii* with $\alpha$ equal to 16, we detected that the first migration is postponed, which results in a larger final time when compared with lower values of $\alpha$. With $\alpha$ 4 for instance, we have more calls for process rescheduling with migrations during the first supersteps. This fact will cause a large overhead to be paid during this period. These penalty costs are amortized when the amount of executed supersteps increases. Thus, the configuration with $\alpha$ 4 outperforms other studied values of $\alpha$ when 2000 supersteps are evaluated. Figure 10 illustrates the frequency of process rescheduling calls when testing 25 processes and 2000 supersteps. We can observe that 6 calls are done with $\alpha$ 16, while 8 are performed when initial $\alpha$ changes to 4. Considering scenarios *ii*, we conclude that the greater is $\alpha$, the lower is the model's impact if migrations are not applied (situation in which migration viability is false).

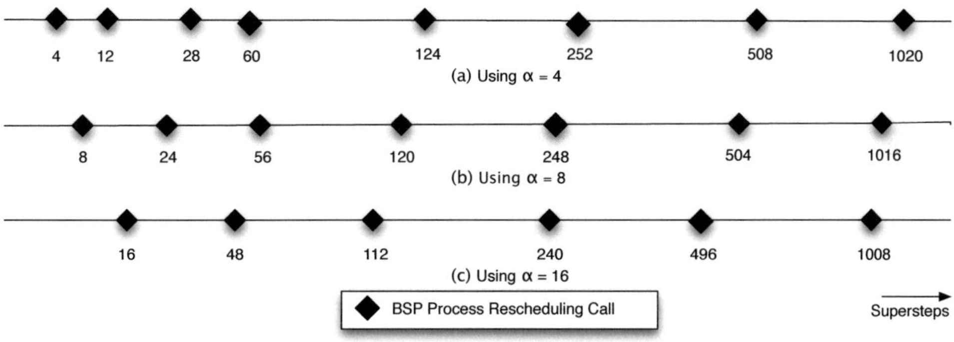


Fig. 10. Number of rescheduling calls when 25 processes and 2000 supersteps are evaluated

Table 3 shows the results when the number of processes is increased to 50. The processes are considered balanced and $\alpha$ increases at each rescheduling call. In this manner, we have the same configuration of calls when testing 25 processes (see Figure 10). We achieved 8 migrations when 2000 supersteps are evaluated: {(p38,a1), (p40,a2), (p42, a3), (p39, a4), (p41, a5), (p37, a6), (p22, a7), (p21, a8)}. MigBSP moves all processes from cluster Frontal to Aquario and transfers two process from Corisco to the fastest cluster. Using $\alpha$ 4, 430.95s and 408.25s were obtained for scenarios *i* and *iii*, respectively. Besides this 5% of gain with $\alpha$ 4, we also achieve a gain when $\alpha$ is equal to 8. However, the final result when changing initial $\alpha$ to 16 in scenario *iii* is worse than scenario *i*, since the migrations are delayed and more supersteps are need to achieve a gain in this situation. Table 4 presents the execution of 100 processes over the tested infrastructure. As the situations with 25 and 50 processes, the environment when 100 processes are evaluated is stable and the processes are balanced among the resources. Thus, $\alpha$ increases at each rescheduling call. The same migrations occurred when testing 50 and 100 processes, since the configuration with 100 just uses more nodes from cluster ICE. In general, the same percentage of gain was achieve with 50 and 100 processes.
The results of scenarios *i*, *ii* and *iii* with 200 processes is shown in Table 5. We have an unstable scenario in this situation, which explains the fact of a large overhead in scenario *ii*. Considering this scenario, $\alpha$ will begin to grow after $\omega$ calls for process rescheduling without migrations. Taking into account scenario *iii* and $\alpha$ equal to 4, 2 migrations are done when executing 10 supersteps: {(p195,a1), (p197,a2)}. Besides these, 10 migrations take place when 50 supersteps were tested: {(p196,a3), (p198,a4), (p199,a5), (p200,a6), (p38,a7), (p39,a8), (p37,a9), (p40,a10), (p41,a11), (p42, a12)}. Despite the happening of these migrations, the processes are still unbalanced with adopted value of $D$ and, then, $\alpha$ does not increase at each superstep.

| Super-steps | Scenario $i$ | $\alpha = 4$ | | $\alpha = 8$ | | $\alpha = 16$ | |
|---|---|---|---|---|---|---|---|
| | | Scen. ii | Scen. iii | Scen. ii | Scen. iii | Scen. ii | Scen. iii |
| 10 | 2.16 | 2.95 | 3.20 | 2.95 | 3.17 | 2.16 | 2.16 |
| 50 | 10.78 | 13.14 | 14.47 | 12.35 | 13.32 | 12.35 | 13.03 |
| 100 | 21.55 | 24.70 | 26.68 | 29.91 | 25.92 | 23.13 | 24.63 |
| 500 | 107.74 | 112.46 | 106.90 | 111.67 | 115.73 | 111.67 | 117.84 |
| 1000 | 215.48 | 220.98 | 199.83 | 220.19 | 207.78 | 219.40 | 226.43 |
| 2000 | 430.95 | 436.79 | 408.25 | 435.88 | 417.56 | 434.68 | 434.30 |

Table 3. Evaluating 50 processes on three considered scenarios (time in seconds)

| Super-steps | Scenario $i$ | $\alpha = 4$ | | $\alpha = 8$ | | $\alpha = 16$ | |
|---|---|---|---|---|---|---|---|
| | | Scen. ii | Scen. iii | Scen. ii | Scen. iii | Scen. ii | Scen. iii |
| 10 | 1.22 | 2.08 | 2.24 | 2.08 | 2.21 | 1.22 | 1.22 |
| 50 | 5.94 | 8.59 | 9.63 | 7.71 | 8.48 | 7.71 | 8.19 |
| 100 | 11.86 | 15.40 | 16.99 | 14.52 | 16.24 | 13.63 | 14.94 |
| 500 | 59.25 | 64.57 | 62.55 | 63.68 | 67.25 | 63.68 | 69.37 |
| 1000 | 118.48 | 124.69 | 113.87 | 123.80 | 119.06 | 122.92 | 129.46 |
| 2000 | 236.96 | 243.70 | 224.48 | 241.12 | 232.87 | 239.23 | 241.52 |

Table 4. Evaluating 100 processes on three considered scenarios (time in seconds)

| Super-steps | Scenario $i$ | $\alpha = 4$ | | $\alpha = 8$ | | $\alpha = 16$ | |
|---|---|---|---|---|---|---|---|
| | | Scen. ii | Scen. iii | Scen. ii | Scen. iii | Scen. ii | Scen. iii |
| 10 | 1.04 | 2.86 | 3.06 | 1.95 | 2.11 | 1.04 | 1.04 |
| 50 | 5.09 | 10.56 | 17.14 | 9.65 | 11.06 | 7.82 | 8.15 |
| 100 | 10.15 | 16.53 | 25.43 | 15.62 | 21.97 | 14.71 | 16.04 |
| 500 | 50.66 | 57.84 | 68.44 | 56.93 | 71.42 | 55.92 | 77.05 |
| 1000 | 101.29 | 108.78 | 102.59 | 107.84 | 106.89 | 105.25 | 117.57 |
| 2000 | 200.43 | 209.46 | 194.87 | 208.13 | 202.22 | 204.69 | 211.69 |

Table 5. Evaluating 200 processes on three considered scenarios (time in seconds)

After these migrations, MigBSP does not indicate the viability of other ones. Thus, after $\omega$ calls without migrations, MigBSP enlarges the value of $D$ and $\alpha$ begins to increase following adaptation 2 (see Subsection 3.2 for details).

| Processes | Scenario $i$ - Without process migration | Scenario $iii$ - With process migration |
|---|---|---|
| 10 | 0.005380s | 0.005380s |
| 25 | 0.023943s | 0.010765s |
| 50 | 0.033487s | 0.025360s |
| 100 | 0.036126s | 0.028337s |
| 200 | 0.043247s | 0.031440s |

Table 6. Barrier times on two situations

Table 6 presents the barrier times captured when 2000 supersteps were tested. More especially, the time is captured when the last superstep is executed. We implemented a centralized master-slave approach for barrier, where process 1 receives and sends a scheduling message from/to other BSP processes. Thus, the barrier time is captured on process 1. The times shown in the third column of Table 6 do not include both scheduling messages and computation. Our idea is to demonstrate that the remapping of processes decreases the time to compute the BSP supersteps. Therefore, process 1 can reduce the waiting time for barrier computation since the processes reach this moment faster. Analyzing such table, we observed that a gain of 22% in time was achieved when comparing barrier time on scenarios *i* and *iii* with 50 processes. The gain was reduced when 100 processes were tested. This occurs because we just include more nodes from cluster ICE with 100 processes if compared with the execution of 50 processes.

### 5.2 Smith-Waterman Application

Our second application is based on dynamic programming (DP), which is a popular algorithm design technique for optimization problems (Low et al., 2007). DP algorithms can be classified according to the matrix size and the dependency relationship of each matrix cell. An algorithm for a problem of size $n$ is called a $tD/eD$ algorithm if its matrix size is $O(n^t)$ and each matrix cell depends on $O(n^e)$ other cells. $2D/1D$ algorithms are all irregular with changes on load computation density along the matrix's cells. In particular, we observed the Smith-Waterman algorithm that is a well-known 2D/1D algorithm for local sequence alignment (Smith, 1988).

#### 5.2.1 Modeling the Problem

Smith-Waterman algorithm proceeds in a series of wavefronts diagonally across the matrix. Figure 11 (a) illustrates the concept of the algorithm for a 4×4 matrix with a column-based processes allocation. The more intense the shading, the greater is the load computation density of the cell. Each wavefront corresponds to a BSP superstep. For instance, Figure 11 (b) shows a 4×4 matrix that presents 7 supersteps. The computation load is uniform inside a particular superstep, growing up when the number of the superstep increases. Both organizations of diagonal-based supersteps mapping and column-based processes mapping bring the following conclusions: (i) $2n-1$ supersteps are crossed to compute a square matrix with order $n$ and; (ii) each process will be involved on $n$ supersteps. Figures 11 (b) and (c) show the communication actions among the processes. Considering that cell $x, y$ ($x$ means a matrix' line, while $y$ is a matrix' column) needs data from the $x, y-1$ and $x-1, y$ other ones, we will have an interaction from process $py$ to process $py+1$. We do not have communication inside the same matrix column, since it corresponds to the same process.
The configuration of scenarios *ii* and *iii* depends on the Computation Pattern $P_{comp}(i)$ of each process $i$ (see Subsection 3.3 for more details) . $P_{comp}(i)$ increases or decreases depending on the prediction of the amount of performed instructions at each superstep. We consider a specific process as regular if the forecast is within a $\delta$ margin of fluctuation from the amount of instructions performed actually. In our experiments, we are using $10^6$ as the amount of instructions for the first superstep and $10^9$ for the last one. The increase of load computational density among the supersteps is uniform. In other words, we take the difference between $10^9$ and $10^6$ and divide by the number of involved supersteps in a specific execution. Considering this, we applied $\delta$ equal to 0.01 (1%) and 0.50 (50%) to scenarios *ii* and *iii*, respectively. This last value was used because $I_2(1)$ is $565.10^5$ and $PI_2(1)$ is $287.10^5$ when a 10×10 matrix is tested (see details about the notations in Subsection 3.3). The percentage of 50% enforces instruction regularity in the system. Both values of $\delta$ will influence the Computation metric, and consequently the choosing of candidates for migration. Scenario ii tends to obtain negatives values for $PM$ since the Computation Metric will be close to 0. Consequently, no migrations will

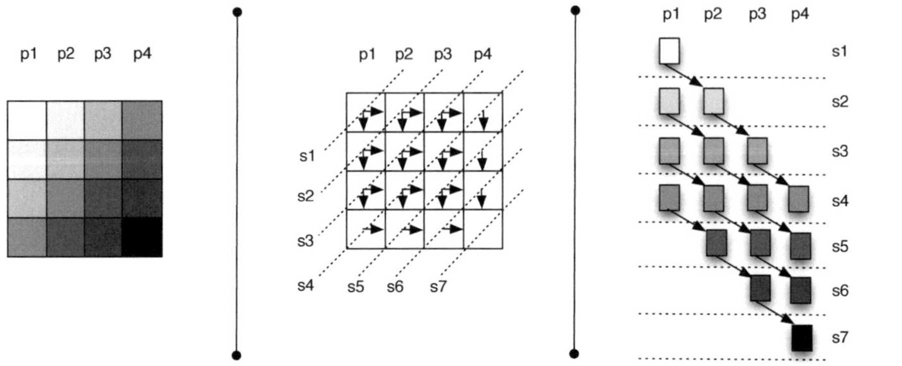


(a) Computational load density along the matrix (b) Mapping of supersteps and communications among the cells (c) Emphasizing that the load is uniform inside a particular superstep

Fig. 11. Different views of Smith-Waterman irregular application

happen on this scenario. We tested the behavior of square matrixes of order 10, 25, 50, 100 and 200. Each cell of a 10×10 matrix needs to communicate 500 Kbytes and each process occupies 1.2 Mbyte in memory (700 Kbytes comprise other application data). The cell of 25×25 matrix communicates 200 Kbytes and each process occupies 900 Kbytes in memory and so on.

### 5.2.2 Results and Discussions

Table 7 presents the application evaluation. Nineteen supersteps were crossed when a 10×10 matrix was tested. Adopting this size of matrix and $\alpha$ 2, 13.34s and 14.15s were obtained for scenarios *i* and *ii* which represents a cost of 8%. The higher is the value of $\alpha$, the lower is the MigBSP overhead on application execution. This occurs because the system is stable (processes are balanced) and $\alpha$ always increases at each rescheduling call. Three calls for process relocation were done when testing $\alpha$ 2 (at supersteps 2, 6 and 14). The rescheduling call at superstep 2 does not produce migrations. At this step, the load computational density is not enough to overlap the consider migration costs involved on process transferring operation. The same occurred on the next call at superstep 6. The last call happened at superstep 14, which resulted on 6 migrations: {(p5,a1), (p6,a2), (p7,a3), (p8,a4), (p9,a5), (p10,a6)}. MigBSP indicated the migration of processes that are responsible to compute the final supersteps. The execution with $\alpha$ equal to 4 implies in a shorter overhead since two calls were done (at supersteps 4 and 12). Observing scenario *iii*, we do not have migrations in the first call, but eight occurred in the other one. Processes 3 up to 10 migrated in this last call to cluster Aquario. $\alpha$ 4 outperforms $\alpha$ 2 for two reasons: (i) it does less rescheduling calls and; (ii) the call that causes process migration was done at a specific superstep in which MigBSP takes better decisions.
The system stays stable when the 25×25 matrix was tested. $\alpha$ 2 produces a gain of 11% in performance when considering 25×25 matrix and scenario *iii*. This configuration presents four calls for process rescheduling, where two of them produce migrations. No migrations are indicated at supersteps 2 and 6. Nevertheless, processes 1 up to 12 are migrated at superstep 14 while processes 21 up to 25 are transferred at superstep 30. These transferring operations occurred to the fastest cluster. In this last call, the remaining execution presents 19 supersteps (from 31 to 49) to amortize the migration costs and to get better performance. The execution when considering $\alpha$ 8 and scenario *iii* brings an overhead if compared with scenario *i*. Two calls for migrations were done, at supersteps 8 and 24. The first call causes

| Scenarios | | Order of considered matrices | | | | |
|---|---|---|---|---|---|---|
| | | 10×10 | 25×25 | 50×50 | 100×100 | 200×200 |
| Scenario i | | 13.34s | 40.74s | 92.59s | 162.66s | 389.91s |
| Scenario ii | $\alpha = 2$ | 14.15s | 43.05s | 95.70s | 166.57s | 394.68s |
| | $\alpha = 4$ | 14.71s | 42.24s | 94.84s | 165.66s | 393.75s |
| | $\alpha = 8$ | 13.78s | 41.63s | 94.03s | 164.80s | 392.85s |
| | $\alpha = 16$ | 13.42s | 41.28s | 93.36s | 164.04s | 392.01s |
| Scenario iii | $\alpha = 2$ | 13.09s | 35.97s | 85.95s | 150.57 | 374.62s |
| | $\alpha = 4$ | 11.94s | 34.82s | 84.65s | 148.89s | 375.53s |
| | $\alpha = 8$ | 13.82s | 41.64s | 83.00s | 146.55s | 374.38s |
| | $\alpha = 16$ | 12.40s | 40.64s | 85.21s | 162.49s | 374.40s |

Table 7. Evaluation of scenarios *i*, *ii* and *iii* when varying the matrix size

the migration of just one process (number 1) to a1 and the second one produces three migrations: {(p21,a2),(p22,a3),(p23,a4)}. We observed that processes p24 and p25 stayed on cluster Corisco. Despite performed migrations, these two processes compromise the supersteps that include them. Both are executing on a slower cluster and the barrier waits for the slowest process. Maintaining the matrix size and adopting $\alpha$ 16, we have two calls: at supersteps 16 and 48. This last call migrates p24 an p25 to cluster Aquario. Although this movement is pertinent to get performance, just one superstep is executed before ending the application.

Fifty processes were evaluated when the 50×50 matrix was considered. In this context, $\alpha$ also increases at each call for process rescheduling. We observed that an overhead of 3% was found when scenario *i* and *ii* were compared (using $\alpha$ 2). In addition, we observed that all values of $\alpha$ achieved a gain of performance in scenario *iii*. Especially when $\alpha$ 2 was used, five calls for process rescheduling were done (at supersteps 2, 6, 14, 30 and 62). No migrations are indicated in the first three calls. The greater is the matrix size, the greater is the amount of supersteps needed to make migrations viable. This happens because our total load is fixed (independent of the matrix size) but the load partition increases uniformly along the supersteps (see Section 4 for details). Process 21 up to 29 are migrated to cluster Aquario at superstep 30, while process 37 up to 42 are migrated to this cluster at superstep 62. Using $\alpha$ equal to 4, 84.65s were obtained for scenario *iii* which results a gain of 9%. This gain is greater than that achieved with $\alpha$ 2 because now the last rescheduling call is done at superstep 60. The same processes were migrated at this point. However, there are two more supersteps to execute using $\alpha$ equal to 4. Three rescheduling calls were done with $\alpha$8 (at supersteps 8, 24 and 56). Only the last two produce migration. Three processes are migrated at superstep 24: {(p21,a1),(p22,a2),(p23,a3)}. Process 37 up to 42 are migrated to cluster Aquario at superstep 56. This last call is efficient since it transfers all processes from cluster Frontal to Aquario.

The execution with a 100×100 matrix shows good results with process migration. Six rescheduling calls were done when using $\alpha$ 2. Migrations did not occur at the first three supersteps (2, 6 and 14). Process 21 up to 29 are migrated to cluster Aquario after superstep 30. In addition, process 37 to 42 are migrated to cluster Aquario at superstep 62. Finally, superstep 126 indicates 7 migrations, but just 5 occurred: p30 up to p36 to cluster Aquario. These migrations complete one process per node on cluster Aquario. MigBSP selected for migration those processes that belonged to cluster Corisco and Frontal, which are the slowest clusters on our infrastructure testbed. $\alpha$ equal to 16 produced 3 attempts for migration when a 100×100 matrix is evaluated (at supersteps 16, 48 and 112). All of them triggered migrations. In the first call, the $11^{th}$ first processes are migrated to cluster Aquario. All process from cluster Frontal are migrated to Aquario at superstep 48. Finally, 15 processes are selected as candidates for migration after crossing 112 supersteps. They are: p21 to p36. This spectrum of candidates is equal to the processes that are running on Frontal. Considering this, only 3 processes were migrated actually: {(p34,a18),(p35a19),(p36,a20)}.

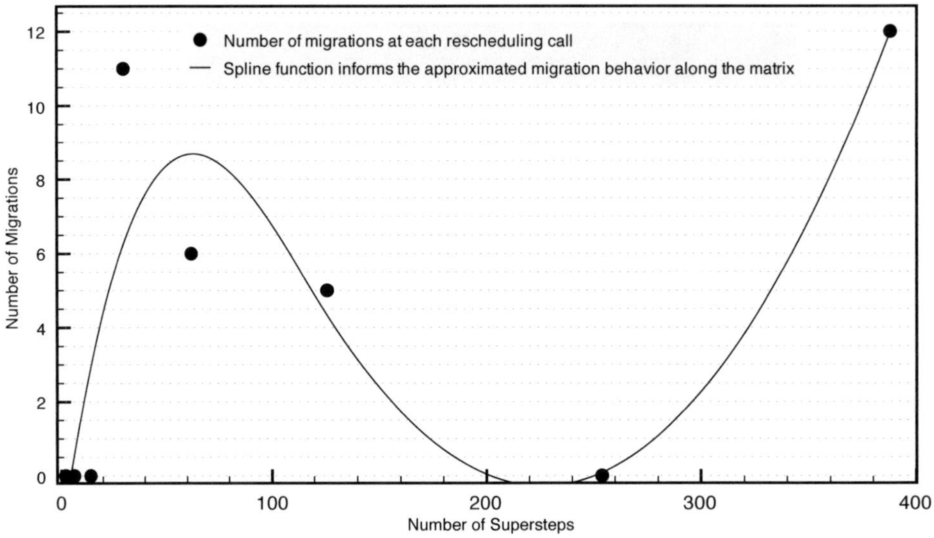


Fig. 12. Migration behavior when testing a 200 $\times$ 200 matrix with initial $\alpha$ equal to 2

Table 7 also shows the application performance when the 200×200 matrix was tested. Satisfactory results were obtained with process migration. The system stays stable during all application execution. Despite having more than one process mapped to one processor, sometimes just a portion of them is responsible for computation at a specific moment. This occurs because the processes are mapped to matrix columns, while supersteps comprise the anti-diagonals of the matrix. Figure 12 illustrates the migrations behavior along the execution with $\alpha$ 2. Using $\alpha$ 2 and considering scenario *iii*, 8 calls for process rescheduling were done. Migrations were not done at supersteps 2, 6 and 14. Processes 21 up to 31 are migrated to cluster Aquario at superstep 30. Moreover, all processes from cluster Frontal are migrated to Aquario at superstep 62. Six processes are candidates for migration at superstep 126: p30 to p36. However, only p31 up to p36 are migrated to cluster Aquario. These migrations happen because the processes initially mapped to cluster Aquario do not collaborate yet with BSP computation. Migrations are not viable at superstep 254. Finally, 12 processes (p189 to p200) are migrated to cluster Aquario when superstep 388 was crossed. At this time, all previous processes allocated to Aquario are inactive and the migrations are viable. However, just 10 remaining supersteps are executed to amortize the process migration costs.

### 5.3 LU Decomposition Application

Consider a system of linear equations $A.x = b$, where $A$ is a given $n \times n$ non singular matrix, $b$ a given vector of length $n$, and $x$ the unknown solution vector of length $n$. One method for solving this system is by using the LU Decomposition technique. It comprises the decomposition of the matrix $A$ into a lower triangular matrix $L$ and an upper triangular matrix $U$ such that $A = LU$. A $n \times n$ matrix $L$ is called unit lower triangular if $l_{i,i} = 1$ for all $i, 0 \le i < n$, and $l_{i,j} = 0$ for all $i, j$ where $0 \le i < j < n$. An $n \times n$ matrix $U$ is called upper triangular if $u_{i,j} = 0$ for all $i, j$ with $0 \le j < i < n$.

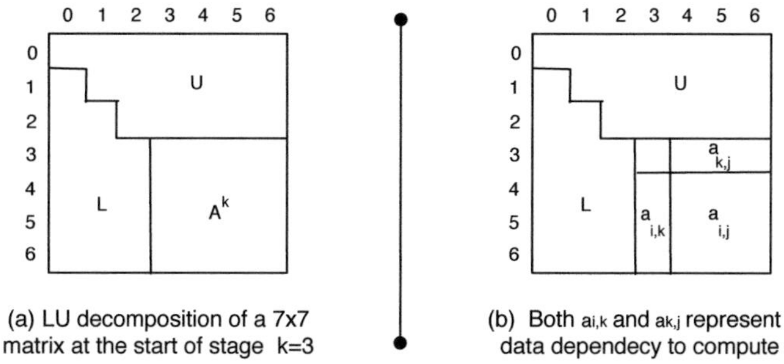


(a) LU decomposition of a 7x7 matrix at the start of stage k=3

(b) Both $a_{i,k}$ and $a_{k,j}$ represent the data dependecy to compute $a_{i,j}$

Fig. 13. L and U matrices with the same memory space of the original matrix $A^0$

**for** $k$ from 0 to $n-1$ **do**
  **for** $j$ from $k$ to $n-1$ **do**
    $u_{k,j} = a^k_{k,j}$
  **endfor**
  **for** $i$ from $k+1$ to $n-1$ **do**
    $l^k_{i,k} = \frac{a^k_{i,k}}{u_{k,k}}$
  **endfor**
  **for** $i$ from $k+1$ to $n-1$ **do**
    **for** $j$ from $k+1$ to $n-1$ **do**
      $a^{k+1}_{i,j} = a^k_{i,j} - l_{i,k} \cdot u_{k,j}$
    **endfor**
  **endfor**
**endfor**

**for** $k$ from 0 to $n-1$ **do**
  **for** $i$ from $k+1$ to $n-1$ **do**
    $a_{i,k} = \frac{a_{i,k}}{a_{k,k}}$
  **endfor**
  **for** $i$ from $k+1$ to $n-1$ **do**
    **for** $j$ from $k+1$ to $n-1$ **do**
      $a_{i,j} = a_{i,j} - a_{i,k} \cdot a_{k,j}$
    **endfor**
  **endfor**
**endfor**

Fig. 14. Two algorithms to solve the LU Decomposition problem

On input, $A$ contains the original matrix $A^0$, whereas on output it contains the values of $L$ below the diagonal and the values of $U$ above and on the diagonal such that $LU = A^0$. Figure 13 illustrates the organization of LU computation. The values of $L$ and $U$ computed so far and the computed sub-matrix $A^k$ may be stored in the same memory space of $A^0$. Figure 14 presents the sequential algorithm for producing $L$ and $U$ in stages. Stage $k$ first computes the elements $u_{k,j}$, $j \ge k$, of row $k$ of $U$ and the elements $l_{i,k}$, $i > k$, of column $k$ of $L$. Then, it computes $A^{k+1}$ in preparation for the next stage. Figure 14 also presents in the right side the functioning of the previous algorithm using just the elements from matrix $A$. Figure 13 (b) presents the data that is necessary to compute $a_{i,j}$. Besides its own value, $a_{i,j}$ is updated using a value from the same line and another from the same column.

### 5.3.1 Modeling the Problem

This section explains how we modeled the LU sequential application on a BSP-based parallel one. Firstly, the bulk of the computational work in stage $k$ of the sequential algorithm is the

modification of the matrix elements $a_{i,j}$ with $i, j \geq k+1$. Aiming to prevent communication of large amounts of data, the update of $a_{i,j} = a_{i,j} + a_{i,k}.a_{k,j}$ must be performed by the process whose contains $a_{i,j}$. This implies that only elements of column $k$ and row $k$ of $A$ need to be communicated in stage $k$ in order to compute the new sub-matrix $A^k$. An important observation is that the modification of the elements in row $A(i, k+1 : n-1)$ uses only one value of column $k$ of $A$, namely $a_{i,k}$. The provided notation $A(i, k+1 : n-1)$ denotes the cells of line $i$ varying from column $k+1$ to $n-1$. If we distribute each matrix row over a limit set of $N$ processes, then the communication of an element from column $k$ can be restricted to a multicast to $N$ processes. Similarly, the change of the elements in $A(k+1 : n-1, j)$ uses only one value from row $k$ of $A$, namely $a_{k,j}$. If we divide each column over a set of $M$ processes, the communication of an element of row $k$ can be restricted to a multicast to $M$ processes.
We are using a Cartesian scheme for the distribution of matrices. The square cyclic distribution is used since it is particularly suitable for matrix computations (Bisseling, 2004). Thus, it is natural to organize the processes by two-dimensional identifiers $P(s,t)$ with $0 \leq s < M$ and $0 \leq t < N$, where the number of processes $p = M.N$. Figure 15 depicts a $6 \times 6$ matrix mapped to 6 processes, where $M = 2$ and $N = 3$. Assuming that $M$ and $N$ are factors of $n$, each process will store $nc$ (number of cells) cells in memory (see Equation 10).

$$nc = \frac{n}{M}.\frac{n}{N} \tag{10}$$

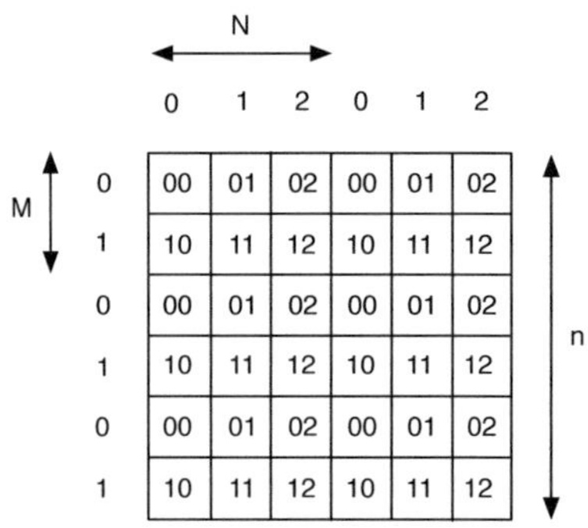


Fig. 15. Cartesian distribution of a matrix over $2 \times 3$ ($M \times N$) processes

A parallel algorithm uses data parallelism for computations and the need-to-know principle to design the communication phase of each superstep. Following the concepts of BSP, all communication performed during a superstep will be completed when finishing it and the data will be available at the beginning of the next superstep (Bonorden, 2007). Concerning this, we modeled our algorithm using three kinds of supersteps. They are explained in Table 8. The element $a_{k,k}$ is passed to the process that computes $a_{i,k}$ in the first kind of superstep.
The computation of $a_{i,k}$ is expressed in the beginning of the second superstep. This superstep is also responsible for sending the elements $a_{i,k}$ and $a_{k,j}$ to $a_{i,j}$. First of all, we pass the element $a_{i,k}$, $k+1 \leq i < n$, to the $N-1$ processes that execute on the respective row $i$. This kind of superstep also comprises the passing of $a_{k,j}$, $k+1 \leq j < n$, to $M-1$ processes which execute on the respective column $j$. The superstep 3 considers the computation of $a_{i,j}$, the increase of $k$ (next stage of the algorithm) and the transmission of $a_{k,k}$ to $a_{i,k}$ elements ($k+1 \leq i < n$). The application will execute one superstep of type 1 and will follow with the interleaving of supersteps 2 and 3. Thus, a $n \times n$ matrix will trigger $2n+1$ supersteps in our LU modeling. We

| Type of superstep | Steps and explanation |
|---|---|
| First | Step 1.1 : $k = 0$ |
| | Step 1.2 - Pass the element $a_{k,k}$ to cells which will compute $a_{i,k}$ $(k+1 \le i < n)$ |
| Second | Step 2.1 : Computation of $a_{i,k}$ $(k+1 \le i < n)$ by cells which own them |
| | Step 2.2 : For each $i$ $(k+1 \le i < n)$, pass the element $a_{i,k}$ to other $a_{i,j}$ elements in the same line $(k+1 \le j < n)$ |
| | Step 2.3 : For each $j$ $(k+1 \le j < n)$, pass the element $a_{k,j}$ to other $a_{i,j}$ elements in the same column $(k+1 \le i < n)$ |
| Third | Step 3.1 : For each $i$ and $j$ $(k+1 \le i,j < n)$, calculate $a_{i,j}$ as $a_{i,j} + a_{i,k}.a_{k,j}$ |
| | Step 3.2 : $k = k+1$ |
| | Step 3.3 : Pass the element $a_{k,k}$ to cells which will compute $a_{i,k}$ $(k+1 \le i < n)$ |

Table 8. Modeling three types of supersteps for LU computation

modeled the Cartesian distribution $M \times N$ in the following manner: $5 \times 5$, $10 \times 5$, $10 \times 10$ and $20 \times 10$ for 25, 50, 100 and 200 processes, respectively. Moreover, we are applying simulation over square matrices with orders 500, 1000, 2000 and 5000. Lastly, the tests were executed using $\alpha = 4$, $\omega = 3$, $D = 0.5$ and $x = 80\%$.

### 5.3.2 Results and Discussions

Table 9 presents the results when evaluating LU application. The tests with the first matrix size show the worst results. Formerly, the higher the number of processes, the worse the performance, as we can observe in scenario *i*. The reasons for the observed times are the overheads related to communication and synchronization. Secondly, MigBSP indicated that all migration attempts were not viable due to low computing and communication loads when compared to migration costs. Considering this, both scenarios *ii* and *iii* have the same results.

| Processes | 500×500 matrix | | | 1000×1000 matrix | | | 2000×2000 matrix | | |
|---|---|---|---|---|---|---|---|---|---|
| | *i* | *ii* | *iii* | *i* | *ii* | *iii* | *i* | *ii* | *iii* |
| 25 | 1.68 | 2.42 | 2.42 | 11.65 | 13.13 | 10.24 | 90.11 | 91.26 | 76.20 |
| 50 | 2.59 | 3.54 | 3.34 | 10.10 | 11.18 | 9.63 | 60.23 | 61.98 | 54.18 |
| 100 | 6.70 | 7.81 | 7.65 | 15.22 | 16.21 | 16.21 | 48.79 | 50.25 | 46.87 |
| 200 | 13.23 | 14.89 | 14.89 | 28.21 | 30.46 | 30.46 | 74.14 | 76.97 | 76.97 |

Table 9. First results when executing LU linked to MigBSP (time in seconds)

When testing a $1000 \times 1000$ matrix with 25 processes, the first rescheduling call does not cause migrations. After this call at superstep 4, the next one at superstep 11 informs the migration of 5 processes from cluster Corisco. They were all transferred to cluster Aquario, which has the highest computation power. MigBSP does not point migrations in the future calls. $\alpha$ always increases its value at each rescheduling call since the processes are balanced after the mentioned relocations. MigBSP obtained a gain of 12% of performance with 25 processes when comparing scenarios *i* and *iii*. With the same size of matrix and 50 processes, 6 processes from Frontal were migrated to Aquario at superstep 9. Although these migrations are profitable,

they do not provide stability to the system and the processes remain unbalanced among the resources. Migrations are not viable in the next 3 calls at supersteps 15, 21 and 27. After that, MigBSP launches our second adaptation on rescheduling frequency in order to alleviate its impact and $\alpha$ begins to grow until the end of the application. The tests with 50 processes obtained gains of just 5% with process migration. This is explained by the fact that the computational load is decreased in this configuration when compared to the one with 25 processes. In addition, the bigger the number of the superstep, the smaller the computational load required by it. Therefore, the more advanced the execution, the lesser the gain with migrations. The tests with 100 and 200 processes do not present migrations owing to the forces that act in favor of migration are weaker than the Memory metric in all rescheduling calls.
The execution with a $2000 \times 2000$ matrix presents good results because the computational load is increased. We observed a gain of 15% with process relocation when testing 25 processes. All processes from cluster Corisco were migrated to Aquario in the first rescheduling call (at superstep 4). Thus, the application can take profit from this relocation in its beginning, when it demands more computations. The time for concluding the LU application is reduced when passing from 25 to 50 processes as we can see in scenario *i*. However, the use of MigBSP resulted in lower gains. Scenario *i* presented 60.23s while scenario *iii* achieved 56.18s (9% of profit). When considering 50 processes, 6 processes were transferred from cluster Frontal to Aquario at superstep 4. The next call occurs at superstep 9, where 16 processes from cluster Corisco were elected as migration candidates to Aquario. However, MigBSP indicated the migration of 14 processes, since there were only 14 unoccupied processors in the target cluster.

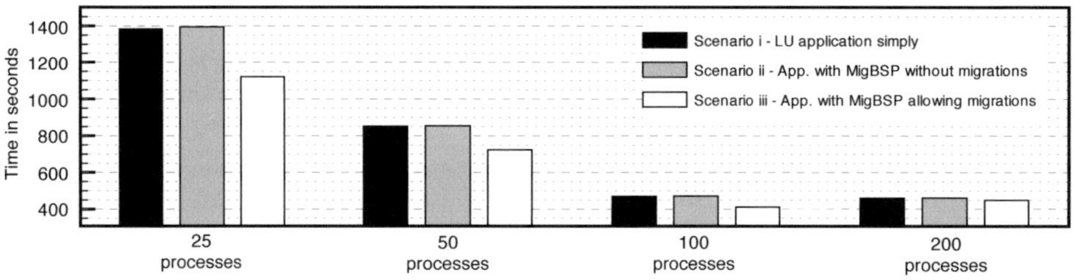


Fig. 16. Performance graph with our three scenarios for a $5000 \times 5000$ matrix

We observed that the higher the matrix order, the better the results with process migration. Considering this, the evaluation of a $5000 \times 5000$ matrix can be seen in the Figure 16. The simple movement of all processes from cluster Corisco to Aquario represented a gain of 19% when executing 25 processes. The tests with 50 processes obtained 852.31s and 723.64s for scenario *i* and *iii*, respectively. The same migration behavior found on the tests with the $2000 \times 2000$ matrix was achieved in Scenario *iii* However, the increase of matrix order represented a gain of 15% (order 5000) instead of 10% (order 2000). This analysis helps us to verify our previous hypothesis about performance gains when enlarging the matrix. Finally, the tests with 200 processes indicated the migration of 6 processes (p195 up to p200) from cluster Corisco to Aquario at superstep 4. Thus, the nodes that belong to Corisco just execute one BSP process while the nodes from Aquario begin to treat 2 processes. The remaining rescheduling calls inform the processes from Labtec as those with the higher values of *PM*. However, their migrations are not considered profitable. The final execution with 200 processes achieved 460.85s and 450.33s for scenarios *i* and *iii*, respectively.

## 6. Conclusion

Scheduling schemes for multi-programmed parallel systems can be viewed in two levels (Frachtenberg & Schwiegelshohn, 2008). In the first level processors are allocated to a job. In the second level processes from a job are (re)scheduled using this pool of processors. MigBSP can be included in this last scheme, offering algorithms for load (BSP processes) rebalancing among the resources during the application runtime. In the best of our knowledge, MigBSP is the pioneer model on treating BSP process rescheduling with three metrics and adaptations on remapping frequency. These features are enabled by MigBSP at middleware level, without changing the application code.

Considering the spectrum of the three tested applications, we can take the following conclusions in a nutshell: (i) the larger the computing grain, the better the gain with processes migration; (ii) MigBSP does not indicate the migration of those processes that have high migration costs when compared to computation and communication loads; (iii) MigBSP presented a low overhead on application execution when migrations are not applied; (v) our tests prioritizes migrations to cluster Aquario since it is the fastest one among considered clusters and tested applications are CPU-bound and; (vi) MigBSP does not work with previous knowledge about application. Considering this last topic, MigBSP indicates migrations even when the application is close to finish. In this situation, these migrations bring an overhead since the remaining time for application conclusion is too short to amortize their costs.

The results showed that MigBSP presented a low overhead on application execution. The calculus of the PM (Potential of Migration) as well as our efficient adaptations were responsible for this feature. PM considers processes and Sets (different sites), not performing all processes-resources tests at the rescheduling moment. Meanwhile, our adaptations were crucial to enable MigBSP as a viable scheduler. Instead of performing the rescheduling call at each fixed interval, they manage a flexible interval between calls based on the behavior of the processes. The concepts of the adaptations are: (*i*) to postpone the rescheduling call if the system is stable (processes are balanced) or to turn it more frequent, otherwise; (*ii*) to delay this call if a pattern without migrations in $\omega$ calls is observed.

## 7. References

Bhandarkar, M. A., Brunner, R. & Kale, L. V. (2000). Run-time support for adaptive load balancing, *IPDPS '00: Proceedings of the 15 IPDPS 2000 Workshops on Parallel and Distributed Processing*, Springer-Verlag, London, UK, pp. 1152–1159.

Bisseling, R. H. (2004). *Parallel Scientific Computation: A Structured Approach Using BSP and MPI*, Oxford University Press.

Bonorden, O. (2007). Load balancing in the bulk-synchronous-parallel setting using process migrations., *21th International Parallel and Distributed Processing Symposium (IPDPS 2007)*, IEEE, pp. 1–9.

Bonorden, O., Gehweiler, J. & auf der Heide, F. M. (2005). Load balancing strategies in a web computing environment, *Proceeedings of International Conference on Parallel Processing and Applied Mathematics (PPAM)*, Poznan, Poland, pp. 839–846.

Casanova, H., Legrand, A. & Quinson, M. (2008). Simgrid: A generic framework for large-scale distributed experiments, *Tenth International Conference on Computer Modeling and Simulation (uksim)*, IEEE Computer Society, Los Alamitos, CA, USA, pp. 126–131.

Casavant, T. L. & Kuhl, J. G. (1988). A taxonomy of scheduling in general-purpose distributed computing systems, *IEEE Trans. Softw. Eng.* **14**(2): 141–154.

Chen, L., Wang, C.-L. & Lau, F. (2008). Process reassignment with reduced migration cost in grid load rebalancing, *Parallel and Distributed Processing, 2008. IPDPS 2008. IEEE International Symposium on* pp. 1–13.

Du, C., Ghosh, S., Shankar, S. & Sun, X.-H. (2004). A runtime system for autonomic rescheduling of mpi programs, *ICPP '04: Proceedings of the 2004 International Conference on Parallel Processing*, IEEE Computer Society, Washington, DC, USA, pp. 4–11.

Du, C., Sun, X.-H. & Wu, M. (2007). Dynamic scheduling with process migration, *CCGRID '07: Proceedings of the Seventh IEEE International Symposium on Cluster Computing and the Grid*, IEEE Computer Society, Washington, DC, USA, pp. 92–99.

Frachtenberg, E. & Schwiegelshohn, U. (2008). New Challenges of Parallel Job Scheduling, *Job Scheduling Strategies for Parallel Processing* **4942**: 1–23.

Heiss, H.-U. & Schmitz, M. (1995). Decentralized dynamic load balancing: the particles approach, *Inf. Sci. Inf. Comput. Sci.* **84**(1-2): 115–128.

Hernandez, I. & Cole, M. (2007). Scheduling dags on grids with copying and migration., *in* R. Wyrzykowski, J. Dongarra, K. Karczewski & J. Wasniewski (eds), *PPAM*, Vol. 4967 of *Lecture Notes in Computer Science*, Springer, pp. 1019–1028.

Huang, C., Zheng, G., Kalé, L. & Kumar, S. (2006). Performance evaluation of adaptive mpi, *PPoPP '06: Proceedings of the eleventh ACM SIGPLAN symposium on Principles and practice of parallel programming*, ACM Press, New York, NY, USA, pp. 12–21.

Kondo, D., Casanova, H., Wing, E. & Berman, F. (2002). Models and scheduling mechanisms for global computing applications, *IPDPS '02: Proceedings of the 16th International Symposium on Parallel and Distributed Processing*, IEEE Computer Society, Washington, DC, USA, p. 79.2.

Low, M. Y.-H., Liu, W. & Schmidt, B. (2007). A parallel bsp algorithm for irregular dynamic programming, *Advanced Parallel Processing Technologies, 7th International Symposium*, Vol. 4847 of *Lecture Notes in Computer Science*, Springer, pp. 151–160.

Milanés, A., Rodriguez, N. & Schulze, B. (2008). State of the art in heterogeneous strong migration of computations, *Concurr. Comput. : Pract. Exper.* **20**(13): 1485–1508.

Moreno-Vozmediano, R. & Alonso-Conde, A. B. (2005). Influence of grid economic factors on scheduling and migration., *High Performance Computing for Computational Science - VECPAR*, Vol. 3402 of *Lecture Notes in Computer Science*, Springer, pp. 274–287.

Sánchez, A., Pérez, M. S., Montes, J. & Cortes, T. (2010). A high performance suite of data services for grids, *Future Gener. Comput. Syst.* **26**(4): 622–632.

Schepke, Claudio; Maillard, N. (2007). Performance improvement of the parallel lattice boltzmann method through blocked data distributions, *19th International Symposium on Computer Architecture and High Performance Computing, 2007. SBAC-PAD 2007*, pp. 71–78.

Smith, J. M. (1988). A survey of process migration mechanisms, *SIGOPS Oper. Syst. Rev.* **22**(3): 28–40.

Tanenbaum, A. (2003). *Computer Networks*, 4th edn, Prentice Hall PTR, Upper Saddle River, New Jersey.

Utrera, G., Corbalan, J. & Labarta, J. (2005). Dynamic load balancing in mpi jobs, *The 6th International Symposium on High Performance Computing*.

Vadhiyar, S. S. & Dongarra, J. J. (2005). Self adaptivity in grid computing: Research articles, *Concurr. Comput. : Pract. Exper.* **17**(2-4): 235–257.

Valiant, L. G. (1990). A bridging model for parallel computation, *Commun. ACM* **33**(8): 103–111.

# Reliability modeling and analysis of flexible manufacturing cells

Mehmet Savsar
*Kuwait University*
*Kuwait*

## 1. Introduction

During the past half century, market competition has been very intense and production companies have been trying to find more efficient ways to manufacture their products. While during the period 1960 to 1970 manufacturing cost was the primary concern, later it was followed by product quality and delivery speed. New strategies had to be formulated by companies to adapt to the environment in which they operate, to be more flexible in their operations, and to satisfy different market segments. As a result of these efforts, a new manufacturing technology, called Flexible Manufacturing Systems (FMS), was innovated. FMS is a philosophy, in which "systems" is the key concept. A system view is incorporated into manufacturing. FMS is also one way that manufacturers are able to achieve agility, to have fastest response to the market, to operate with the lowest total cost, and to gain the greatest skills in satisfying the customers.

Today flexibility means to produce reasonably priced customized products of high quality that can be quickly delivered to customers. With respect to manufacturing, flexibility could mean the capability of producing different products without major retooling; ability to change an old line to produce new products; or the ability to change a production schedule to handle multiple parts. From customer's point of view, flexibility is the ability to have flexible speed of delivery. With respect to operations, flexibility means the ability to efficiently produce highly customized and unique products. With respect to capacity, flexibility means the ability to rapidly increase or decrease production levels or to shift capacity quickly from one product or service to another. Finally, strategically flexibility means the ability of a company to offer a wide variety of products to its customers. In a manufacturing system, machine flexibility, material handling flexibility, and operation flexibility are the important aspects considered.

The idea of an FMS was proposed in England in 1960s under the name "System 24", which was a flexible machining system that could operate without human operators 24 hours a day under computer control. Initial emphasis was on automation rather than the reorganization of workflow. Initial FMS were very large and complex, consisting of many Computer Numerical Controlled (CNC) machines and sophisticated material handling systems, such as robots, automated guided vehicles (AGV) and automated pallets, all controlled by complex software. Part and tool handling robots could handle any family of parts for which the system had been designed and developed. Only a limited number of industries could afford investing in a highly expensive FMS as described above. However, the current trend is toward small versions of the traditional FMS, called flexible manufacturing cells (FMC) or flexible manufacturing modules (FMM). Today one or more CNC machines served by one or more robots and a pallet system are considered a flexible cell and two ore more cells are considered as a flexible manufacturing system. Other related systems are Flexible Assembly Cells (FAC), Flexible Manufacturing Groups (FMG), Flexible Production Systems (FPS), and Flexible Manufacturing Lines (FML).

A basic FMC consists of a robot, one or more flexible machines including inspection, and an external material handling system such as an automated pallet for moving blanks and finished parts into and out of the cell. The robot is utilized for internal material handling which includes machine loading and unloading. The FMC is capable of doing different operations on a variety of parts, which usually form a part family with selection by a group technology approach. Chan and Bedworth (1990) indicated that the most feasible approach to automate a production system with flexibility is to initially incorporate small FMC into the system. This approach requires lower investment, less risk, and also satisfies many of the benefits gained through larger and more costly structures, such as flexible manufacturing systems (FMS). While FMS are very expensive and generally require investments in millions of dollars, FMC are less costly, smaller and less complex systems. Therefore, for smaller companies with restricted capital resources, a gradual integration is initiated with limited investment in a small FMC, which facilitates subsequent integration into a larger system, such as an FMS.

Machining cells are widely used in industry to process a variety of parts to achieve high productivity in production environments with rapidly changing product structures and customer demand. They offer flexibility to be adapted to the changes in operational requirements. There are various types of Flexible Manufacturing Cells (FMC) incorporated into Flexible Manufacturing Systems (FMS) with a variety of flexible machines for discrete part machining. In addition to discrete part machining systems, there are different types of assembly machines and CNC punching press systems, which are also configured as flexible cells. FMS and FMC performance depends on several operational and system characteristics, which may include part scheduling and system operational characteristics. In the past, most of the FMC related research has been in the areas of part scheduling and system control. Scheduling algorithms are developed to determine the best processing sequence of parts to optimize FMC performance and equipment utilization. It has also been realized that system characteristics, such as design configuration and operation of an FMC have significant effect on its performance. Machining rate, pallet capacity, robot and pallet speed, and equipment failure and repair rates are important system characteristics affecting FMC performance. Several models have been developed for FMS and FMC in relation to the effects of different parameters on system performance. Wang and Wan (1993) studied the dynamic reliability of a FMS based on fuzzy information. Yuanidis et al. (1994) used a heuristic procedure called group method of data handling to asses FMS reliability with minimal data available. Han et al. (2006) analyzed FMC reliability through the method of fuzzy fault tree based on triangular fuzzy membership. Khodabandehloo and Sayles (2007) investigated the applicability of fault tree analysis and event tree analysis to production reliability in FMS and concluded that event tree analysis was more effective in solving this problem. Henneke and Choi (1990), Savsar and Cogun (1993), and Cogun and Savsar (1996) have presented stochastic and simulation models for evaluating the performance of FMC and FMS with respect to system configuration and component speeds, such as machining rate, robot and pallet speeds. Koulamas (1992) and Savsar (2000) have looked into the reliability and maintenance aspects and presented stochastic models for the FMC, which operate under stochastic environment with tool failure and replacement consideration. They developed Markov models to study the effects of tool failures on system performance measures for a FMC with a single machine served by a robot for part loading/unloading and a pallet for part transfers. There are several other studies related to the reliability analysis of manufacturing systems. Butler and Rao (1993) use symbolic logic to analyze reliability of complex systems. Their heuristic approach is based on artificial intelligence and expert systems. Black and Mejabi (1995) have used object oriented simulation modeling to study reliability of complex manufacturing equipment. They presented a hierarchical approach to model complex systems.

Simeu-Abazi, et. al. (1997) uses decomposition and iterative analysis of Markov chains to obtain numerical solutions for the reliability and dependability of manufacturing systems. Adamyan and He (2002) presented a methodology to identify the sequences of failures and probability of their occurrences in an automated manufacturing system. They used Petri nets and reachability trees to develop a model for sequential failure analysis in manufacturing systems. Aldaihani and Savsar (2005a) and Savsar (2008) presented a stochastic model and numerical solutions for a reliable FMC with two machines served by a single robot. Savsar and Aldaihani (2004) and Savsar and Aldaihani (2008) have developed stochastic models for unreliable FMC systems with two unreliable machines served by a robot and a pallet system. Aldaihani and Savsar (2005b) and Aldaihani and Savsar (2008) have presented stochastic models and numerical solutions for performance analysis of an unreliable FMC with two unreliable machines served by two robots and a pallet. These performance measures are compared to the previous results obtained for the FMC with a single robot. Abdulmalek, Savsar, and Aldaihani (2004) presented a simulation model and analysis for tool change policies in a FMC with two machines and a robot, based on ARENA simulation software. Closed form analytical solutions are obtained and FMC analysis is performed for different performance measures and selected cell operations. The results are also compared to reliable FMC system.

This chapter summarizes several stochastic models and results for reliability analysis of FMC systems with single machines and multiple machines served by one or two robots for loading and unloading of parts; and a pallet handling device for moving batch of parts into and out of the cell. Because flexible manufacturing cells are designed to process a wide variety of parts, they have relatively high utilizations compared to traditional machining systems. As a result of high utilizations, these systems are subject to failures more than traditional systems. Therefore, reliability and availability analysis of FMC systems are extremely important for flexible manufacturing systems. The model and the results presented in this chapter can be useful for design engineers as well as operational managers in production and maintenance planning.

## 2. Operation of a Flexible Manufacturing Cell

Operations of two FMC systems are illustrated in Figure 1. In case of two machine FMC system, operation sequence is as follows: An automated pallet handling system delivers n blanks consisting of different parts into the cell. The robot reaches to the pallet, grips a blank, moves to the first machine and loads the blank. While the machine starts operation on the part, the robot reaches the pallet, grips the second part and moves to the second machine and loads it. Next, robot reaches to the machine which finishes its operation first, unloads the finished part and loads a new part. The loading/unloading operation continues in this way with the preference given to the machine which finishes its operation first. After the machining operations of all parts on the pallet are completed, the pallet with n finished parts moves out and a new pallet with n blanks is delivered into the cell by the pallet handling system automatically. In case of the FMC with a single machine, robot loads the machine with a blank and waits until the part is completed; then unloads the finished part and loads a new blank. The operation sequence continues in this manner. Machines are assumed to be unreliable and fail during the operations. Time to failure and time to repair are assumed to follow exponential distribution. Due to the introduction of different parts into the FMC, failures of machines, and random characteristics of system operation, processing times as well as loading/unloading times are random, which present a complication in studying and modeling FMC operations. If there were no randomness in system parameters and the pallet exchange times were neglected, the problem could be analyzed by a man-machine assignment chart for non-identical machines, and by a symbolic formulation for identical machines. However, because of random operations the system needs to be modeled by a stochastic process.

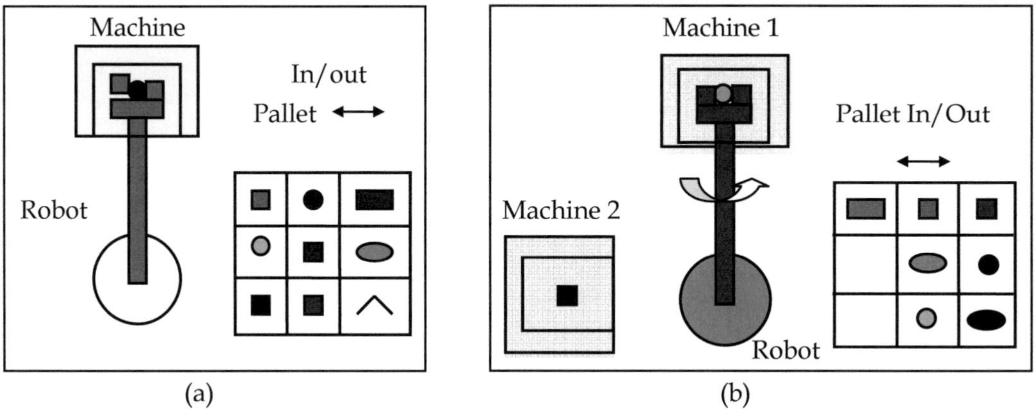


Fig. 1. Flexible manufacturing cells: (a) one machine, a robot and a pallet; and (b) two machines, a robot and a pallet

## 3. Reliability Modeling of a FMC with a Single Machine

In this section, stochastic models are presented for the FMC system with a single machine as discussed above and illustrated in Figure 1a. A reliability model and related analysis are presented for the FMC system with a single machine and a robot. Processing times on the machine, robot loading and unloading times, pallet transfer times and the equipment up and down times are all assumed as random quantities that follow exponential distribution. The model is applied to a case example and the results are presented in graphical forms.

### 3.1 A Stochastic Model for a FMC with a Single Machine

In order to model FMC operations, the following system states and notations are defined:

$S_{ijk}(t)$ = state of the FMC at time t

$P_{ijk}(t)$ = probability that the system will be in state $S_{ijk}(t)$

i = number of blanks in FMC (on the pallet and on the machine or the robot gripper)

j = state of the production machine (j=0 if the machine is idle; j=1 if the machine is operating on a part; and j=d if the machine is down under repair)

k= state of the robot (k=1 if the robot is loading/unloading the machine; k=0 if the robot is not engaged in loading/unloading the machine; and k=d if the robot is down under repair).

$\iota$ = loading rate of the robot (parts/unit time)

u = unloading rate of the robot (parts/unit time)

z = combined loading/unloading rate of the robot (parts/unit time)

$\omega$ = pallet transfer rate (pallets/unit time)

$\lambda$ = failure rate of the production machine ($1/\lambda$ = mean time between machine failures)

$\mu$ = repair rate of the production machine ($1/\mu$ = mean machine repair time)

$\alpha$ = failure rate of the robot

$\beta$ = repair rate of the robot

$\nu$ = machining rate (or production rate) of the machine (parts/unit time)

n = pallet capacity (number of parts/pallet)

$Q_c$ = production output rate of the cell in terms of parts/unit time

Using the state probability definitions and the notations above, the stochastic transition flow diagram of the unreliable FMC operation, with machine tool and robot failures, is shown in Figure 2. Using the fact that the *net flow* rate at each state is equal to the difference between the rates of *flow in* and *flow out*, the following system of differential equations are constructed for the unreliable FMC with machine and robot failures. While robot failures are not as significant as machine failures, they are incorporated into the model in the last column of figure 2 as $S_{ijd}$.

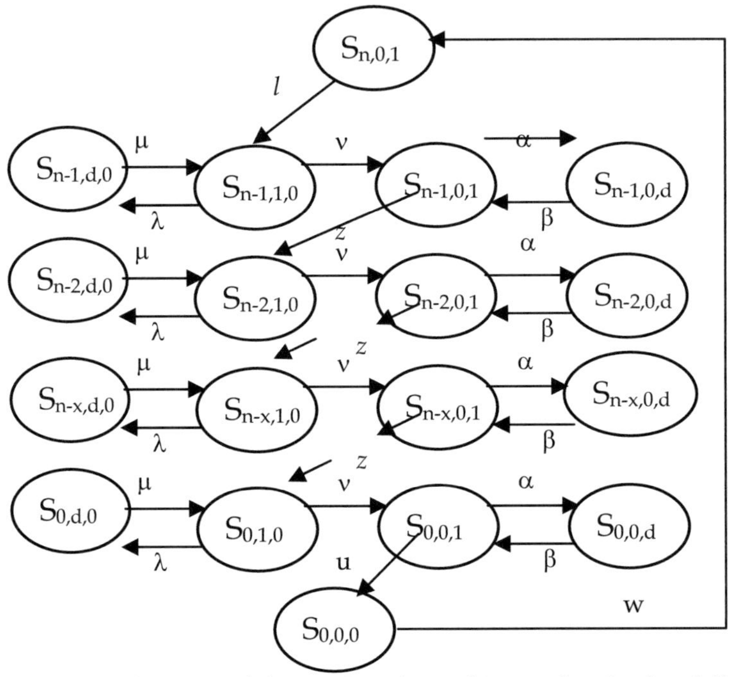


Fig. 2. Stochastic transition diagram of the FMC with machine tool and robot failures

$$dP_{n,0,0}(t)/dt = \omega P_{0,0,0} - \iota P_{n,0,0}$$
$$dP_{n-1,d,0}(t)/dt = \lambda P_{n-1,1,0} - \mu P_{n-1,d,0}$$
$$dP_{n-1,1,0}(t)/dt = \iota P_{n,0,0} + \mu P_{n-1,d,0} - (\lambda + \nu) P_{n-1,1,0}$$
$$dP_{n-1,0,1}(t)/dt = \nu P_{n-1,1,0} + \beta P_{n-1,0,d} - (\alpha + z) P_{n-1,0,1}$$
$$dP_{n-1,0,d}(t)/dt = \alpha P_{n-1,0,1} - \beta P_{n-1,0,d}$$

......................................................

$$dP_{n-x,d,0}(t)/dt = \lambda P_{n-x,1,0} - \mu P_{n-x,d,0}$$
$$dP_{n-x,1,0}(t)/dt = z P_{n-x+1,0,1} + \mu P_{n-x,d,0} - (\lambda + \nu) P_{n-x,1,0}$$
$$dP_{n-x,0,1}(t)/dt = \nu P_{n-x,1,0} + \beta P_{n-x,0,d} - (\alpha + z) P_{n-x,0,1}$$
$$dP_{n-x,0,d}(t)/dt = \alpha P_{n-x,0,1} - \beta P_{n-x,0,d}$$

......................................................

$$dP_{0,d,0}(t)/dt = \lambda P_{0,1,0} - \mu P_{0,d,0}$$
$$dP_{0,1,0}(t)/dt = z P_{1,0,1} + \mu P_{0,d,0} - (\lambda + \nu) P_{0,1,0}$$
$$dP_{0,0,1}(t)/dt = \nu P_{0,1,0} + \beta P_{0.0.d} - (\alpha + u) P_{0,0,1}$$
$$dP_{0,0,d}(t)/dt = \alpha P_{0,0,1} - \beta P_{0,0,d}$$
$$dP_{0,0,0}(t)/dt = u P_{0,0,1} - \omega P_{0,0,0} \tag{1}$$

For the steady state solution, we let $t \to \infty$ and thus $dP(t)/dt \to 0$ in the equation set (1) above. The resulting set of difference equations are solved by using the fact that sum of all state probabilities is 1;

$$\sum_{i=0}^{n}\sum_{j=0}^{d}\sum_{k=0}^{d} P_{ijk} = 1 \tag{2}$$

The following general solution set given by equation (3) is obtained for the state probabilities.

$$\begin{aligned}
P_{i,1,0} &= (\omega / \nu) P_{0,0,0}, \quad && i=0,\dots,n-1;\\
P_{i,d,0} &= (\lambda\omega / \mu\nu) P_{0,0,0}, \quad && i=0,\dots,n-1;\\
P_{i,0,1} &= (\omega / z) P_{0,0,0}, \quad && i=1,\dots,n-1;\\
P_{0,0,1} &= (\omega / u) P_{0,0,0};\\
P_{i,0,d} &= (\alpha\omega / \beta z) P_{0,0,0}, \quad && i=1,\dots,n-1;\\
P_{0,0,d} &= (\alpha\omega / \beta u) P_{0,0,0}\\
P_{n,0,0} &= (\omega / \iota) P_{0,0,0};
\end{aligned} \tag{3}$$

where, $s=\lambda/\mu$; $r=\alpha/\beta$ and,

$$P_{0,0,0} = 1/\{n\omega\nu^{-1}(1+s) + \omega(1+r)(nz^{-1} + u^{-1} - z^{-1}) + \omega\iota^{-1} + 1\} \tag{4}$$

System performance is measured by the utilization rate of the production machine ($L_m$), utilization rate of the robot ($L_r$) and utilization rate of the pallet handling system ($L_h$). These measures are calculated by using the system state probabilities determined above. $P_{0,0,0}$ represents the utilization rate of the pallet handling system. It is fraction of the time that handling system is loading/unloading a pallet at a rate of $\omega$ pallets/unit time or $n\omega$ parts/unit time. Thus,

$$L_h = P_{0,0,0} \tag{5}$$

Similarly, utilization rate of the machine is the fraction of time that the machine is operational, and is given by:

$$L_m = \sum_{i=0}^{n-1} P_{i,1,0} = (n\omega/\nu)P_{0,0,0} \tag{6}$$

and utilization rate of the robot is the fraction of time that the robot is operational given by:

$$L_r = P_{n,0,0} + \sum_{i=1}^{n-1} P_{i,0,1} + P_{0,0,1} = [\omega/\iota + (n-1)\omega/z + \omega/u]P_{0,0,0} \quad (7)$$

The above model is for the unreliable cell with machine tool and robot failures. For the reliable FMC without machine and robot failures, system states corresponding to $S_{i,d,0}$ and $S_{i,0,d}$, where i=0,1,...,n-1, are not applicable. A procedure similar to the above could be applied to the rest of the transition diagram and the utilization rates of the reliable FMC components could be obtained. However, an easier way is to use the fact that a reliable FMC is a system with no failures, i.e. $\lambda=0$ and $\alpha=0$. Thus, setting $s=\lambda/\mu=0$ and $r=\alpha/\beta=0$ in Equations 5-7, the following set of equations (8-10) are easily obtained for the reliable FMC.

$$L_h = P_{0,0,0} = 1/[1+n\omega/\nu+(n-1)\,\omega/z + (u+\iota)\omega/u\iota] \quad (8)$$

$$L_m = \frac{n\omega}{\nu} P_{0,0,0} \quad (9)$$

$$L_r = [(n-1)\omega/z + \omega/\iota + \omega/u]\, P_{0,0,0} \quad (10)$$

Production output rate of the cell, $Q_c$, is defined as the number of parts processed by the machine per unit time. It is obtained for both, reliable and unreliable cells as follows:

$$Q_c = L_m\nu = (n\omega/\nu)P_{0,0,0}\nu = n\omega P_{0,0,0} \quad (11)$$

Equations (5-11) are easily used to determine the utilization rates of the cell components, as well as the production output rate of the cell, for both, reliable and unreliable cell systems. It is interesting to note that the ratios $L_m/L_h$ and $L_r/L_h$ are the same for reliable and unreliable cells. This can be easily verified by substituting the corresponding values and determining the ratios:

$L_m/L_h = n\omega/\nu$ is the same for both reliable and unreliable cells. Similarly,

$L_r/L_h = (n-1)\omega/z + \omega/\iota + \omega/u$ is also the same for reliable and unreliable cells.

The implications of these results are that failures of system components have no effects on the two ratios given above. The functional relationships or the proportionality rates are the same regardless of the cell reliability. In other words, *relative utilization rates* of the machine and the robot remain constant regardless of the degree of reliability introduced. In order to illustrate application of the stochastic model, a case example is solved with the model and the results are presented in the next section.

### 3.2 Case Example for a Single-Machine FMC

A case example has been selected with the following FMC parameters in order to illustrate the application of the model. The results are presented in graphical forms.

Followings are the assumed mean values for various cell parameters:

Operation time per part = $\nu^{-1}$ = 4 time units

Robot loading time (for the first part) = $\iota^{-1}$ = 1/6 time units

Robot loading/unloading time for subsequent parts = $z^{-1}$ = 1/3 time units
Robot unloading time for the last part = $u^{-1}$ = 1/6 time units
Time between machine tool failures = $\lambda^{-1}$ = 100 time units
Repair time (down time) of the machine tool = $\mu^{-1}$ = 10 time units
Time between robot failures = $\alpha^{-1}$ = Assumed to be zero for this case.
Repair time (down time) of the robot = $\beta^{-1}$ = Assumed to be zero for this case.
Pallet transfer time = $\omega^{-1}$ = 4 time units per pallet
Pallet capacity, n, has been varied from 1 to 20 parts/pallet.

Utilization rates of the production machine, the robot, and the pallet handling systems are compared for the reliable and unreliable FMC with component failures in order to visualize the effects of these failures on the utilization of different components for different pallet capacities. Figure 3 illustrates the utilization rate of the production machine. As it can be seen from this figure, machine tool utilization is highly affected by the pallet capacity up to a certain level and stabilizes thereafter. However, there is significant gap between fully reliable cell and the unreliable cell, with specified component hazard rates. Decrease in machine tool utilization is directly reflected in cell productivity. The mentioned gap increases with increasing pallet capacity. Production output rate of the cell, $Q_c$, is obtained by multiplying the machine tool utilization with the average production output rate. For example, in case of the pallet capacity of 20 parts/pallet, production output rate of the fully reliable cell would be about $Q_c = L_m\nu$ =(0.88)(1/4)=0.22 parts/time unit, while the production output rate of the unreliable cell would be about (0.75)(1/4) = 0.19 parts/time unit. Note that, since the average processing time is 4 time units, the average output rate is 1/4 = 0.25 parts/time unit if the machine is fully utilized. Figure 4 shows the percentage of time the machine would be down due to failures. Reliable cell has zero percentage in this case. Figure 5 shows the percent of time machine is idle with respect to pallet capacity. Reliable cell has slightly higher idle time as compared to unreliable cell, but the trend is very similar. Figure 6 shows robot utilization for both reliable and unreliable cases. Robot utilization for reliable cell is much higher than that for unreliable cell due to low utilization of the machine. Figure 7 shows the pallet utilizations, which is almost the same for reliable and unreliable cell. Figure 8 shows the production output rate of the FMC as a function of pallet capacity. There is a significant difference in production rates between the reliable and unreliable cells. The results that are shown in these figures with respect to the effects of pallet capacity on various FMC performance measures, may seem to be obvious; however, exact effects of specific parameters on various FMC performance measures can not be predicted without formulation and solution of the models presented in this chapter. These models are useful for design engineers and operational managers for analysis of FMC systems operating under different conditions.

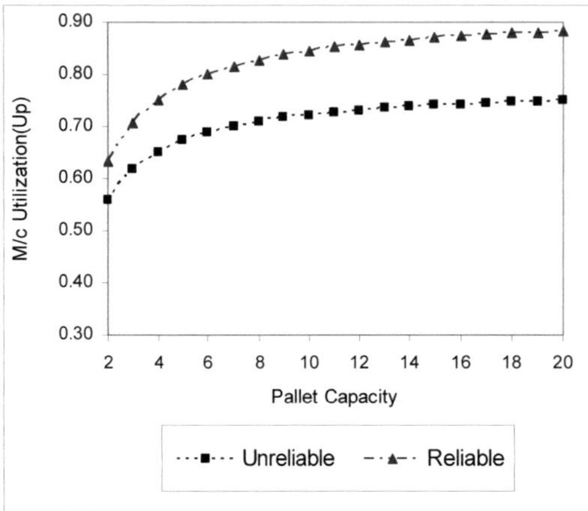


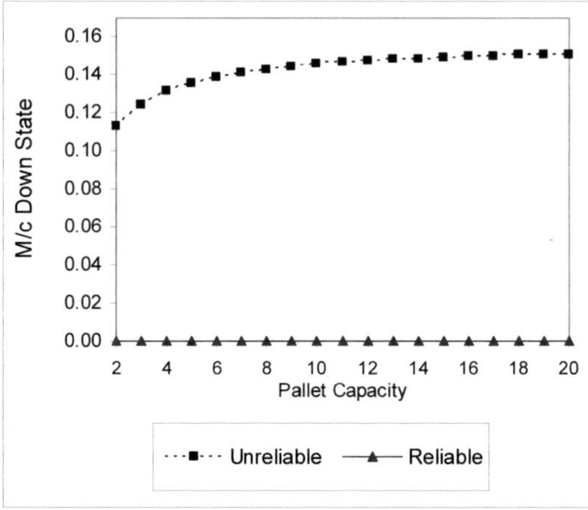


Fig. 3. Effects of pallet capacity on machine utilization
Fig. 4. Effects of pallet capacity on machine down state

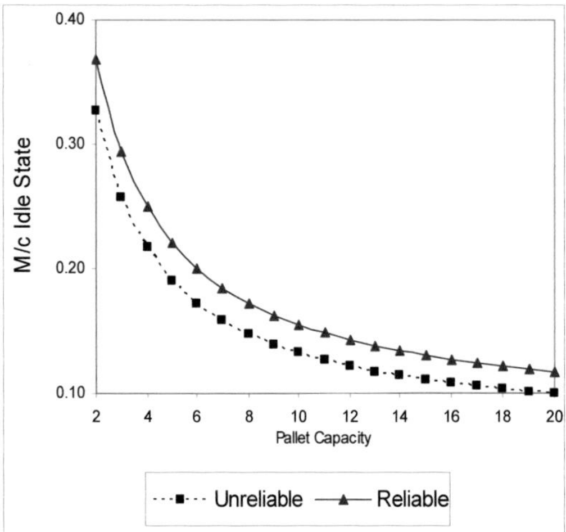


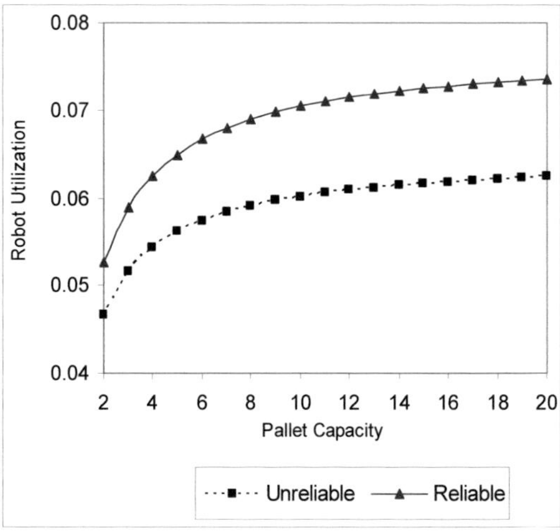


Fig. 5. Effects of pallet capacity on machine idle state
Fig. 6. Effects of pallet capacity on robot utilization

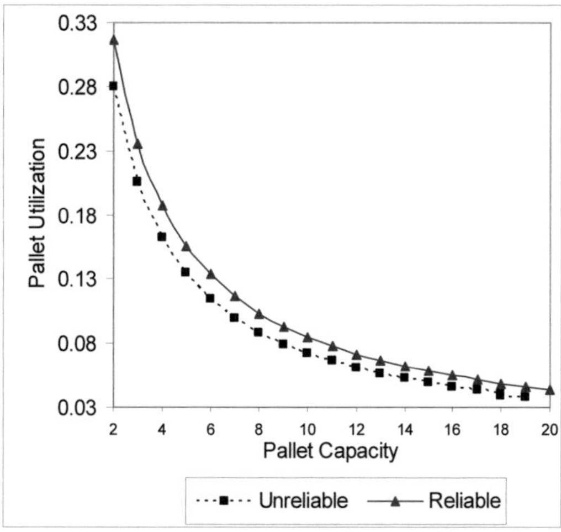


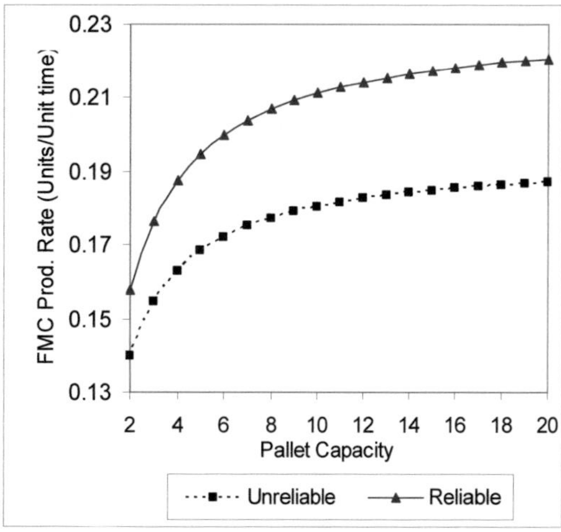


Fig. 7. Effects of pallet capacity on pallet utilization
Fig. 8. Effects of pallet capacity on FMC Production rate

### 3.3 Economic Analysis of a Single-Machine FMC

In order to demonstrate application of the stochastic model for single-machine FMC to cost analysis and optimization of the system, the following notations and cost equations are developed and a case example is solved to illustrate the results.

$C_m$=Total machine cost per unit time; $C_{mf}$=Fixed machine cost per unit of time; $C_{mv}$= Variable machine cost per unit time; $C_r$= Total robot cost per unit time; $C_{rf}$= Fixed robot cost per unit time; $C_{rv}$ = Variable robot cost per unit time;
$C_p$= Total pallet cost per unit time; $C_{pf}$= Fixed pallet cost per unit time; $C_{pv}$ =Variable pallet cost per unit time.

$$C_m = C_{mf} + C_{mv} * v_1 \tag{12}$$

$$C_r = C_{rn} + C_{rv} * z_i \tag{13}$$

$$C_p = C_{pf} C_{pv} * n \tag{14}$$

Total FMC cost per unit of production, TC, is given by the following equation, where $Q_c$ is production rate (units produced per unit time).

$$TC = (C_m + C_r + C_p)/Q_c \tag{15}$$

In order to illustrate behavior of the system with respect to various cost measures, a case problem with specified cost and speed parameters are selected as follows: z = 3; $C_{mf}$ = 1.0; $C_{mv}$ = 0.2; $C_{rf}$ = 0.108; $C_{pv}$ = 0.054; $C_{pf}$= 0.108; $C_{rv}$ = 0.054. Other parameters are as given in section 2. 2. Figure 9 shows the behavior of FMC cost per unit of production as function of pallet capacity for the reliable and unreliable FMC operations. Optimum occurs at n=4 and n=3 for the reliable and unreliable FMC systems respectively. The trend in cost is almost the same in both cases. Total costs show a decreasing pattern with increasing pallet capacity with optimum pallet capacity ranging between 3 and 4 units depending on the FMC operational conditions. It is possible to include other cost parameters related to lot sizes (pallet capacity) and develop cost models that could be utilized in real life applications. These results show the usefulness of the stochastic model presented with respect to cost optimization of FMC systems.

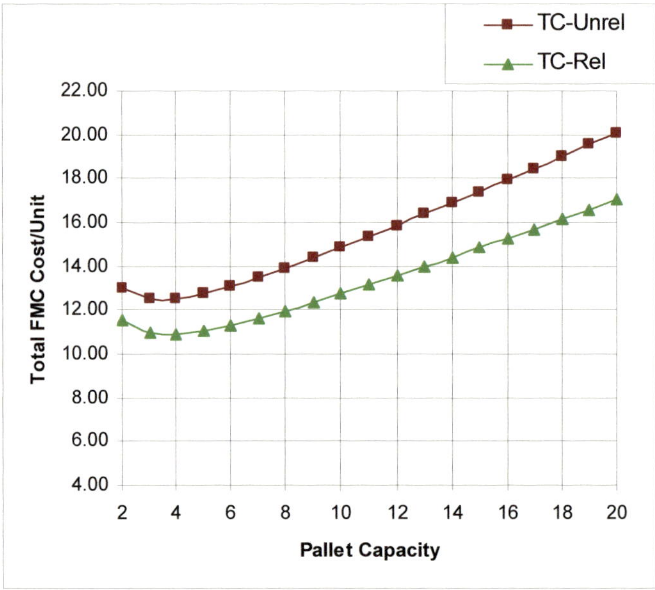


Fig. 9. Total FMC cost as a function of pallet capacity.

## 4. Reliability Modeling of a FMC with Multiple Machines

In this section, reliability modeling of a FMC with two machines and a single robot is presented. Processing times on the machines, robot loading and unloading times, pallet transfer times, machine operation times, as well as machine failure and repair times are all assumed as random quantities that follow exponential distribution. Development of the model follows the same procedure as it was done for the single-machine FMC system with necessary modifications for multiple machines.

### 4.1 A Stochastic Model for a FMC with Two Machines and a Robot

In order to analyze FMC operations with stochastic parameters, stochastic model is developed using Markov chains as in the previous section. First, the following system states and notations are defined:

$S_{ijkl}(t)$ = state of the FMC at time t, with subscripts i, j, k, and l as described below.

$P_{ijkl}(t)$ = probability that the system will be in state $S_{ijkl}(t)$

i =number of blanks in FMC (on the pallet, the machine, or the robot gripper)

j = state of the production machine 1 (j=0 if the machine is idle; j=1 if the machine is operating on a part;

and j=2 if the machine is waiting for the robot; j=3 if the machine is under repair)

k = state of the production machine 2 (j=0 if the M/C is idle; j=1 if the machine is operating on a part;

and j=2 if the machine is waiting for the robot, j=3 if the machine is under repair)

l = state of the robot (l=0 if the robot is idle; l=1 if the robot is loading/unloading machine 1 ; k=2 if the robot is loading/unloading machine 2)

$l_m$ = loading rate of the robot for machine m (m=1,2) (parts/unit time)

$u_m$ = unloading rate of the robot for machine m (m=1,2) (parts/unit time)

$z_m$ = combined loading/unloading rate of the robot for machine m (m=1,2)

w = pallet transfer rate (pallets/unit time)

$\lambda_m$= failure rate of production machine m ($1/\lambda_m$ = mean time between failures)

$\mu_m$= repair rate of the production machine m ($1/\mu_m$ = mean machine repair time)

$v_m$= machining rate (or production rate) of machine m (parts/unit time)

n = pallet capacity (number of parts/pallet)

$Q_c$ = production output rate of the cell in terms of parts/unit time

In order to analyze the FMC system with two machines, state equations that describe the rate of flow between the states are developed and presented here. Because of its large size, the transition flow diagram has not been shown here. Using the fact that the *net flow* rate at each state is equal to the difference between the rates of *flow in* and *flow out*, a set of differential equations are obtained for the system. For example, for the state (n,001), rate of change with respect to time t is given by:

$$dP_{n,001}(t)/dt = (w)P_{0,\,000}-(l_1)P_{n,\,001}$$

At steady state, $t\to\infty$; $dP_{n,001}(t)/dt\to 0$ and the differential equation changes into a difference equation. The resulting difference equations for all states are given by the equation sets (16a), (16b) and (16c) below. The whole set of equations is divided into three subsets: The first subset (16a) includes the equations for the loading of initial parts; the last subset (16c) includes the equations for the unloading of the final parts; and the subset (16b) includes all the general equations for intermediate parts. These equations must be solved to obtain the state probabilities and FMC system performance measures.

$$\begin{gathered}
w\, P_{0,\,000} - l_1\, P_{n,\,001} = 0 \\
l_1\, P_{n,\,001} + \mu_1\, P_{n-1,\,302} - (v_1 + l_2 + \lambda_1)\, P_{n-1,\,102} = 0 \\
\lambda_1\, P_{n-1,\,102} - (l_2 + \mu_1)\, P_{n-1,\,302} = 0 \\
v_1\, P_{n-1,\,102} - l_2\, P_{n-1,\,202} = 0 \\
\lambda_2\, P_{n-2,\,110} + \mu_1\, P_{n-2,\,330} - (v_1 + \lambda_1 + \mu_2)\, P_{n-2,\,130} = 0 \\
l_2\, P_{n-1,\,102} + \mu_1\, P_{n-2,\,310} + \mu_2\, P_{n-2,\,130} - (v_1 + v_2 + \lambda_1 + \lambda_2)\, P_{n-2,\,110} = 0 \\
l_2\, P_{n-1,\,302} + \lambda_1\, P_{n-2,\,110} + \mu_2\, P_{n-2,\,330} - (v_2 + \lambda_2 + \mu_1)\, P_{n-2,\,310} = 0 \\
\lambda_1\, P_{n-2,\,130} + \lambda_2\, P_{n-2,\,310} - (\mu_1 + \mu_2)\, P_{n-2,\,330} = 0 \\
v_2\, P_{n-2,\,011} - z_1\, P_{n-2,\,021} = 0 \\
v_1\, P_{n-2,\,110} + l_2\, P_{n-1,\,202} + \mu_2\, P_{n-2,031} - (v_2 + z_1 + \lambda_2)\, P_{n-2,\,011} = 0 \\
v_1\, P_{n-2,\,130} + \lambda_2\, P_{n-2,\,011} - (z_1 + \mu_2)\, P_{n-2,\,031} = 0 \\
v_2\, P_{n-2,\,310} + \lambda_1\, P_{n-2,\,102} - (z_2 + \mu_1)\, P_{n-2,\,302} = 0 \\
v_2\, P_{n-2,\,110} + \mu_1\, P_{n-2,\,302} - (v_1 + z_2 + \lambda_1)\, P_{n-2,\,102} = 0 \\
v_1\, P_{n-2,\,102} - z_2\, P_{n-2,\,202} = 0
\end{gathered} \tag{16a}$$

$$z_1 P_{x+1,031} + \lambda_2 P_{x,110} + \mu_1 P_{x,330} - (v_1 + \lambda_1 + \mu_2) P_{x,130} = 0$$
$$z_1 P_{x+1,011} + z_2 P_{x+1,102} + \mu_1 P_{x,310} + \mu_2 P_{x,130} - (v_1 + v_2 + \lambda_1 + \lambda_2) P_{x,110} = 0$$
$$z_2 P_{x+1,302} + \lambda_1 P_{x,110} + \mu_2 P_{x,330} - (\lambda_2 + \mu_1 + v_2) P_{x,310} = 0$$
$$\lambda_1 P_{x,130} + \lambda_2 P_{x,310} - (\mu_1 + \mu_2) P_{x,330} = 0$$
$$v_1 P_{x,102} - z_2 P_{x,202} = 0$$
$$v_2 P_{x,110} + z_1 P_{x+1,021} + \mu_1 P_{x,302} - (v_1 + z_2 + \lambda_1) P_{x,102} = 0$$
$$v_2 P_{x,310} + \lambda_1 P_{x,102} - (z_2 + \mu_1) P_{x,302} = 0$$
$$v_1 P_{x,130} + \lambda_2 P_{x,011} - (z_1 + \mu_2) P_{x,031} = 0$$
$$v_1 P_{x,110} + z_2 P_{x+1,202} + \mu_2 P_{x,031} - (v_2 + z_1 + \lambda_2) P_{x,011} = 0$$
$$v_2 P_{x,011} - z_1 P_{x,021} = 0 \qquad \text{for } x = 1,2,\ldots, n-3 \tag{16b}$$

……………………………………

$$z_1 P_{1,031} + \lambda_2 P_{0,110} + \mu_1 P_{0,330} - (v_1 + \lambda_1 + \mu_2) P_{0,130} = 0$$
$$\mu_1 P_{0,310} + \mu_2 P_{0,130} - (v_1 + v_2 + \lambda_1 + \lambda_2) P_{0,110} + z_1 P_{1,011} + z_2 P_{1,102} = 0$$
$$z_2 P_{1,302} + \lambda_1 P_{0,110} + \mu_2 P_{0,330} - (v_2 + \lambda_2 + \mu_1) P_{0,310} = 0$$
$$\lambda_1 P_{0,130} + \lambda_2 P_{0,310} - (\mu_1 + \mu_2) P_{0,330} = 0$$
$$v_1 P_{0,102} - u_2 P_{0,202} = 0$$
$$v_2 P_{0,110} + z_1 P_{1,021} + \mu_1 P_{0,302} - (v_1 + u_2 + \lambda_1) P_{0,102} = 0$$
$$v_2 P_{0,310} + \lambda_1 P_{0,102} - (u_2 + \mu_1) P_{0,302} = 0$$
$$v_1 P_{0,130} + \lambda_2 P_{0,011} - (u_1 + \mu_2) P_{0,031} = 0$$
$$v_1 P_{0,110} + z_2 P_{1,202} + \mu_2 P_{0,031} - (v_2 + u_1 + \lambda_2) P_{0,011} = 0$$
$$v_2 P_{0,011} - u_1 P_{0,021} = 0$$
$$u_2 P_{0,302} + \lambda_1 P_{0,100} - \mu_1 P_{0,300} = 0$$
$$u_2 P_{0,102} + \mu_1 P_{0,300} - (v_1 + \lambda_1) P_{0,100} = 0$$
$$u_1 P_{0,011} + \mu_2 P_{0,030} - (v_2 + \lambda_2) P_{0,010} = 0$$
$$u_1 P_{0,031} + \lambda_2 P_{0,010} - \mu_2 P_{0,030} = 0$$
$$v_1 P_{0,100} + u_2 P_{0,202} - u_1 P_{0,001} = 0$$
$$v_2 P_{0,010} + u_1 P_{0,021} - u_2 P_{0,002} = 0$$
$$u_1 P_{0,001} + u_2 P_{0,002} - w P_{0,000} = 0 \tag{16c}$$

The system consists of 10n+1 equations and equal number of unknowns. For example, for n=4, number of system states, as well as number of equations, is 10(4) +1=41 and for n=10, it is 10(10) + 1= 101. It is possible to obtain an exact solution for this system of equations given by PT=0, where P is the state probabilities vector to be determined and T is the probability transition rate matrix. It is known that all the equations in PT=0 are not linearly independent and thus the matrix T is singular, which does not have an inverse. We must add the normalizing condition given by equation (17) below, which assures that sum of all state probabilities, is 1, to the three sets of equations above by eliminating one of them.

$$\sum_{i=0}^{n}\sum_{j=0}^{2}\sum_{k=0}^{2}\sum_{l=0}^{2} P_{ijkl} = 1 \tag{17}$$

Exact numerical solutions can be obtained for all state probabilities. However, for large values of n, exact numerical solution becomes tedious and one has to resort to software in order to obtain a solution for a given system. Therefore, it is preferable to have a closed form solution for the state probabilities, as well as system performance measures for applications of the model in system design and analysis. Savsar and Aldaihani (2008) have obtained a closed form solution for this problem, which involved sophisticated algebraic analysis and manipulations. In the following section, we present the results obtained for the closed form solution.

### 4.2 A Closed Form Solution for Two-Machine FMC Model

After a systematic procedure and comprehensive algebraic manipulations, equation sets (16a, 16b and 16c) are solved for the unknown probabilities. Equation set (16a) consists of 14 equations and involves n, n-1, and n-2; equation set (16b) consists of 10 equations with x, for x=1,…,n-3; equation set (16c) consists of 17 equations involving with n=0. In order to present the solution, a set of intermediary variables are defined based on the system parameters as given in tables 1-3.

Based on the definitions given in tables 1-3, algebraic equation sets 16a, 16b, and 16c are solved step by step for the unknown state probabilities. The solution results are summarized in Table 4 for $n \geq 3$. In the case of n<3, a solution will be obtained only for n=2, since FMC system has two machines and therefore it is physically meaningless to have n=1 part delivered into the system by the pallet. Therefore, a special solution is obtained for n=2 due to the reduction in number of equations in this case. To find $P_{0,000}$ we use the renumbering of the state probabilities as shown in table 4. For example, state probability $P_{n,001}$ is represented by $P_{n,1}$; $P_{n-1,102}$ by $P_{n-1,2}$; $P_{n-1,302}$ by $P_{n-1,3}$; and so on until $P_{0,002}$ by $P_{0,40}$. We have a normalizing condition represented by equation (17) above and the last equation in table 4, in addition to a set of 40 state equations (set 18) in the table. Since the sum of probabilities has to be 1, we need to use the normalizing condition to determine $P_{0,000}$. Substituting the state probabilities into the normalizing condition given by equation 17, we obtain equation 19 given below. Finally, values of state probabilities, $P_{ijkl}$ given in table 4, are substituted into equation 19 to obtain $P_{0,000}$ with respect to known parameters. All state probabilities are then determined with respect to $P_{0,000}$.

| | | |
|---|---|---|
| $a = v_1 + \lambda_1$ | $b = v_2 + \lambda_2$ | $c = v_1 + v_2 + \lambda_1 + \lambda_2$ |
| $D = v_1 + v_2 + \mu_1 + \mu_2$ | $e = z_1 + \mu_2$ | $f = z_2 + \mu_1$ |
| $g = v_1 + l_2 + \lambda_1$ | $h = v_1 + \lambda_1 + \mu_2$ | $k = v_2 + \lambda_2 + \mu_1$ |
| $p = v_1 + z_2 + \lambda_1$ | $q = v_2 + z_1 + \lambda_2$ | $r = l_2 + \mu_1$ |
| $s = \mu_1 + \mu_2$ | $t = \lambda_1 + \lambda_2$ | $x = v_1 + u_2 + \lambda_1$ |
| $y = v_2 + u_1 + \lambda_2$ | $A = u_2 + \mu_1$ | $B = u_1 + \mu_2$ |

Table 1. First Set of Variables

| | | |
|---|---|---|
| $C_1 = \lambda_2/h$ | $C_2 = \mu_1/h$ | $C_3 = l_2 rw/c(gr - \mu_1\lambda_1)$ |
| $C_4 = u_1/c$ | $C_5 = \mu_2/c$ | $C_6 = l_2\lambda_1 w/\mathrm{k}(gr - \mu_1\lambda_1)$ |
| $C_7 = \lambda_1/k$ | $C_8 = \mu_2/k$ | $C_9 = \lambda_1/s$ |
| $C_{10} = \lambda_2/s$ | $C_{11} = qe - \mu_2\lambda_2$ | $C_{12} = gr - \mu_1\lambda_1$ |
| $C_{13} = pf - \mu_1\lambda_1$ | $C_{14} = \mu_1\lambda_1 - kc$ | $C_{15} = \mu_1\lambda_1 - \mu_2\lambda_2$ |
| $C_{16} = \mu_2\lambda_2 - sk$ | | |

Table 2. Second Set of Variables

| | | | |
|---|---|---|---|
| $F_1 = w/l_1$ | $F_2 = rw/C_{12}$ | $F_3 = \lambda_1 w/C_{12}$ | $F_4 = v_1 rw/l_2 C_{12}$ |
| $F_5 = \dfrac{C_2C_{10}\left(C_6 + C_3C_7\right) + C_1C_3\left(1 - C_8C_{10}\right) + C_1C_4C_6}{\left(C_5C_{10} - C_4C_9\right)\left(C_1C_8 - C_2C_7\right) - C_1C_5 - C_2C_9 - C_4C_7 - C_8C_{10} + 1}$ | | | |
| $F_6 = \left[C_3(1 - C_8C_{10}) + C_4C_6 - F_5C_8\left(C_5C_{10} - C_4C_9\right) - F_5C_5\right]/\left(1 - C_4C_7 - C_8C_{10}\right)$ | | | |
| $F_7 = \left[F_6C_{10} - (C_5C_{10} - C_4C_9)F_5 - C_3C_{10}\right]/C_4$ | $F_8 = \left[F_6 - C_5F_5 - C_3\right]/C_4$ | | |
| $F_9 = ev_1F_6/C_{11} + ev_1rw/C_{11}C_{12} + v_1\mu_2F_5/C_{11}$ | $F_{10} = v_2F_9/z_1$ | | |
| $F_{11} = \left(v_1F_5 + \lambda_2F_9\right)/e$ | $F_{12} = v_2F_6f + v_2\mu_1F_8/C_{13}$ | | |
| $F_{13} = v_2F_8/f + \lambda_1F_{12}/f$ | $F_{14} = v_1F_{12}/z_1$ | | |
| $A_1 = \left(z_1F_9 + z_2F_{12}\right)/c$ | $A_2 = A_1 + \mu_1 z_2 F_{13}/kc$ | | |
| $A_3 = \left[C_{14}C_{15}/k\lambda_2c^2\right] - \mu_2/c$ | $A_4 = \mu_1\mu_2/kc + s\mu_1C_{14}/\left(\lambda_2kc^2\right)$ | | |
| $A_5 = A_1 C_{14}/kc$ | $A_6 = (\mu_1\mu_2/kc) - \mu_1 C_{14}/kc\lambda_2$ | | |
| $A_7 = -z_1C_{14}(v_1F_5 + \lambda_2F_9)/kc\lambda_2e$ | $A_8 = -hC_{14}/(kc\lambda_2) - \mu_2/c$ | | |
| $R_1 = z_1\left(v_1D_8 + \lambda_2D_{14}\right)/e$ | $R_2 = z_1D_{14} + z_2D_{12}$ | $R_3 = z_2D_{13}$ | |
| $G_1 = \left[sk\lambda_1\mu_1/C_{16}\right] + C_{15}$ | $G_2 = \left[s\lambda_1\lambda_2\mu_1/C_{16}\right] + c\lambda_2$ | $G_3 = \lambda_2R_2 - \left[s\lambda_2R_3\mu_1/C_{16}\right]$ | |
| $G_4 = -\left[\mu_1k\lambda_1/C_{16} + h\right]$ | $G_5 = \lambda_2 - \mu_1\lambda_1\lambda_2/C_{16}$ | $G_6 = \mu_1\lambda_2R_3/C_{16} - R_1$ | |
| $F_{15} = \left[A_4\left(A_2 + A_7\right) - A_6\left(A_2 + A_5\right)\right]/\left(A_4A_8 - A_3A_6\right)$ | $F_{16} = \left[A_3F_{15} - \left(A_2 + A_5\right)\right]/A_4$ | | |
| $F_{17} = -\left[kcA_2 + \mu_2kF_{15} + \mu_1\mu_2F_{16}\right]/\left(\mu_1\lambda_1 - kc\right)$ | $F_{18} = \left[cF_{17} - \mu_2F_{15} - cA_1\right]/\mu_1$ | | |
| $F_{19} = v_2\left[fF_{17} + fF_9 + \mu_1F_{18}\right]/\left(C_{13}\right)$ | $F_{20} = \left(v_2F_{18} + \lambda_1F_{19}\right)/f$ | | |
| $F_{21} = v_1F_{19}/z_2$ | $F_{22} = v_1e\left(z_2F_{12}/z_1 + \mu_2F_{15}/e + F_{17}\right)/C_{11}$ | | |
| $F_{23} = v_2F_{22}/z_1$ | $F_{24} = \left(v_1F_{15} + \lambda_2F_{22}\right)/e$ | | |
| $F_{25} = \left(G_2G_6 - G_3G_5\right)/\left(G_2G_4 - G_1G_5\right)$ | $F_{26} = \left(G_3 - G_1F_{25}\right)/G_2$ | | |
| $F_{27} = \left(-\lambda_2R_3 - k\lambda_1F_{25} - \lambda_1\lambda_2F_{26}\right)/C_{16}$ | $F_{28} = \left(-\lambda_1F_{25} + sF_{27}\right)/\lambda_2$ | | |
| $F_{29} = \left(v_2xF_{28} + v_2\lambda_1F_{26} + \lambda_1v_2D_{14}\right)/\left(Ax - \mu_1\lambda_1\right)$ | $F_{30} = \left(v_2F_{26} + v_2D_{14} + \mu_1F_{29}\right)/x$ | | |
| $F_{31} = v_1F_{30}/u_2$ | $F_{32} = \left(Bv_1F_{26} + Bv_1D_{12} + \mu_2v_1F_{25}\right)/\left(By - \mu_2\lambda_2\right)$ | | |
| $F_{33} = \left(v_1F_{25} + \lambda_2F_{32}\right)/B$ | $F_{34} = v_2F_{32}/u_1$ | $F_{35} = u_2\left(F_{30} + F_{29}\right)/\left(a - \lambda_1\right)$ | |
| $F_{36} = \left(u_2F_{29} + \lambda_1F_{35}\right)/\mu_1$ | $F_{37} = \left(bu_1F_{33} + u_1\lambda_2F_{32}\right)/\mu_2\left(b - \lambda_2\right)$ | | |
| $F_{38} = \left(u_1F_{32} + \mu_2F_{37}\right)/b$ | $F_{39} = v_1\left(F_{35} + F_{30}\right)/u_1$ | $F_{40} = v_2\left(F_{38} + F_{32}\right)/u_2$ | |

Table 3. Third Set of Variables

| | | |
|---|---|---|
| $P_{n,001} = F_1P_{0,000}$ (1) | $P_{n\text{-}1,102} = F_2P_{0,000}$ (2) | $P_{n\text{-}1,302} = F_3P_{0,000}$ (3) |
| $P_{n\text{-}1,202} = F_4P_{0,000}$ (4) | $P_{n\text{-}2,130} = F_5P_{0,000}$ (5) | $P_{n\text{-}2,110} = F_6P_{0,000}$ (6) |
| $P_{n\text{-}2,330} = F_7P_{0,000}$ (7) | $P_{n\text{-}2,310} = F_8P_{0,000}$ (8) | $P_{n\text{-}2,011} = F_9P_{0,000}$ (9) |
| $P_{n\text{-}2,021} = F_{10}P_{0,000}$ (10) | $P_{n\text{-}2,031} = F_{11}P_{0,000}$ (11) | $P_{n\text{-}2,102} = F_{12}P_{0,000}$ (12) |
| $P_{n\text{-}2,302} = F_{13}P_{0,000}$ (13) | $P_{n\text{-}2,202} = F_{14}P_{0,000}$ (14) | **$P_{x,130} = F_{15}P_{0,000}$ (15)** |
| **$P_{x,110} = F_{17}P_{0,000}$ (16)** | **$P_{x,310} = F_{18}P_{0,000}$ (17)** | **$P_{x,330} = F_{16}P_{0,000}$ (18)** |
| **$P_{x,302} = F_{20}P_{0,000}$ (19)** | **$P_{x,102} = F_{19}P_{0,000}$ (20)** | **$P_{x,202} = F_{21}P_{0,000}$ (21)** |
| **$P_{x,021} = F_{23}P_{0,000}$ (22)** | **$P_{x,011} = F_{22}P_{0,000}$ (23)** | **$P_{x,031} = F_{24}P_{0,000}$ (24)** **$x = 1, \ldots\ldots, n-3$** |
| $P_{0,130} = F_{25}P_{0,000}$ (25) | $P_{0,110} = F_{26}P_{0,000}$ (26) | $P_{0,330} = F_{27}P_{0,000}$ (27) |
| $P_{0,310} = F_{28}P_{0,000}$ (28) | $P_{0,302} = F_{29}P_{0,000}$ (29) | $P_{0,102} = F_{30}P_{0,000}$ (30) |
| $P_{0,202} = F_{31}P_{0,000}$ (31) | $P_{0,031} = F_{33}P_{0,000}$ (32) | $P_{0,011} = F_{32}P_{0,000}$ (33) |
| $P_{0,021} = F_{34}P_{0,000}$ (34) | $P_{0,300} = F_{35}P_{0,000}$ (35) | $P_{0,100} = F_{36}P_{0,000}$ (36) |
| $P_{0,010} = F_{37}P_{0,000}$ (37) | $P_{0,030} = F_{38}P_{0,000}$ (38) | $P_{0,001} = F_{39}P_{0,000}$ (39) |
| $P_{0,002} = F_{40}P_{0,000}$ (40) | $\sum_j P_j = 1$ (41) | **Equation Set (18)** |

Table 4. Equation set (18); summary of solutions (for $n \geq 3$)

From equation (17) and substitution of the values results in the following equation:

$$\sum_{i=0}^{n}\sum_{j=0}^{2}\sum_{k=0}^{2}\sum_{l=0}^{2} P_{ijkl} = \left[ P_{n,000} + \sum_{j=2}^{4} P_{n-1,j} + \sum_{j=5}^{14} P_{n-2,j} + (n-3)\sum_{j=15}^{24} P_{x,j} + \sum_{j=25}^{40} P_{0,j} \right] + P_{0,000} = 1 \quad (19)$$

$$= \left[ F_1 + \sum_{j=2}^{4} F_j + \sum_{j=5}^{14} F_j + \{(n-3)\sum_{j=15}^{24} F_j\} + \sum_{j=25}^{40} F_j + 1 \right] P_{0,000} = 1$$

which is solved to obtain $P_{0,000}$ as:

$$P_{0,000} = 1 \Big/ \left[ F_1 + \sum_{j=2}^{4} F_j + \sum_{j=5}^{14} F_j + (n-3)\sum_{j=15}^{24} F_j + \sum_{j=25}^{40} F_j + 1 \right]$$

Finally $P_{0,000}$ is substituted back into the equations in table 4 to obtain values of state probabilities. Equation (19) and the solution obtained for the state probabilities are valid for $n \geq 3$. For n=2, special modifications need to be made since none of the equations in set 16b are applicable in the case of n=2. Only the first 4 equations in set 16a and all 17 equations in set 16c are applicable for the case n=2. In the set 16a, n is replaced by 2. Table 5 shows the modified parameter equations for n=2, while all other equations are the same as those given for n=3 in tables 1-3. In addition to the changes given in table 5, there are slight modifications in some of the equations in the set 16c. These changes are listed below:

- Delete all the terms involving $z_1$ including, ($z_1P_{1,031}$) from the first equation, ($z_1P_{1,011}$) from the second equation, and ($z_1P_{1,021}$) from the sixth equation of the set 16c; replace $z_2$ by $l_2$ in the second equation of the set 16c to obtain the following equations respectively.

$$\lambda_2 P_{0,110} + \mu_1 P_{0,330} - (v_1 + \lambda_1 + \mu_2) P_{0,130} = 0$$
$$\mu_1 P_{0,310} + \mu_2 P_{0,130} - (v_1 + v_2 + \lambda_1 + \lambda_2) P_{0,110} + l_2 P_{1,102} = 0$$
$$v_2 P_{0,110} + \mu_1 P_{0,302} - (v_1 + u_2 + \lambda_1) P_{0,102} = 0$$

- Replace $z_2$ with $l_2$ in the third and in the ninth equations of the set 16c to obtain the following equations respectively.

$$l_2 P_{1,302} + \lambda_1 P_{0,110} + \mu_2 P_{0,330} - (v_2 + \lambda_2 + \mu_1) P_{0,310} = 0$$
$$v_1 P_{0,110} + l_2 P_{1,202} + \mu_2 P_{0,031} - (v_2 + u_1 + \lambda_2) P_{0,011} = 0$$

| | |
|---|---|
| $G_3 = \lambda_2 l_2 F_2 - \dfrac{s\mu_1\lambda_2 F_3}{C_{16}}$ | $G_6 = \dfrac{\mu_1\lambda_2 l_2 F_3}{C_{16}}$ |
| $F_{27} = \dfrac{-\lambda_2 l_2 F_3 - k\lambda_1 F_{25} - \lambda_1\lambda_2 F_{26}}{\lambda_2\mu_2 - ks}$ | $F_{29} = \dfrac{v_2\left(xF_{28} + \lambda_1 F_{26}\right)}{Ax - \lambda_1\mu_1}$ |
| $F_{30} = \dfrac{v_2 F_{26} + \mu_1 F_{29}}{x}$ | $F_{31} = \dfrac{v_1}{u_2} F_{30}$ |
| $F_{32} = \dfrac{Bv_1 F_{26} + Bv_1 F_2 + v_1\mu_2 F_{25}}{By - \mu_2\lambda_2}$ | All other values are as given in table 4. |

Table 5. Definition of variables for $n = 2$

All the remaining solution steps are similar to the general case of $n \geq 3$. Thus, for n=2, equations 1-4 of the set (18) in table 4 remain the same, equations 5-24 of the set (18) are dropped, and equations 25-40 of the set (18) are changed according to the modifications mentioned above. Once the state probabilities are determined, it is then possible to determine various system and subsystem performance measures.

$$\text{M1 idle} = P_{0,000}\left[1 + F_1 + \sum_{i=1}^{11} F_i + (n-3)\sum_{i=22}^{24} F_i + \sum_{i=32}^{34} F_i + \sum_{i=37}^{40} F_i\right] \tag{20}$$

$$\text{M1 busy} = P_{0,000}\left[F_2 + F_5 + F_6 + F_{12} + (n-3)(F_{15} + F_{17} + F_{19}) + F_{25} + F_{26} + F_{30} + F_{36}\right] \quad (21)$$

$$\text{M1 waiting} = P_{0,000}\left[F_4 + F_{14} + (n-3)F_{21} + F_{31}\right] \quad (22)$$

$$\text{M1 repair} = P_{0,000}\left[F_3 + F_7 + F_8 + F_{13} + (n-3)(F_{16} + F_{18} + F_{20}) + \sum_{i=27}^{29} F_i + F_{35}\right] \quad (23)$$

$$\text{M2 idle} = P_{0,000}\left[1 + \sum_{i=1}^{4} F_i + \sum_{i=12}^{14} F_i + (n-3)\sum_{i=19}^{21} F_i + \sum_{i=29}^{31} F_i + F_{35} + F_{36} + F_{39} + F_{40}\right] \quad (24)$$

$$\text{M2 busy} = P_{0,000}\left[F_6 + F_8 + F_9 + (n-3)(F_{17} + F_{18} + F_{22}) + F_{26} + F_{28} + F_{32} + F_{37}\right] \quad (25)$$

$$\text{M2 waiting} = P_{0,000}\left[F_{10} + (n-3)F_{23} + F_{34}\right] \quad (26)$$

$$\text{M2 repair} = P_{0,000}\left[F_5 + F_7 + F_{11} + (n-3)(F_{15} + F_{16} + F_{24}) + F_{25} + F_{27} + F_{33} + F_{38}\right] \quad (27)$$

$$\text{Robot idle} = P_{0,000}\left[1 + \sum_{i=5}^{8} F_i + (n-3)\sum_{i=15}^{18} F_i + \sum_{i=25}^{28} F_i + \sum_{i=35}^{38} F_i\right] \quad (28)$$

$$\text{Robot busy with M1} = P_{0,000}\left[F_1 + \sum_{i=5}^{8} F_i + (n-3)\sum_{i=22}^{24} F_i + \sum_{i=32}^{34} F_i + F_{39}\right] \quad (29)$$

$$\text{Robot busy with M2} = P_{0,000}\left[\sum_{i=2}^{4} F_i + \sum_{i=12}^{14} F_i + (n-3)\sum_{i=19}^{21} F_i + \sum_{i=29}^{31} F_i + F_{40}\right] \quad (30)$$

### 4.3 Case Example for a Two-Machine FMC

In this section, we present some numerical results for the two-machine unreliable FMC problem with different parameters. The results are also compared to the reliable FMC results in order to see the effects of equipment failures on system performance measures. The parameter values for the two-machine unreliable FMC system are shown in table 6. The parameters for the reliable system are the same with the exception that there are no failures and repairs in the reliable system. Values given in the table are the mean values for various parameters and the mean is the inverse of the rate in each case. Figure 10 shows the production output rate as a function of the pallet capacity (n) at different robot loading/unloading rates, z. As it is seen from the figure, production rate increases with increasing pallet capacity and robot loading rates. While the rate of increase is higher initially, it levels off at higher values of n.

| | |
|---|---|
| Operation time per part | $1/v_m$ = 1 time unit, m=1, 2 for machines 1 and 2 |
| Robot loading time for the first part | $1/l_m$ = 0.25 time units, for machines m=1, 2 |
| Robot load/unload time for subsequent parts | $1/z_m$ = 0.5 time units, m=1, 2 |
| Robot unloading time for the last part | $1/u_m$ = 0.25 time units, m=1, 2 |
| Mean time to failure for machine m | $1/\lambda_m$ = 100 time units |
| Mean time to repair the machine m | $1/\mu_m$ = 10 time units |
| Pallet transfer time | 1/w = 1,...,10 time units per pallet |
| Pallet capacity | n=4 units |

Table 6. Parameter values for the unreliable FMC system

Figure 11 shows the production rate of the FMC system as a function of pallet capacity (n) and four different machine repair rates (μ=0.05, 0.1, 0.5, 1). Other parameters are kept constant as before. As the machine repair rates are increased, FMC production rate increases, but the increase is marginal when μ is changed from 0.5 to 1 repair per unit time. While it is not shown here, the effects of pallet capacity (n) and pallet loading rate (w) on FMC production rate has a similar trend. Production rate increases with increasing w up to a certain level and levels off after that. Effects of various parameters on FMC component utilizations, including the machines, the robot, and the pallet handling system, can also be obtained by the equations presented above.

Unreliable FMC system performance results obtained from the stochastic model presented in this paper are compared to the reliable FMC system performance results, whose model is reported elsewhere in Savsar and Aldaihani (2004) and Aldaihani and Savsar (2005a). Figure 12 shows the production output results for both the reliable and unreliable FMC system for a pallet capacity of n=4 units at different pallet transfer rates. Production rate increases with respect to pallet transfer rates for both, reliable and unreliable systems. Reliable FMC has 8-10% higher production rate than the unreliable FMC in this case. Production rate significantly increases with increasing pallet transfer rate up to the rate of 4 pallets per time unit. After this rate, production rate increases at a slower pace. Figure 13 compares machine utilizations for reliable and unreliable cells. Machine 1 has higher utilizations in either case because priority is given to machine 1 during initial loading when a pallet moves into the cell. Machine utilizations are significantly higher for reliable FMC than unreliable FMC. The utilization rates also increase sharply with respect to pallet transfer rates up to the rate of 4 pallets per time unit. After this, utilization increases are not as significant. The best pallet transfer rate must be established for each particular FMC using similar analysis as presented here. Other performance measures can be compared by using the models presented.

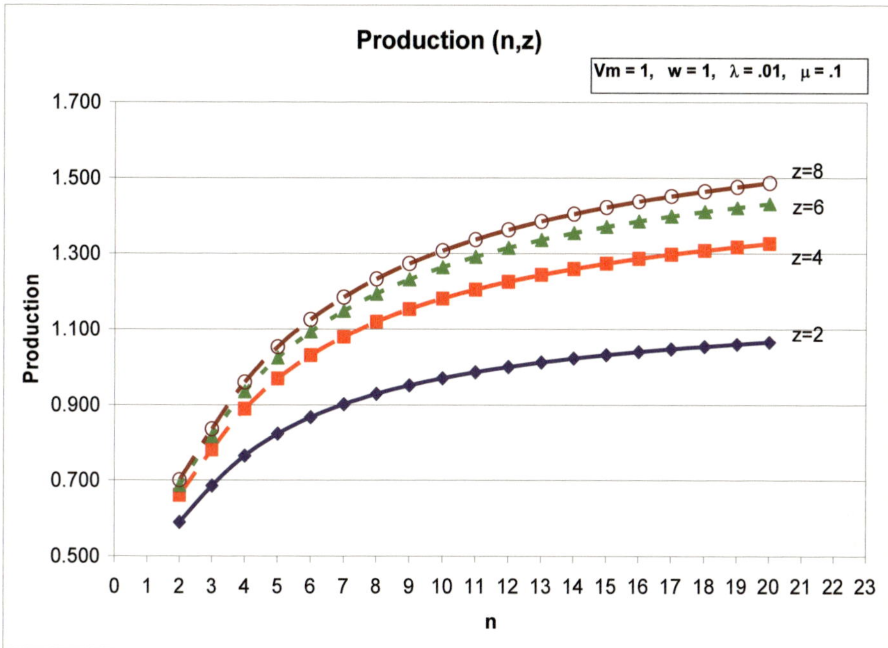


Fig. 10. Effects of pallet capacity (n) and robot loading rata (z)on FMC production rate.

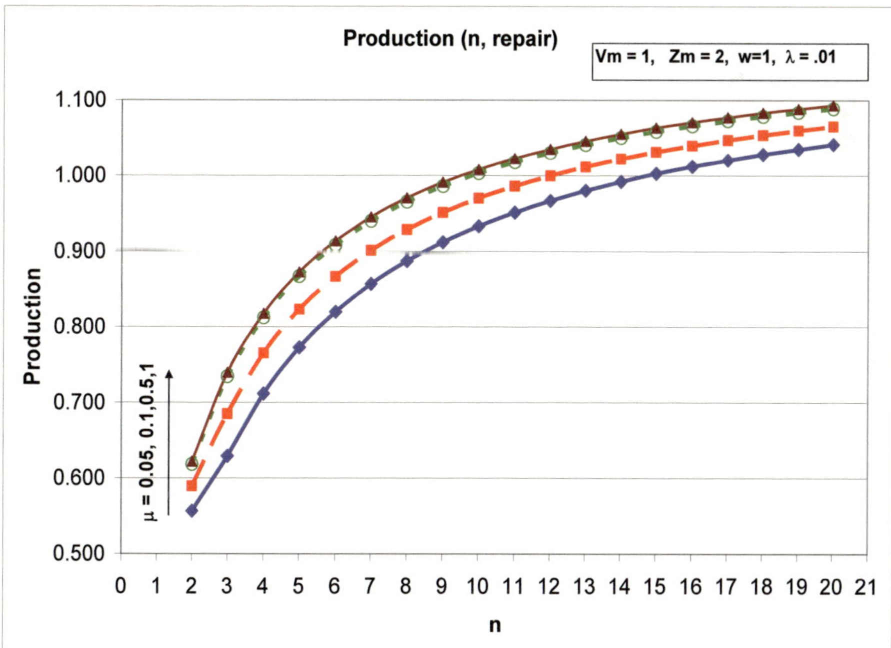


Fig. 11. Effects of pallet capacity (n) and repair rate (μ) on FMC production rate.

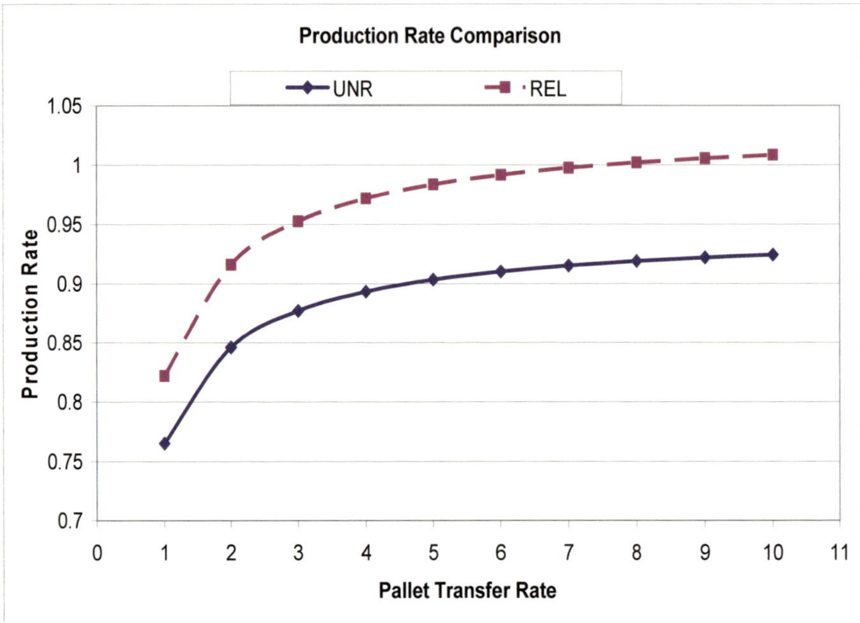


Fig. 12. Effects of pallet transfer rates on FMC production rates for unreliable and reliable FMC

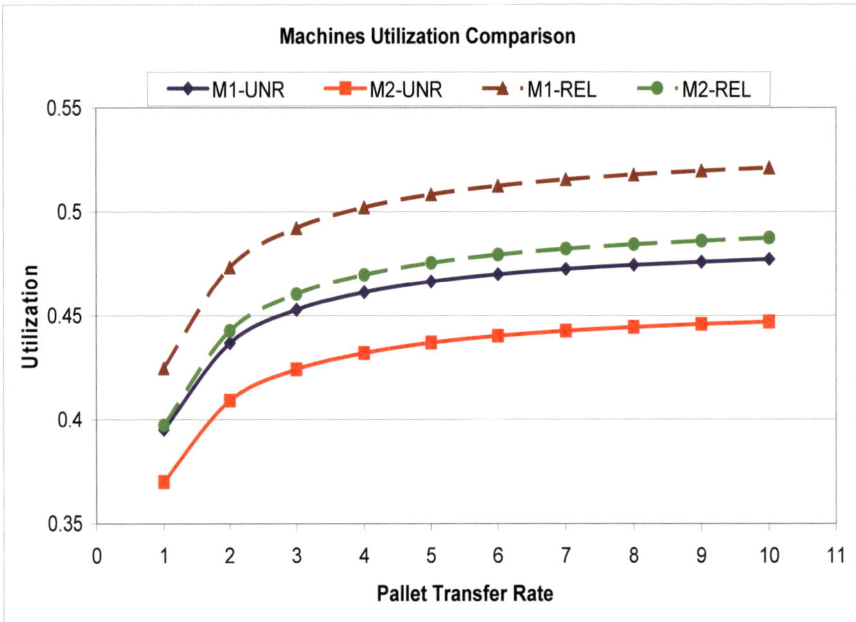


Fig. 13. Effects of pallet transfer rates on FMC Machine utilizations for unreliable and reliable FMC

## 5. Reliability Modeling of FMC with Multiple Robots

FMC systems with more than one machine can be served with more than one robot depending on the requirements for loading and unloading operations. Figure 14 shows a FMC with two unreliable machines and two robots. Operation of the cell is similar to the FMCs discussed in section 1. The only difference here is that each machine is attended by a specific robot for loading and unloading operations, which are carried out according to a scheduling program. The robots, the machines, and the pallet must be coordinated by a common control system, which controls all the operations in the cell. When the parts of various shapes and types are delivered into the system, the control system can direct the robots for necessary loading and unloading activities.

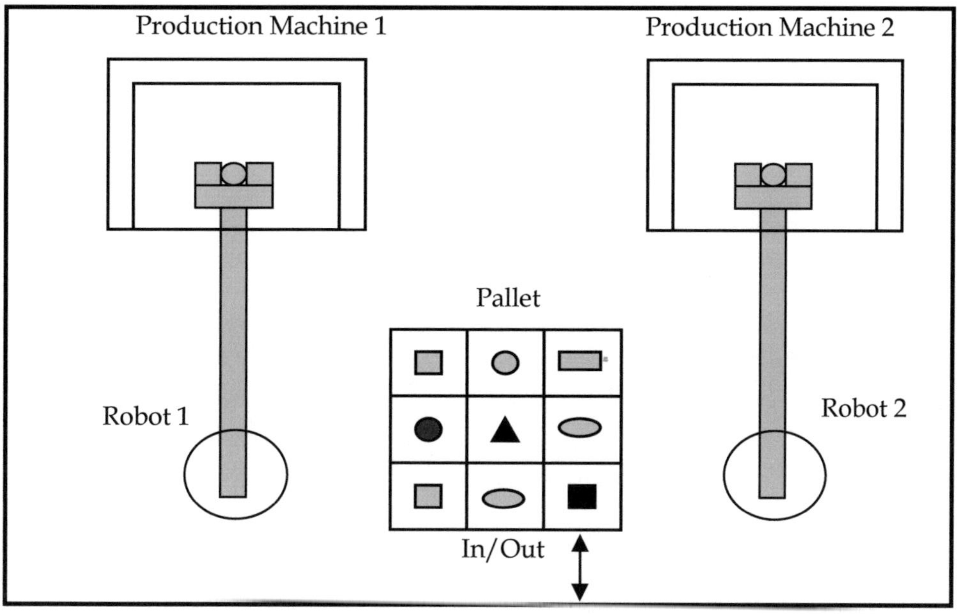


Fig. 14. An FMC with two machines served by two robots and a pallet handling system

### 5.1 A Stochastic Model for a FMC with Two Machines and Two Robots

In order to analyze reliability and performance of this FMC system, a stochastic model similar to the model given in section 3 has been developed by Aldaihani and Savsar (2008). This model is an extension of the model developed for the FMC with two machines and a single robot. The only addition here is another subscript in the state definition for the system due to one additional robot, which results in several additional state equations. In particular, the state of the system is described by $S_{i,jklm}(t)$ and probability of the system being in this state is given by $P_{i,jklm}(t)$, where

i = number of blanks in FMC (on the pallet, the machine, or the robot gripper)

j = state of the production machine 1 (j=0 if the machine is idle; j=1 if the machine is operating on a part; and j=2 if the machine is under repair)

k = state of the production machine 2 (k=0 if the M/C is idle; k=1 if the machine is operating on a part; and k=2 if the machine is under repair)

l = state of the robot 1 (l=0 if the robot is idle; l=1 if the robot is loading/unloading)
m= state of the robot 2 (l=0 if the robot is idle; l=1 if the robot is loading/unloading)
System parameters change slightly due to additional robot and the following notations are used:
$l_r$ = loading rate of robot r for machine r (r=1,2) (parts/unit time)
$u_r$ = unloading rate of robot r for machine r (r=1,2) (parts/unit time)
$z_r$ = combined loading/unloading rate of robot r for machine r (r=1,2)
w = pallet transfer rate (pallets/unit time)
$\lambda_r$= failure rate of the production machine r ($1/\lambda_r$ = mean time between failures)
$\mu_r$= repair rate of the production machine r ($1/\mu_r$ = mean machine repair time)
$v_r$= machining rate (or production rate) of machine r (parts/unit time)
n = pallet capacity (number of parts/pallet)
$Q_c$ = production output rate of the cell in terms of parts/unit time

A set of differential equations are obtained for the two-machine two-robot stochastic FMC in a similar way as it was done for the single robot case by using the fact that the *net flow rate* at each state is equal to the difference between the rates of *flow in* and *flow out*. For example, for the state (n,0011), rate of change with respect to time t is given by:

$$dP_{n,0011}(t)/dt = (w)P_{0,0000} - (l_1+l_2)P_{n,0011}$$

At steady state, $t\to\infty$; $dP(t)/dt\to 0$ and the differential equation changes into a difference equation. The resulting difference equations for all states, which govern the FMC behavior at steady state, are given by equation set 31 below. These equations must be solved to obtain the state probabilities and system performance measures. There are 9n+4 equations and equal number of unknowns. It is difficult to obtain a closed form solution for these equations due to large number of states. However, exact numerical solutions have been obtained by using the MAPLE equation solver with symbolic manipulation.

$$w\,P_{0,0000} - (l_1 + l_2)\,P_{n,0011} = 0$$
$$v_1 P_{n-1,1001} + v_2 P_{n-1,0110} - (z_1 + z_2)\,P_{n-1,0011} = 0$$
$$l_2 P_{n,0011} + \mu_2 P_{n-1,0210} - (v_2 + l_1 + \lambda_2)\,P_{n-1,0110} = 0$$
$$l_1 P_{n,0011} + \mu_1 P_{n-1,2001} - (v_1 + l_2 + \lambda_1)\,P_{n-1,1001} = 0$$
$$\lambda_1 P_{n-1,1001} - (l_2 + \mu_1)\,P_{n-1,2001} = 0$$
$$\lambda_2 P_{n-1,0110} - (l_1 + \mu_2)\,P_{n-1,0210} = 0$$
$$l_1 P_{n-1,0110} + l_2 P_{n-1,1001} + \mu_1 P_{n-2,2100} + \mu_2 P_{n-2,1200} - (v_1 + v_2 + \lambda_1 + \lambda_2)\,P_{n-2,1100} = 0$$
$$l_2 P_{n-1,2001} + \lambda_1 P_{n-2,1100} + \mu_2 P_{n-2,2200} - (v_2 + \lambda_2 + \mu_1)\,P_{n-2,2100} = 0$$
$$\lambda_1 P_{n-2,1200} + \lambda_2 P_{n-2,2100} - (\mu_1 + \mu_2)\,P_{n-2,2200} = 0$$
$$l_1 P_{n-1,0210} + \lambda_2 P_{n-2,1100} + \mu_1 P_{n-2,2200} - (v_1 + \lambda_1 + \mu_2)\,P_{n-2,1200} = 0$$
$$v_1 P_{n-2,1001} + v_2 P_{n-2,0110} - (z_1 + z_2)\,P_{n-2,0011} = 0$$
$$v_1 P_{n-2,1100} + z_2 P_{n-1,0011} + \mu_2 P_{n-2,0210} - (v_2 + z_1 + \lambda_2)\,P_{n-2,0110} = 0$$
$$v_2 P_{n-2,1100} + z_1 P_{n-1,0011} + \mu_1 P_{n-2,2001} - (v_1 + z_2 + \lambda_1)\,P_{n-2,1001} = 0$$
$$v_2 P_{n-2,2100} + \lambda_1 P_{n-2,1001} - (z_2 + \mu_1)\,P_{n-2,2001} = 0$$
$$v_1 P_{n-2,1200} + \lambda_2 P_{n-2,0110} - (z_1 + \mu_2)\,P_{n-2,0210} = 0$$

$$z_1 P_{n-2,0110} + z_2 P_{n-2,1001} + \mu_1 P_{n-x,2100} + \mu_2 P_{n-x,1200} - (v_1 + v_2 + \lambda_1 + \lambda_2) P_{n-x,1100} = 0$$
$$z_2 P_{n-2,2001} + \lambda_1 P_{n-x,1100} + \mu_2 P_{n-x,2200} - (v_2 + \lambda_2 + \mu_1) P_{n-x,2100} = 0$$
$$\lambda_1 P_{n-x,1200} + \lambda_2 P_{n-x,2100} - (\mu_1 + \mu_2) P_{n-x,2200} = 0$$
$$z_1 P_{n-2,0210} + \lambda_2 P_{n-x,1100} + \mu_1 P_{n-x,2200} - (v_1 + \lambda_1 + \mu_2) P_{n-x,1200} = 0$$
$$v_1 P_{n-x,1001} + v_2 P_{n-x,0110} - (z_1 + z_2) P_{n-x,0011} = 0$$
$$v_1 P_{n-x,1100} + z_2 P_{n-2,0011} + \mu_2 P_{n-x,0210} - (v_2 + z_1 + \lambda_2) P_{n-x,0110} = 0$$
$$v_2 P_{n-x,1100} + z_1 P_{n-2,0011} + \mu_1 P_{n-x,2001} - (v_1 + z_2 + \lambda_1) P_{n-x,1001} = 0$$
$$v_2 P_{n-x,2100} + \lambda_1 P_{n-x,1001} - (z_2 + \mu_1) P_{n-x,2001} = 0$$
$$v_1 P_{n-x,1200} + \lambda_2 P_{n-x,0110} - (z_1 + \mu_2) P_{n-x,0210} = 0$$

..........................................................

$$z_1 P_{2,0110} + z_2 P_{2,1001} + \mu_1 P_{1,2100} + \mu_2 P_{1,1200} - (v_1 + v_2 + \lambda_1 + \lambda_2) P_{1,1100} = 0$$
$$z_2 P_{2,2001} + \lambda_1 P_{1,1100} + \mu_2 P_{1,2200} - (v_2 + \lambda_2 + \mu_1) P_{1,2100} = 0$$
$$\lambda_1 P_{1,1200} + \lambda_2 P_{1,2100} - (\mu_1 + \mu_2) P_{1,2200} = 0$$
$$z_1 P_{2,0210} + \lambda_2 P_{1,1100} + \mu_1 P_{1,2200} - (v_1 + \lambda_1 + \mu_2) P_{1,1200} = 0$$
$$v_1 P_{1,1001} + v_2 P_{1,0110} - (z_1 + z_2) P_{1,0011} = 0$$
$$v_1 P_{1,1100} + z_2 P_{2,0011} + \mu_2 P_{1,0210} - (v_2 + z_1 + \lambda_2) P_{1,0110} = 0$$
$$v_2 P_{1,1100} + z_1 P_{2,0011} + \mu_1 P_{1,2001} - (v_1 + z_2 + \lambda_1) P_{1,1001} = 0$$
$$v_2 P_{1,2100} + \lambda_1 P_{1,1001} - (z_2 + \mu_1) P_{1,2001} = 0$$
$$v_1 P_{1,1200} + \lambda_2 P_{1,0110} - (z_1 + \mu_2) P_{1,0210} = 0$$
$$z_1 P_{1,0110} + z_2 P_{1,1001} + \mu_1 P_{0,2100} + \mu_2 P_{0,1200} - (v_1 + v_2 + \lambda_1 + \lambda_2) P_{0,1100} = 0$$
$$z_2 P_{1,2001} + \lambda_1 P_{0,1100} + \mu_2 P_{0,2200} - (v_2 + \lambda_2 + \mu_1) P_{0,2100} = 0$$
$$\lambda_1 P_{0,1200} + \lambda_2 P_{0,2100} - (\mu_1 + \mu_2) P_{0,2200} = 0$$
$$z_1 P_{1,0210} + \lambda_2 P_{0,1100} + \mu_1 P_{0,2200} - (v_1 + \lambda_1 + \mu_1) P_{0,1200} = 0$$
$$v_1 P_{0,1100} + z_2 P_{1,0011} + \mu_2 P_{0,0210} - (v_2 + u_1 + \lambda_2) P_{0,0110} = 0$$
$$v_2 P_{0,1100} + z_1 P_{1,0011} + \mu_1 P_{0,2001} - (v_1 + u_2 + \lambda_1) P_{0,1001} = 0$$
$$v_2 P_{0,2100} + \lambda_1 P_{0,1001} - (u_2 + \mu_1) P_{0,2001} = 0$$
$$v_1 P_{0,1200} + \lambda_2 P_{0,0110} - (u_1 + \mu_2) P_{0,0210} = 0$$
$$u_1 P_{0,0110} + \mu_2 P_{0,0200} - (v_2 + \lambda_2) P_{0,0100} = 0$$
$$v_1 P_{0,1001} + v_2 P_{0,0110} - (u_1 + u_2) P_{0,0011} = 0$$
$$u_2 P_{0,1001} + \mu_1 P_{0,2000} - (v_1 + \lambda_1) P_{0,1000} = 0$$
$$u_2 P_{0,2001} + \lambda_1 P_{0,1000} - \mu_1 P_{0,2000} = 0$$
$$u_1 P_{0,0210} + \lambda_2 P_{0,0100} - \mu_2 P_{0,0200} = 0$$
$$v_2 P_{0,0100} + u_1 P_{0,0011} - u_2 P_{0,0001} = 0$$
$$v_1 P_{0,1000} + u_2 P_{0,0011} - u_1 P_{0,0010} = 0$$
$$u_1 P_{0,0010} + u_2 P_{0,0001} - w P_{0,0000} = 0 \qquad (31)$$

For example, for n=6, number of system states, as well as number of equations is 9(6)+4=58. In order to determine numerical solutions, for 58 state probabilities represented by the vector, P, the set of equations given by PT=0 must be solved for P, where T is the probability transition rate matrix, which is the matrix of the coefficients in equation set (31). It is known that all the equations in PT=0 are not linearly independent and thus matrix T is singular with no inverse. We must add the normalizing condition given by equation (32) below, which assures that sum of all state probabilities is 1, to the set of 58 equations given for the FMC case above by eliminating one of them.

$$\sum_{i=0}^{n}\sum_{j=0}^{2}\sum_{k=0}^{2}\sum_{l=0}^{1}\sum_{m=0}^{1} P_{i,jklm} = 1 \quad (32)$$

Exact numerical solutions are obtained for each state probability. In the following section, several numerical solutions are obtained and the results are presented.

### 5.2 Case Example For Multi-Robot FMC System

In this section, we present numerical results of a case example for the FMC with two machines and two robots. The case example is solved by the proposed model and the results are presented in tabular form. Table 7 shows the operational parameters of the FMC system considered. The time related parameters are mean values in each case.

| Operational Parameters | Parameter Values |
|---|---|
| Mean processing time per part, $1/v_r$ | 1.00 time unit, r = 1 and 2 (for machines 1 and 2) |
| Mean robot loading time for the first part, $1/l_r$ | 0.25 time units for machines and robots r = 1,2 |
| Mean robot loading/unloading time for subsequent parts, $1/z_r$ | 0.50 time units for both robots and machines, r = 1,2 |
| Mean robot unloading time for last part, $1/u_r$ | 0.25 time units for robots and machines, r = 1,2 |
| Mean time to failure for machine r, $1/\lambda_r$ | 100 time units for both machines, r = 1,2 |
| Mean repair time of machine r, $1/\mu_r$ | 10 time units for both machines, r = 1,2 |
| Mean pallet transfer time, $1/w$ | Varied between 1,...,10 time units per pallet |
| Pallet capacity | n = 4 units. |

Table 7. Multi-Robot FMC operational parameters considered in the analysis.

Results of analysis for the double robot FMC are shown in tabular form in table 8, which summarizes the percentages of times for system states, particularly the states in which each FMC component is idle, busy or under repair. One can use this table to evaluate performance of each system component, such as the fraction of time that each machine would be idle, under repair, or operational; or the fraction of time that each robot and the pallet would be idle or operational in the steady state. Figure 15 shows the production output rate of the FMC system with respect to pallet transfer rates at three different machine repair rates. As it is seen in the figure, when the repair rate is doubled from 0.1 to 0.2, the production rate increases twice more than the rate of increase when the repair rate is doubled from 0.2 to 0.4. The increase in the production rate as a function of the pallet rate is also shown in this figure. The trend in the production rate is similar to previous cases, that is, a pallet transfer rate of 3 or more units/unit time is effective for increasing production output rate under any possible repair policy. Such analysis are extremely useful for operational engineers and maintenance managers for effective operation of an FMC and for planning of maintenance/repair activities as well as for production planning.

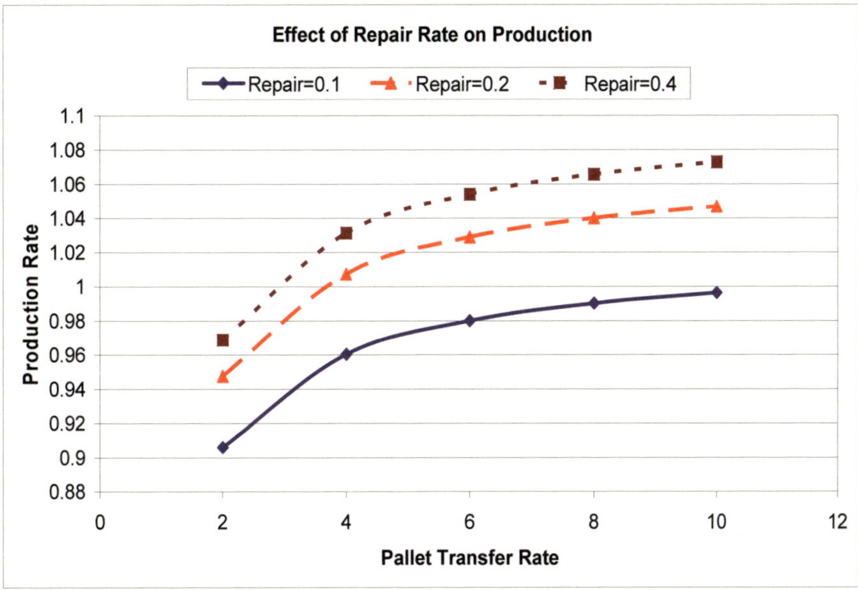


Fig. 15. FMC production rate as a function of pallet transfer rate at three different machine repair rates

## 6. Conclusions

The demand for customized products has been continuously increasing in recent years and a great deal of attention has been given to automation of manufacturing, specifically to flexible manufacturing systems (FMS) and flexible manufacturing cells (FMC). FMC are less costly, smaller, and less complex systems than FMS. In order to get full benefit from these systems, they have to be analyzed in detail before implementation as well as during their operations. While modeling and analysis of traditional machines and production systems have been subject of extensive research over the past several years, not as many studies can be seen on FMC systems. Also, there are many books written about conventional manufacturing systems and very few on FMS and FMC systems.

In this chapter, we have presented a method for developing stochastic models to be used in the design and analysis of unreliable FMC systems with one or more machines that are served by one or more robots and a common pallet handling system. The models are used to determine system performance measures, such as production output rate and system component utilizations, under different parametric conditions. Exact solutions were obtained for the stochastic models either in closed form or numerically and case problems were solved by these models to illustrate their applications and the results obtained. As shown by various results and graphs, the models could be a useful tool in the design as well as in the operational phases of FMC. It was observed that the FMC production rate was significantly affected by cell parameters such as pallet capacity, pallet transfer rate, robot loading/unloading rates, or the repair rates of the machines. Equipment utilizations are also analyzed with respect to different parameters by using the models presented. These models could be useful for system designers, FMC manufacturers, and the operation engineers.

| P.T.R. | | 1 | 2 | 3 | 4 | 5 | 6 | 7 | 8 | 9 | 10 |
|---|---|---|---|---|---|---|---|---|---|---|---|
| Machine 1 | Busy | 0.407 | 0.453 | 0.471 | 0.480 | 0.486 | 0.490 | 0.493 | 0.495 | 0.497 | 0.498 |
| | Idle | 0.552 | 0.502 | 0.482 | 0.472 | 0.465 | 0.461 | 0.458 | 0.455 | 0.454 | 0.452 |
| | Repair | 0.041 | 0.045 | 0.047 | 0.048 | 0.049 | 0.049 | 0.049 | 0.050 | 0.050 | 0.050 |
| Machine 2 | Busy | 0.407 | 0.453 | 0.471 | 0.480 | 0.486 | 0.490 | 0.493 | 0.495 | 0.497 | 0.498 |
| | Idle | 0.552 | 0.502 | 0.482 | 0.472 | 0.465 | 0.461 | 0.458 | 0.455 | 0.454 | 0.452 |
| | Repair | 0.041 | 0.045 | 0.047 | 0.048 | 0.486 | 0.049 | 0.049 | 0.050 | 0.050 | 0.050 |
| Robot 1 | Busy | 0.209 | 0.232 | 0.241 | 0.246 | 0.249 | 0.251 | 0.253 | 0.254 | 0.255 | 0.255 |
| | Idle | 0.792 | 0.768 | 0.759 | 0.754 | 0.751 | 0.749 | 0.747 | 0.746 | 0.745 | 0.745 |
| Robot 2 | Busy | 0.209 | 0.232 | 0.241 | 0.246 | 0.249 | 0.251 | 0.253 | 0.254 | 0.255 | 0.255 |
| | Idle | 0.792 | 0.768 | 0.759 | 0.754 | 0.751 | 0.749 | 0.747 | 0.746 | 0.745 | 0.745 |
| Pallet | Busy | 0.204 | 0.113 | 0.079 | 0.060 | 0.049 | 0.041 | 0.035 | 0.031 | 0.028 | 0.025 |
| | Idle | 0.797 | 0.887 | 0.922 | 0.940 | 0.951 | 0.959 | 0.965 | 0.969 | 0.972 | 0.975 |
| Production | | 0.814 | 0.906 | 0.942 | 0.960 | 0.972 | 0.980 | 0.986 | 0.990 | 0.994 | 0.996 |

Table 8. Percentages of time in which system components are in different states at different pallet transfer rates (P.T.R.) for pallet capacity of n=4 and the FMC parameters of table 7.

Stochastic models and the closed form solution formulas obtained in this paper could be used to analyze and optimize the productivity and other performance measures of a FMC under different machine, robot, and pallet operational characteristics. Using the models presented in this paper, best parameter combinations can be determined for a given FMC system. In particular, best machining rates, robot loading and unloading rates, pallet capacity, and pallet transfer rates can be determined for a given set of FMC machine characteristics. Furthermore, reliability and availability analysis of the FMC system can be determined based on different failure/repair characteristics of the machines in the system. It is possible to optimize machine repair rates, based on other system parameters, to achieve maximum production output rates and other performance measures.

## Acknowledgement

This chapter was prepared with a support from Kuwait University Research Administration under the research grant number EI01/08.

## 7. References

Abdulmalek, F., Savsar, M., and Aldaihani, M. (2004). "Simulation of Tool Change Policies in a Flexible Manufacturing Cell", *WSEAS Transactions on Systems, Vol.* 7, No. 3, pp. 2546-2552.

Adamyan, A. and He, D. (2002). "Analysis of sequential failures for assessment of reliability and safety of manufacturing systems", *Reliability Engineering & System Safety,* No. 76, pp. 227-236.

Aldaihani, M. and Savsar, M. (2005a). "Stochastic Modeling and Analysis of a Two-Machine Flexible Manufacturing Cell", *Computers and Industrial Engineering*, No. 49, pp. 600-610.

Aldaihani, M. and Savsar, M. (2005b). "Modeling of a Flexible Manufacturing Cell with Two Unreliable Machines Served by Two Robots", *INFORMS Annual Conference*, November 13-16, San Francisco, CA, USA.

Aldaihani, M. and Savsar, M. (2008). "Stochastic models for reliable and unreliable flexible manufacturing cells with two machines and two robots", *International journal of Industrial and Systems Engineering, Vol. 3, No. 5, pp. 610-624.*

Black, J. J. and Mejabi, O. O. (1995). "Simulation of Complex Manufacturing Equipment Reliability Using Object Oriented Methods", *Reliability Engineering & System Safety*, No. 48, pp. 11-18.

Chan, D. and Bedworth, D.D. (1990). "Design of a Scheduling System for Flexible Manufacturing Cells" *International Journal of Production Research, No.* 28, pp. 2037-2049.

Butler, A. C. and Rao, S. S. (1993). "Reliability Analysis of Complex Systems Using Symbolic Logic", *Reliability Engineering & System Safety*, No. 40, pp. 49-60.

Cogun, C. and Savsar, M. (1996). "Performance Evaluation of a Flexible Manufacturing Cell (FMC) by Computer Simulation", *Modeling Measurement, & Control B*, Vol. 62, No. 2, pp. 31-44.

Han, B. T., Zhang, C. B., Sun, C. S., and Xu, C. J. (2006), "Reliability Analysis of Flexible Manufacturing Cells Based on Triangular Fuzzy Number", *Communications in Statistics-Theory and Methods,* 35 (10), 1897-1907.

Henneke, M. J. and Choi, R.H. (1990). "Evaluation of FMS Parameters on Overall System Parameters" *Computers an Industrial Engineering*, No. 18, pp.105-110.

Khodabandehloo, K. and Sayles, R. S. (2007). "Production Reliability in Flexible Manufacturing Systems", *Quality and Reliability Engineering International, Vol.* 2, No. 3, pp. 171-182.

Koulamas, C.P. (1992). "A Stochastic Model for a Machining Cell with Tool Failure and Tool Replacement Considerations", *Computers and Operations Research,* No. 19, pp. 717-729.

Savsar, M. and Cogun, C. (1993). "Stochastic Modeling and Comparisons of Two Flexible Manufacturing Cells with Single and Double Gripper Robots", *International Journal of Production Research*, No. 31, pp. 633-645.

Savsar, M. (2000), "Reliability analysis of a flexible manufacturing cell", *Reliability Eng. & System Safety*, No. 67, pp. 147-132.

Savsar, M. and Aldaihani, M. (2004). Modeling and Analysis of a Flexible Manufacturing Cell with two Machines Served by a Robot" IMS'2004: *The 4th International Symposium on Intelligent Manufacturing,* September 4-6, Sakarya, Turkey.

Savsar, M. and Aldaihani, M. (2008). "Modeling of Machine Failures in a Flexible Manufacturing Cell with Two Machines Served by a Robot", *Journal of Reliability and System Safety, Vol. 93, No. 10, pp. 1551-1562.*

Savsar, M. (2008). "Calculating Production Rate of a Flexible Manufacturing Module", *International Journal of Advanced Manufacturing Technology*, Vol. 37, pp. 760-769.

Simeu-Abazi, Z., Daniel, O., and Descotes-Genon, B. (1997). "Analytical Method to Evaluate the Dependability of Manufacturing Systems", *Reliability Engineering & System Safety*, Vol. 55, pp. 125-130.

Wang, K. and Wan, E. (1993), "Reliability Consideration of a Flexible Manufacturing System from Fuzzy Information", *International Journal of Quality and Reliability Management*, Vol. 10, No. 7, pp. 366-376.

Yuanidis, P., Styblinski, M. A., Smith, D. R., and Singh, C. (1994). "Reliability Modeling of Flexible Manufacturing Systems", *Microelectronics and Reliability*, Vol. 34, No. 7, pp. 1203-1220.

5

# Multi agent and holonic manufacturing control

Sugimura Nobuhiro[1], Tehrani Nik Nejad Hossein[2] and Iwamura Koji[1]
[1]*Graduate School of Engineering, Osaka Prefecture University, Osaka, Japan*
[2]*Information and communication Engineering, Toyota Technological Institute, Japan*

## 1. Introduction

The manufacturing industry will continue to be in the future one of the main wealth generators of the world economy (CMV, 1998). In the last decades world has moved towards a global economy, with markets demanding for products with high quality at lower costs, highly customized and with short life cycles, imposing new requirements on manufacturing enterprises, namely in terms of quality, response, agility and flexibility, that are crucial for an enterprise staying in the business.
The traditional manufacturing control systems are not designed to exhibit these capabilities of responsiveness, flexibility, robustness and re-configurability, since they are built upon centralized and hierarchical control structures. They present good production optimization, but a weak response to adopt due to the rigidity and centralization of their control structures. Such centralized hierarchical organization normally leads to situations where the whole system is shutting down by single failures at one point of the system hierarchy (Colombo et al., 2006). The current challenge is to develop collaborative and reconfigurable manufacturing control systems that support efficiently small batches, product diversity, high quality and low costs, by introducing innovative characteristics of adaptation, agility and modularization.
Information and communication technologies, and artificial intelligence techniques, have been used for more than two decades addressing this challenge. Namely, agent-based and holonic manufacturing control seem to be suitable to face these requirements such as modularity, scalability, autonomy and re-usability, since they present decentralization of control over distributed structures. When properly designed and implemented, agent-based control systems result in a performance that is flexible, robust, adaptive and fully tolerant, which are key factors for manufacturing success in the increasingly global marketplace.
In this chapter, we review the different manufacturing control architecture including centralized and distributed. We discuss about intelligent and distributed manufacturing control systems using emerging paradigms, such as multi-agent systems and holonic manufacturing systems (HMSs), and present two case studies about the applications of agent-based manufacturing control systems for process planning and scheduling. The objective of this chapter is to provide an overview about the application of multi-agent systems and holonic manufacturing principles to manufacturing environment, but to focus on the manufacturing control applications.
The chapter is organized as follows. Section 2 reviews the concepts associated with manufacturing control systems, describing the traditional approaches and the distributed

and intelligent ones, namely the agent-based and holonic manufacturing control system. In section 3, we present two case studied including real-time scheduling method for holonic manufacturing system and agent-based dynamic integrated process planning and scheduling in flexible manufacturing system. Finally, we briefly discuss about realizing the agent based manufacturing system by applying the ORiN (Open Robot Interface Network) architecture which recently has been developed for manufacturing automation.

## 2. Manufacturing control systems

### 2.1 Traditional approach to manufacturing control problem

The manufacturing control is concerned with managing and controlling the physical activities in the factory aiming to execute the manufacturing plans, provided by the manufacturing planning activity, and to monitor the progress of the product as it is being processed, assembled, moved, and inspected in the factory. Algorithms at this level are used to decide what to produce, how much to produce, when production is to be finished, how and when to use the resources or make them available, when to release jobs into the factory, which jobs to release, job routing, and job/operation sequencing (Baker, 1998).

Due to its complexity, especially the high number of interactions between the different components and the variety of functions executed, manufacturing control systems are traditionally implemented using centralized or hierarchical control approaches, comprising, the following main components: planning, scheduling, execution (i.e. dispatching, monitoring, diagnosis and error recovery) and machine/device control. Each one of these components operates in a specific temporal horizon, ranging from weeks at the strategic level to seconds at the shop floor.

The traditional approach to manufacturing control systems based on centralized or hierarchical control structures, presents good characteristics in terms of productivity, essentially due to its intrinsic optimization capabilities. However, dynamic and adaptive response to change is, currently, the key to competitiveness, and the traditional approaches to manufacturing control typically fall into large monolithic and centralized software packages that are developed and adapted case by case, requiring a huge and expensive effort to implement, maintain or re-configure. In conclusion, they are not adequate because they do not support efficiently the current requirements imposed to manufacturing systems, namely in terms of flexibility, expansibility, agility and re-configurability.

### 2.2 Agent-based manufacturing control

The multi-agent system paradigm derives from the distributed artificial intelligence (DAI) field, being characterized by decentralization and parallel execution of activities based on autonomous entities, called agents. The definition of agent concept is neither unique nor consensual (Russel & Norvig, 1995; Wooldridge & Jennings, 1995;Wooldridge, 2002). Despite some definitions and interpretations for agents, a suitable definition is: "An autonomous component that represents physical or logical objects in the system, capable to act in order to achieve its goals, and being able to interact with other agents, when it does not possess knowledge and skills to reach alone its objectives". The most important properties of an agent are the autonomy, intelligence, adaptation and co-operation.

There are several agent architectures, ranging from reactive agents, operating in a stimulus-response manner, to deliberative agents characterized by their pro-active reasoning and goal-oriented behaviour. A well-known deliberative and cognitive agent-type is belief–desire–intention (BDI) architecture, which origin lies in a theory of human practical reasoning, focusing particularly on the role of intentions in practical reasoning (Wooldridge, 2002). In the BDI agents, the decision-making depends on the manipulation of beliefs, desires and intentions of the agents.

A multi-agent system can be defined as a set of agents that represent the objects of a system, capable of interacting, in order to achieve their individual goals, when they have not enough knowledge and/or skills to achieve individually their objectives. Agents organize themselves into a heterarchical structure characterized by the high-level of autonomy and co-operation, being the client–server structure with fixed relations no more applied (Diltis et al., 1991). These features allow a high performance against disturbances, but the global optimization reduced, because the decision-making is local and autonomous, without a global view of the system. The expansibility of the system is easier, and only enough to modify the functioning of some agents or add new agents to the control system.

In the automation and manufacturing domains, an agent can represent physical resources, such as machine tools, robots, auto-guided vehicles (AGVs) and products or logical objects, such as the schedulers and orders (Sepehri & Tehrani, 2005). Using the appropriate distributed control algorithms, individual machines and product agents can make their own manufacturing control decisions relating to resource allocation and coordination, using an automated form of "negotiation". The key benefit of such approach is that if production is disrupted or re-organized in some way, the same negotiation process still takes place, with different machines or products making the decisions, and hence the system is relatively robust to change.

### 2.3 Holonic manufacturing control

The Holonic Manufacturing System (HMS) is a paradigm that translates into the manufacturing world the concepts developed by Arthur Koestler from living organisms and social organizations. In middle of sixties, Koestler introduced the word holon to describe the basic unit of organization in living organisms and social organizations, based on Herbert Simon theories and on his observations (Koestler, 1969).

The HMS has been proposed and discussed in the HMS consortium, in order to develop a new autonomous distributed architecture of the manufacturing systems, which are applicable to very small batch productions. The HMS consortium has developed the following definitions to help the common understanding of the HMS (Wyns, 1999).

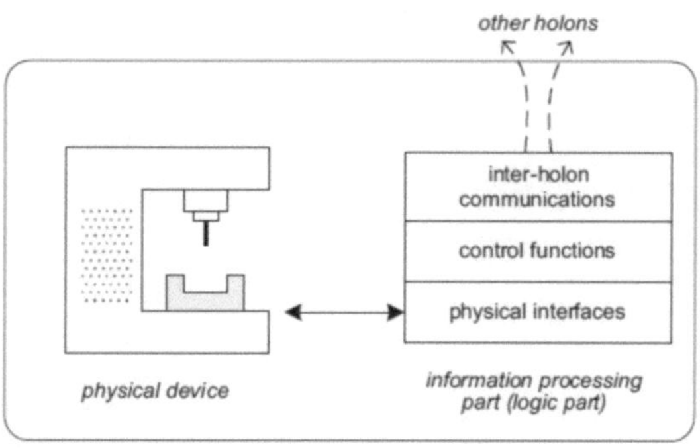


Fig. 1. A physical holon.

- *Holon*: An autonomous and cooperative building block of a manufacturing system for transforming, transporting, storing and/or validating information and physical objects. A holon can represent a physical or logical activity, such as a robot, a machine, an order, a flexible manufacturing system or even an human operator. The holon has information about itself and the environment, containing an information processing part and a physical processing part when the holon represents a physical device (Winkler & Mey, 1994), such as an industrial robot, as illustrated in Fig. 1.
- *Autonomy*: The capability of an entity to create and control the execution of its own plans and strategies.
- *Cooperation*: A process whereby a set of entities develops mutually acceptable plans and executes these plans.
- *Holarchy:* A system of holons that can cooperate to achieve a goal or objective. The holarchy defines the basic rules for cooperation of the holons and thereby limits their autonomy.

Brussel et al. (1998) has proposed PROSA a reference architecture is built around three types of basic holons: order holons, product holons, and resource holons. Each of them is responsible for one aspect of manufacturing control, logistics, technological planning, or resource capabilities respectively. These basic holons are structured using object-oriented concepts like aggregation and specialization. Staff holons can be added to assist the basic holons with expert knowledge. These allow for the use of centralized algorithms and for the incorporation of legacy systems.

## 3. Agent and Holonic Case Studies in Manufacturing Systems

### 3.1 Real-time scheduling method for HMS

In this case study, we discuss about a real-time scheduling method for the manufacturing processes in the HMS. The components in the HMS are basically divided into three classes; they are CNC machine tool (CMT) holons, job holons, and coordination holons. We develop a coordination method for the coordination holon to determine a suitable combination of the CMT holons and the job holons based on the utility values. We verify the effectiveness of the proposed real-time scheduling method from the view point of the objective functions of the individual CMT holons and job holons.

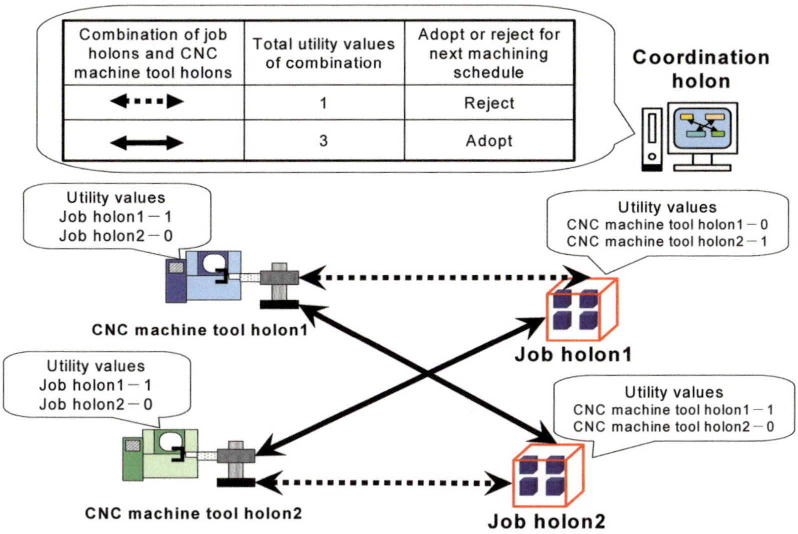

Fig. 2. Real-time scheduling method based on utility values

### 3.1.1 Basic Architecture of HMS

The holons in the HMS are divided into three classes based on their roles in the manufacturing processes and the scheduling processes.

- CNC machine tool (CMT) holons: They transform the job holons in the manufacturing process. In the scheduling process, they evaluate the utility values for the candidate job holons which carry out the machining process in the next time period.
- Job holons: They are transformed by the CMT holons from the blank materials to the final products in the manufacturing process. In the scheduling process, they evaluate the utility values for the candidate CMT holons which carry out the machining process in the next time period.
- Coordination holon: It carries out the coordination among the holons, and selects a most suitable combination of the CMT holons and the job holons for the machining process in the next time period in the scheduling process.

### 3.1.2 Real-time scheduling processes based on utility values

It is assumed here that the individual job holons have the following technological information representing the machining process of the jobs.

$M_{ik}$ : $k$-th machining process of the job holon $i$. ($i = 1,..,\alpha$), ($k = 1,..,\beta$).

$AC_{ik}$ : Required machining accuracy of machining process $M_{ik}$. It is assumed that the machining accuracy is represented by the levels of accuracy indicated by 1, 2, and 3, which mean rough, medium high, and high accuracy, individually.

$R_{ikm}$ : $m$-th candidate of CMT holon, which can carry out the machining process $M_{ik}$. ($m = 1,...,\gamma$).

$T_{ikm}$ : Machining time in the case where the CMT holon $R_{ikm}$ carries out the machining process $M_{ik}$.

$W_i$ : Waiting time until the job holon $i$ becomes idle if it is under machining status.

| Objective functions | | Objective function values |
|---|---|---|
| CMT holon | Efficiency | Σ Machining time / Total time |
| | Machining accuracy | Σ(Machining accuracy of CMTs – Required machining accuracy of jobs) |
| Job holon | Flow-time | Σ(Machining time + Waiting time) |
| | Machining cost | Σ(Machining cost of CMTs) |

Table 1. Objective functions of holons

The individual CMT holons have the following technological information representing the machining capability of the resources for the machining process $M_{ik}$.

$MAC_{ikm}$ : Machining accuracy in the case where the CMT holon $R_{ikm}$ carries out the machining process $M_{ik}$. $MAC_{ikm}$ is also represented by the level of 1, 2 and 3.

$MCO_{ikm}$ : Machining cost in the case where the CMT holon $R_{ikm}$ carries out the machining process $M_{ik}$.

$W_{ikm}$ : Waiting time until CMT holon $R_{ikm}$ becomes idle if it is under machining status.

Based on the information above mentioned, a real-time scheduling method based on the utility values shown in Fig. 2 is proposed, to select a suitable combination between the job holons and the CMT holons which carries out the machining process in the next time period. At the time $t$, all the 'idling' holons have to select their machining schedules in the next time period, as shown in the Fig. 2. The following procedure is proposed for the individual holons to select their machining schedules.

(1) *Retrieval of status data*: The individual 'idling' holons firstly get the status data from the other holons which are 'operating' or 'idling'. The 'idling' holons can start the machining process in the next time period.

(2) *Selection of candidate holons*: The individual 'idling' holons select all the candidate holons for the machining process in the next time period. For instances, the job holon $i$ selects the CMT holons which can carry out the next machining process $M_{ik}$. On the other hand, the CMT holon $j$ select all the candidate job holons which can be machined by the CMT holon $j$.

(3) *Determination of utility values*: The individual holons determine the utility values for the individual candidates selected in the second step. For instances, the job holon determines the utility values, based on its own decision criteria for all the candidate CMT holons which can carry out the next machining process.

(4) *Coordination:* All the job holons and the CMT holons send the selected candidates and the utility values of the candidates to the coordination holon. The coordination holon determine a suitable combination of the job holons and the CMT holons which carry out the machining processes in the next time period, based on the utility values. The decision criteria of the coordination holon is to maximize the total sum of the utility values of all the holons.

### 3.1.3 Evaluation of utility values

It is assumed that the individual holons have one of the objective functions shown in Table 1 for evaluating the utility values.

(1) Efficiency of CMT holons→ ME to be maximized
*ME* is the ratio of the total machining time of the CMT holon and the total time after the CMT holon starts the operations. The total time includes both the machining time and the idling time of the CMT holon.

(2) Machining accuracy of CMT holons→MA to be minimized
*MA* is the difference between the level of machining accuracy of the CMT holons and the required level of accuracy of the machining process.

(3) Flow-time of job holons→JT to be minimized
*JT* is the flow-time of the job holon. *JT* includes the machining time and the idling time of the job holons.

(4) Machining cost of job holons→ JC to be minimized
*JC* is the sum of the machining cost of the job holon, which are evaluated from the machining costs of the CMT holons.

The following procedures are provided for the CMT holons to evaluate the utility values. Let us consider a CMT holon $j$ at a time $t$. It is assumed that $TT_{j \cdot t}$, $ME_{j \cdot t}$, and $MA_{j \cdot t}$ show the total time after the CMT holon $j$ starts its operations, the efficiency, and the evaluated value of machining accuracy of the CMT holon $j$, respectively. If the CMT holon $j$ selects a candidate job holon $i$ for carrying out the machining process $M_{ik}$, the efficiency and the evaluated value of the machining accuracy are estimated by the following equations.

$$ME_{j \cdot t+1}(i) = (ME_{j \cdot t} \cdot TT_{j \cdot t} + T_{ikm}) / (TT_{j \cdot t} + T_{ikm} + W_i) \tag{1}$$

$$MA_{j \cdot t+1}(i) = MA_{j \cdot t} + (MAC_{ikm} - AC_{ik}) \tag{2}$$

where, the CMT holon $j$ can carry out the machining process $M_{ik}$ of job holon $i$ ($j = R_{ikm}$).
As regards the job holons, the following equations are applied to evaluate the flow-time and the machining costs, for the case where a job holon $i$ selects a candidate CMT holon $j$ ($= R_{ikm}$) for carrying out the machining process $M_{ik}$. It is assumed that $JT_{i \cdot t}$ and $JC_{i \cdot t}$ give the total time after the job holon $i$ is inputted to the HMS and the machining cost, respectively.

$$JT_{i \cdot t+1}(j) = JT_{i \cdot t} + T_{ikm} + W_{ikm} \tag{3}$$

$$JC_{i \cdot t+1}(j) = JC_{i \cdot t} + MCO_{ikm} \tag{4}$$

The objective functions mentioned above have different units. Some of them shall be maximized and others shall be minimized. Therefore, the utility values are normalized from 0 to 1, by applying the following equations.

- **Efficiency of CMT holons**

$$RUV_j(i) = 1 - \{\max_{i=1,\cdots\tau} [ME_{j \cdot t+1}(i)] - ME_{j \cdot t+1}(i)\} / \{\max_{i=1,\cdots\tau} [ME_{j \cdot t+1}(i)] - \min_{i=1,\cdots\tau} [ME_{j \cdot t+1}(i)]\} \tag{5}$$

- **Machining accuracy of CMT holons**

$$RUV_j(i) = \{\max_{i=1,\cdots\tau} [MA_{j \cdot t+1}(i)] - MA_{j \cdot t+1}(i)\} / \{\max_{i=1,\cdots\tau} [MA_{j \cdot t+1}(i)] - \min_{i=1,\cdots\tau} [MA_{j \cdot t+1}(i)]\} \tag{6}$$

- **Flow-time of job holons**

$$JUV_i(j) = \{\max_{j=1,\cdots\gamma} [JT_{i \cdot t+1}(j)] - JT_{i \cdot t+1}(j)\} / \{\max_{j=1,\cdots\gamma} [JT_{i \cdot t+1}(j)] - \min_{j=1,\cdots\gamma} [JT_{i \cdot t+1}(j)]\} \tag{7}$$

- **Machining cost of job holons**

$$JUV_i(j) = \{\max_{j=1,\cdots\gamma} [JC_{i \cdot t+1}(j)] - JC_{i \cdot t+1}(j)\} / \{\max_{j=1,\cdots\gamma} [JC_{i \cdot t+1}(j)] - \min_{j=1,\cdots\gamma} [JC_{i \cdot t+1}(j)]\} \tag{8}$$

where, max[$f(x)$] and min[$f(x)$] give the maximum value and the minimum value of $f(x)$

| | Resource 1 | Resource 2 | ... | Resource $\gamma$ |
|---|---|---|---|---|
| Job 1 | $a_{11}$ | $a_{12}$ | ... | $a_{1\gamma}$ |
| Job 2 | $a_{21}$ | $a_{22}$ | ... | $a_{2\gamma}$ |
| ... | ... | ... | ... | ... |
| Job $\tau$ | $a_{\tau 1}$ | $a_{\tau 2}$ | ... | $a_{\tau\gamma}$ |

Table 2. Combination of CMT and job holons

evaluated for all candidates $x$. $\tau$ and $\gamma$ gives the number of the candidate job holons for the CMT holon $j$, and the number of the candidate CMT holons for the job holon $i$, respectively.

### 3.1.4 Coordination among holons

After evaluating the utility values, all the 'idling' holons send all the candidates and their utility values to the coordination holon, and the coordination holon select a most suitable combination of the CMT holons and the job holons, which execute the machining processes in the next time period. The coordination process is summarized in the following, for the case where the coordination holon determines a suitable combination of all the candidates of the job holons $i$ ($i$ = 1,2,..,$\tau$) and the CMT holons $j$ ($j$ =1,2,.., $\gamma$).

The utility value $\delta_{ij}$ of the combination of job holon $i$ and CMT holons $j$ is given by the following equation

$$\delta_{ij} = RUV_j(i) + JUV_i(j) \tag{9}$$

The problem to be solved by the coordination holon is to select a combination of job holons and CMT holons which maximize the total of the utility value, as shown in the following equation.

$$\underset{a_{ij} \in A}{\text{maximize}} \left( \sum_{i=1}^{\tau} \sum_{j=1}^{\gamma} a_{ij} \cdot \delta_{ij} \right) \tag{10}$$

where, $a_{ij}$ (= 0 or 1) are the decision parameters, as shown in Table 2. If $a_{ij}$ =1, the job holon $i$ is machined by the CMT holon $j$ in the next time period. Otherwise, job holon $i$ is not machined by the CMT holon $j$. Only one job holon is machined by one CMT holon, therefore, $A$ is a set of $a_{ij}$ which satisfy the following equation.

$$\sum_{i=1}^{\tau} a_{ij} \leq 1, \quad \sum_{j=1}^{\gamma} a_{ij} \leq 1 \qquad 11)$$

### 3.1.5 Case study

Some case studies have been carried out to verify the effectiveness of the proposed methods. The HMS model consisting of 10 CMT holons is considered for the case study. The individual CMT holons have the different objective functions and the different machining capacities, such as the machining time $T_{ikm}$, the machining accuracy $MAC_{ikm}$, and the machining cost $MCO_{ikm}$.

As regards the job holons, 12 job holons are considered in the case study, which have the different objective functions and the machining process. 12 cases are considered in the case study by changing the machining capacities of the individual CMT holons. Fig. 3 summarizes the comparison between the proposed scheduling method based on utility values and the rule-based method, from the viewpoint of the objective function values of the individual holons. In the Fig. 3, the horizontal axis gives the type of the objective functions of the individual holons, and the vertical axis shows the average number of holons $\lambda$, the objective function values of which are improved by the utility values based methods. The $\lambda$ is calculated by the following equation.

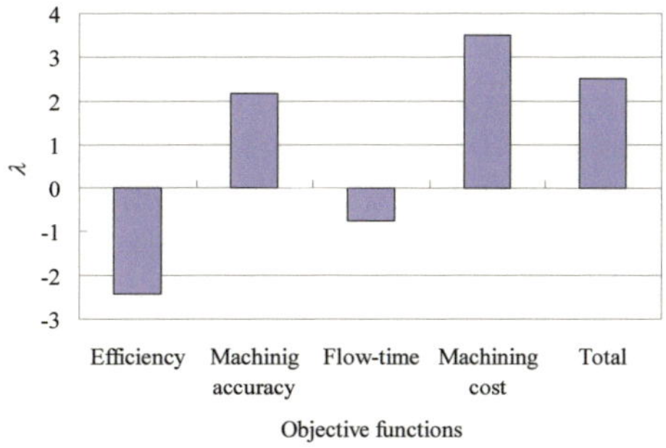


Fig. 3. Results of case study

$$\lambda = \sum_{g=1}^{12} (a_g - b_g) / 12 \tag{12}$$

$a_g$: number of holons, the objective function values of which are improved by the proposed method in the case $g$,
$b_g$: number of holons, the objective function values of which are deteriorated by the proposed method in the case $g$.
As shown in Fig. 3. the proposed scheduling method based on utility values is effective to improve the objective function values of the individual holons from the view point of total number of holons. However, the dispatching rules applied to the rule-based method is very effective to reduce the total make span, therefore some holons with the objective functions of the efficiency or the flow-time do not improve their objective function values by the proposed method.

### 3.2 Agent-based dynamic integrated process planning and scheduling

Process planning and scheduling are important manufacturing planning activities which deal with resource utilization and time span of manufacturing operations. The process planning and scheduling tasks are very complicated and time consuming, if it is applied to the dynamically changing FMSs (Flexible Manufacturing Systems).
PROSA has already proposed a reference architecture for developing distributed manufacturing system including three types of basic holons: resource holons, product holons and order holons (Brussel et al., 1998). Leitao & Restivo (2006) also proposed a control architecture, designated by ADAptive holonic Control aRchitecture (ADACOR) for FMS, intends to contribute to the improvement of the manufacturing control systems performance in term of agile reaction to disturbances and change. Although, there are general architecture and far more detail than is needed for practical applications. We use the elements of PROSA and ADACOR architectures to develop a multi agent architecture for real time integrated process planning and scheduling system in the FMSs, which generate

suitable process plans and schedules based on the status of the FMSs. A comprehensive design and implementation have been done to show the effectiveness of the multi agent approach for dynamic integrated process planning and scheduling of mechanical parts. The methods proposed in the literatures deal mainly with the process planning and scheduling tasks in the static environment in which the jobs specifications and the manufacturing system status are stable (Leitao, 2009). However, it is now required to develop an integrated process planning and scheduling systems applicable to the dynamic environment in which some unforeseen disturbances may occur. In this case study, we propose a multi-agent based integrated system for process planning and scheduling in the dynamic environment, in order to cope with the jobs specification changes and the unforeseen disruptions, such as the malfunction of the machine tools and through the simulation, we are going to illustrate how the agents are able to real timely handle disturbances effectively.
In the literature of agent based manufacturing system, many researches apply simple algorithms such as dispatching rules which are applicable for real time decision making. These methods are simple and applicable, but they do not guarantee the effectiveness for complex problems in the manufacturing systems. As the efficiency becomes more important in the agent based manufacturing (Shen et al., 2006, Wang et al., 2006), we apply coordination agent and a mathematical model for assigning the jobs to the machine tools. The developed model is enough fast to be applicable for the real time agent based manufacturing system when we face unforeseen disturbances such as machine tool breakdown and job specification changes.

### 3.2.1 Multi Agent Architecture

A multi-agent architecture is proposed to carry out the integrated process planning and scheduling. We mainly use the PROSA and ADACOR reference architectures to define the necessary elements such as agents but we have tailored and customized them according to our specific problem. We also define new agents such as machining process agents and production engineering agents to increase the performance of the architecture for disturbance handling and dynamically generating alternative process plans. In the following sections, the system architectures proposed here are discussed from the viewpoints of the agent definitions, negotiation protocols, and coordination mechanism among the agents.

### 3.2.2 Agents Definition

Two types of agents are considered in this research to develop the integrated process planning and scheduling systems. They are, physical agents and information agents. The physical agents represent jobs, the manufacturing resources and the machining processes in the FMSs and they are similar to the product, order and resource holons defined in PROSA and ADACOR reference architectures although the contents of the agents and their communications have been customized. The manufacturing resources considered here are the machine tools, the preparation stations, the AGVs (Automated Guided Vehicles), the fixtures and the cutting tools. The information agents are virtual agents which are responsible for governing the negotiation protocol and decision-making.

#### 3.2.2.1 Physical agents

The UML class diagram of the physical agents including their attributes and relationships

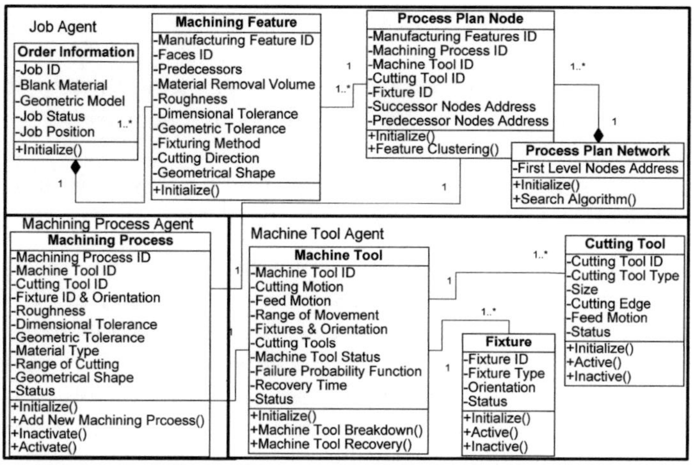


Fig. 4. UML Class diagram of physical agents and their relations.

are summarized in Fig.4. The attributes of the physical agents according to the Fig. 4 are summarized in the following.

*Job agents.* The job agents represent the jobs to be manufactured in the FMSs. The role of the job agents in the process planning is to certify the correct machining processes of the jobs. It is quite similar of product holon in PROSA architecture although we encapsulate the process plan networks in order to dynamically generate alternative process plans which is the key factor for integrating the process planning and scheduling and disturbance handling. The process plan network has essential role for improving the efficiency of the total system and handling the unforeseen disturbances. Generating alternative process plans has not been discussed clearly in the previous architectures and we present a reliable method based on the process plan network to generate process plans according to available manufacturing processes. The job agents include the following information to describe the orders and the machining features.

<u>Job information:</u>

The job information section describes the order information, the locations and the progresses of the machining processes of the jobs.

<u>Machining features:</u>

The machining feature section gives the machining features of the jobs and their technical data such as the types, the tolerances and the roughness. These technical data are required to select appropriate machining processes.

<u>Process plan networks:</u>

The process plan networks represent the generated process plans in non-linear and hierarchical ways. It includes all the alternative process plans that satisfy the technological requirements of the jobs. Although the mechanical parts are complicated but the size of process plan network in practice will remain limited as we are able to group the manufacturing features. Generally in FMSs, the machining centers which include wide range of cutting tools are able to carry out many manufacturing processes of the job at the same time.

Job Status:

We consider the following status for the job agents.

- Idle: The job agent is idle and waiting for the next machining operations.
- Machining operation: The job agent is under machining processes on the machine tools.
- Transportation and re-fixturing: The job agent is transported and/or re-fixtured for its next machining operations.

*Machine tool agents.* The machine tool agents represent the machine tools which are quite similar as the resource holon at PROSA and operational holon at ADACOR reference architectures. PROSA considers the machine tool, cutting tools and fixtures as individual holons. At FMSs generally each machine tools include wide range of the cutting tools for doing different machining processes. To avoid the complexity for practical applications in large manufacturing systems, we do not consider separate agents for cutting tools and fixtures to decrease the total number of agents. We are trying to identify the differences between cutting tools and fixtures by using the information from machining process agent. The agents representing such resources as the preparation stations and the AGV are not considered, at present, since only the machining processes are discussed in the present research. The machine tool agents are responsible for generating proposals to the machining processes required from the job agents. The proposals include the machining time, the transportation time and the re-fixturing time needed to carry out the required machining processes of the job agents. The machine tool agents include the following information to represent the machine tools in the FMSs.

1. Machine tool information:

   The machine tool section specifies the shape generation functions, which are represented by the cutting motions, the spindle directions, the feed motions and the maximum product size.
2. Machine tool status:

   We consider the following status for the machine tool agents in the simulation
   - Idle: The machine tool is idle and negotiating with job agents for next machining operation
   - Machining operation: The machine tool is machining the job agent
   - Breakdown: The machine tool has been broken and is under recovery process
3. Cutting tool:

   The characteristics of the cutting tools are described in the cutting tool section, which includes the information about the cutting tool types, the tool sizes and the cutting edge types.
4. Fixture:

   The fixture section describes the fixture types, and the positions of the fixtures against the spindle axis.

*Machining process agents.* The machining process agents represent the machining processes of machining features of the jobs, which are carried out by the machine tools. It plays a key role for dynamic process planning and disturbance handling by providing the available and suitable machining processes for job agents. This task has mainly handled by the resource holons at previous architectures. In PROSA the resource holon is mainly responsible for selecting the best process by selecting the appropriate tools and cutting speed. Although in the practical applications, selecting the suitable manufacturing process is mainly is done by using manufacturing standards and tables which we encapsulate this knowledge in the machining process agent. The information related to the available machining processes and their capabilities will be updated from design department. The agents include the following information.

1. Machining process ID which is the combination of the ID of the machine tools, the ID of the fixtures and the ID of the cutting tools.
2. Machining process types and machining features types, which can be generated by the machining processes
3. Surface roughness, tolerances and material removal rate of the machining processes.
4. Machining process status:
   The status represents the dynamic status of the machining processes in the FMSs. The machining process agents have two statuses.
   - inactive: if one of the machine tool, the cutting tool and the fixture related to the machining process are broken-down
   - active: otherwise

### 3.2.2.2 Information agents

The information agents are virtual agents for governing the negotiation protocol and decision-making.

*Production engineering agents.* The production engineering agents generate the job agents, the machine tool agents, and the machining process agents to specify the geometric and technological information of the jobs, the machine tools and the machining processes of the FMSs. The agents play a key role for initializing the information of the physical agents. The design department has an important role for initialization and disturbance handling in our architecture. It real timely modifies the job specifications according to the design change orders or customer requests during the manufacturing.

*Job order agents.* The job order agents represent the manufacturing tasks and it is quite similar to order holon at PROSA and task holon at ADACOR reference architectures. They are information agents, which carry out the negotiation processes between the job agents and the machine tool agents to generate suitable process plans. The agents have crucial influence on the system performance by deploying efficient decision-making mechanism to select the appropriate machine tools for the individual machining features of the jobs.

*Coordination agent.* The multi agent architecture proposed here is distributed, and the benefits inherited from distributed architecture include flexibility and robustness. However, absence of higher authority, in general, may result in a lower performance compared with hierarchical systems that are able to achieve global optimization. We introduce a coordination agent to improve the performance which is similar to staff holon at PROSA and supervisor holon at ADACOR reference architectures. The coordination agents are proposed to determine a suitable assignment of the job agents to the machine tool agents at each step of the negotiation. The coordination agents are capable to make mathematical models according the information sent from the machine tool and job agents to find a suitable assignment of the job agents to the machine tool agents at each step of negotiation. There are a few research in the literature about the optimizing the multi agent architecture for integrated process planning and scheduling (Shen et al., 2006). In this case study, we present a method by combining the mathematical modeling and process plan networks to optimize the total system. The optimization method works effectively for real time applications and generates suitable solutions.

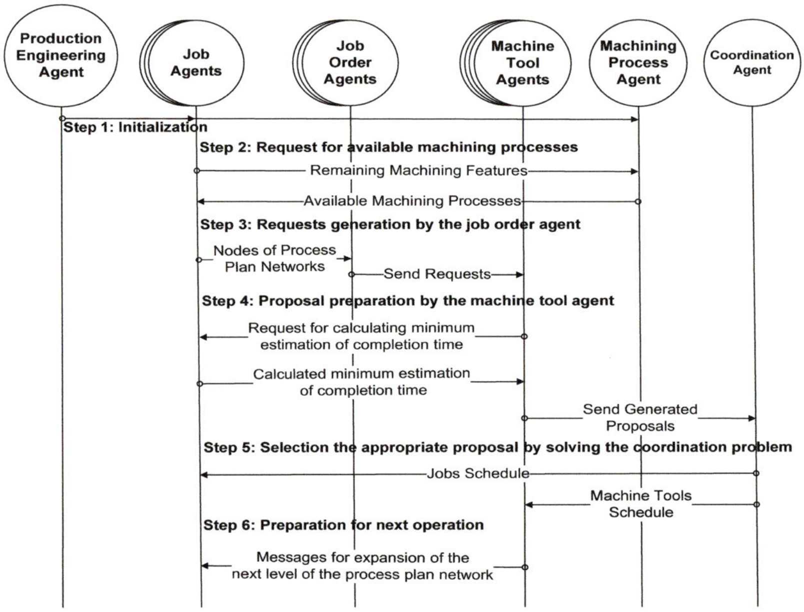


Fig. 5. Negotiation protocol among agents

### 3.2.3 Negotiation Protocol

A negotiation protocol among the agents is required to coordinate the distributed decisions of the individual agents for solving the complex problems in the integrated process planning and scheduling of the FMSs. Here, we define a negotiation protocol to meet the requirements for integrated process planning and scheduling and also handling the unforeseen disturbances. The problems to be solved here are as follows.

1. Selection of candidate machining processes for individual machining features.
2. Selection of suitable combinations of the machine tools, the cutting tools and the fixtures.
3. Selection of suitable machining sequences of the machining features.

In the protocol proposed here, the individual agents have three types of boards named "request boards", "proposal boards" and "status boards" for the communication. The requests from the other agents are firstly sent to the request boards, and the agents scan and read the requests from the boards, every fixed time intervals named RTIP (reading time interval period). The individual agents secondly generate the proposals to the requests, and store them in the proposal boards. The statuses of the agents are changed and stored in the status board, if necessary. Fig. 5. summarizes the negotiation protocol proposed here to generate suitable process plans and schedules by the distributed

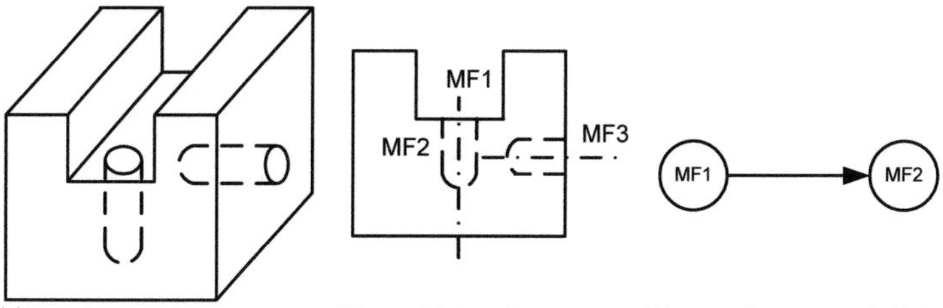


(a) part geometry (b) machining features (c) precedence constraints

Fig. 6. An example of parts

| Machining Features | Machining Process ID | Machine Tool ID | Fixture ID | Cutting Tool ID |
|---|---|---|---|---|
| MF1 | $mp_1$ | $mt_1$ | $fi_1$ | $ct_1$ |
| MF1 | $mp_5$ | $mt_2$ | $fi_2$ | $ct_1$ |
| MF3 | $mp_4$ | $mt_1$ | $fi_2$ | $ct_2$ |
| MF3 | $mp_7$ | $mt_2$ | $fi_1$ | $ct_2$ |

$mt_1$: Vertical machining center, $mt_2$: Horizontal machining center $ct_1$: End Milling, $ct_2$: Drill

Table 3. Alternative machining processes for machining features MF1 and MF3

decision-makings of the individual agents and the negotiations among the agents. The negotiation processes are carried out through the following steps.

**Step 1: Initialization**

The production engineering agents firstly generate all the job agents, the machine tool agents and the machining process agents to initialize the status of the target FMSs. They also define the machining features, which can be generated simultaneously by the same combinations of the machine tools, the cutting tools and the fixtures, and assign them to the job agents.

**Step 2: Requests for available machining processes**

The job agents select a set of the machining features, which can be machined in the next machining process, based on the precedence constraints among the machining features. For example, let us consider a case shown in Fig.6. This part consists of three machining features (see Fig.6 (b)). They are, one slot MF1, and two holes MF2 and MF3. The precedence constraints for this example is shown in Fig. 6(c). In the first step, the machining features MF1 and MF3 are sent to the machining process agents, and they select a set of available alternative machining processes for the individual machining features, based on the specifications of the machining features, such as the geometries, the sizes, the surface roughness, and the tolerances as shown in the Table 3.

The selected machining processes include the information about the machine tools, the cutting tools and the fixtures. The selected machining processes for the individual machining features are sent back to the job agents. The job agents generate a set of groups of the machining features, which can be machined by the same combinations of the machine tools, the cutting tools and the fixtures. Machining features belonging to the same setup have been grouped together for one machine to minimize the setup/fixture time. This means that the grouped machining features are machined by one machining process concurrently, and the job agents generate the nodes representing all the grouped machining features in the process

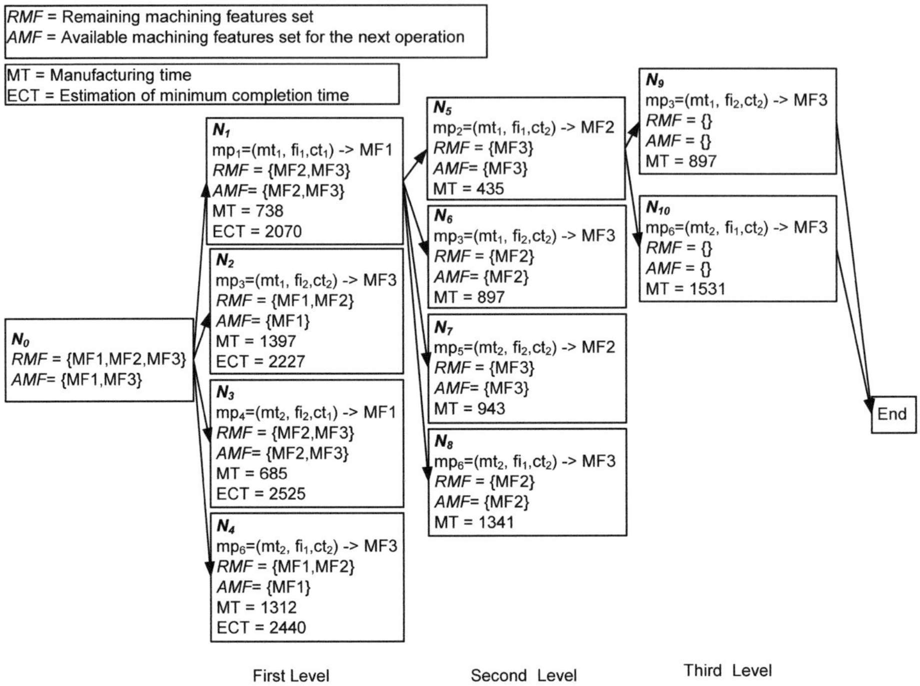


Fig. 7. Process plan network.

plan networks as shown in the first level nodes $N_1$ to $N_4$ in Fig.7. The contents of the process plan networks are described in the previous paper. (Tehrani et al., 2007). The individual nodes in the process plan networks represent a set of the machining features which could be machined by one machining process. An algorithm also generates nodes representing the machining sequences of the machining features based on the precedence constraints. The information of the generated nodes is sent to the job order agents for the negotiations.

**Step 3: Request generation by job order agents**

The job order agents create requests for the machining process execution for the individual nodes of the process plan networks, which are the groups of the machining features that can be generated by same machine tools. The requests for the job agents are generated according to the process plan network which guarantees that we always generate feasible solutions and different operations of one job agents can not be processed simultaneously. The generated requests are sent to the request boards of the corresponding machine tool agents. The content of request includes the machining features and selected machine tool, cutting tool and fixture. As you can see in Fig. 7, there are four nodes $N_1$ to $N_4$ in the first level of the process plan network. For each of them, the requests are generated and sent to the related machine tools MT1 and MT2.

**Step 4: Proposal preparation by machine tool agents**

The machine tool agents read all the requests from the request boards every RTIP (Reading Time Interval Period) if it is idle and each machine can handle only one job at a time. The machine tool agents analyze the request messages, and generate appropriate proposals to all the requests. We consider a heuristic algorithm to estimate minimum completion time for generating appropriate proposal for each request by the machine tool agents, as shown in the followings.

Minimum completion time estimation

The machine tool agents need to estimate the completion time of the remaining machining features of the job agents. A procedure is developed and given to the job agents to estimate the minimal completion time of the remaining machining features, based on the process plan networks shown in Fig. 7. When a machine tool agent requires a job agent to estimate the minimum completion time, the job agent starts the procedures from the start node which is specified by the machine tool agent, and repeat to generate and to select suitable successive nodes with the minimum machining time. When all the machining features are included in the process plan networks, the job agent find both the machining sequences of the machining features and the estimated minimal completion time. Consider a case where we are at node $N_i$ of the process plan network and we are going to estimate the manufacturing time from the node $N_i$ to the end node. The algorithm for calculating the estimated minimum completion time from node $N_i$ to the end node is summarized in the followings.

*Initialization:*

- Set RMF and AMF. The RMF is the set of the remaining machining features. The AMF is the set of the available machining features that do not have any preceding machining features and could be done firstly considering the precedence constraints among the nodes of the process plan networks.(AMF ⊆ RMF).
- Put the node $N_i$ in ECTS set. The ECTS is the set of the nodes in the path from node the $N_i$ to the end of the process plan network, which has the minimum manufacturing time.
- Set a initial node $N_i$. In the $N_i$, RMF ={set remaining machining features}, AMF={set of machining features without any successors}.

**(1)** Generate a set of successor nodes $SN = \{N_t | t = 1,2, \dots |SN|\}$ of the node $N_i$ for all feasible machining processes $mp_r = (mt_i, fx_f, ct_t), r = 1, \dots, R$, ($R$ = total number of available machining processes) by applying the following algorithm.
- Cluster all features of the AMF set of the node $N_i$ that could be machined with the machining process $mp_r$,
- Generate a new node $N_t$ representing a set of machining features which can be machined by the machining process $mp_r$ and put it in the SN set. The links to the nodes $N_t$, which are successor nodes, are stored in the node $N_i$ for further processing,
- Estimate the manufacturing time for node $N_t$ that includes the time of the machining, the transportation and the re-fixturing processes,
- Update the RMF and AMF sets for the node $N_t$,

**(2)** Select a successor node $N_k$ from the SN set which has the minimum machining time for the next step of extension, and move it to the ECTS set.
**(3)** If RMF set of $N_k$ is not empty consider node $N_k$ as node $N_i$ and go to (1).
**(4)** If RMF set of $N_k$ is empty, it means that we are in the end of the process plan network. The sum of the manufacturing time for the nodes in ECTS set is the estimation of the minimum completion time from node $N_i$ to the end.

Let us consider a case where we are going to calculate the estimation of minimum completion time for node $N_1$ at the process plan network shown in Fig. 7. We start with node $N_1$, and there are four successor nodes $N_5$, $N_6$, $N_7$, $N_8$ from the node $N_1$ as shown in Fig. 7. We select the node $N_5$ which has the minimum manufacturing time, and we put it in the ECTS set. We expand the node $N_5$ at the next stage of the algorithm and there are two successor nodes $N_9$ , $N_{10}$ . The node $N_9$ is selected and which has the minimum manufacturing time, we put it in the ECTS set. As you can see in Fig. 7, for the node $N_9$ the RMF set is empty and the algorithm stops. It is because that there are no remaining machining features in the node $N_9$. The sum of the manufacturing time for the nodes in ECTS set is the estimation of the completion time from node $N_1$ until end.
Following this, the job agent returns the estimated completion time to the machine tool agent. As you can see in the Fig.7, the estimation of completion time for all nodes $N_1$, $N_2$, $N_3$, $N_4$ are calculated and these values are returned to the machine tool agent. This procedure can estimate the completion time of all the remaining machining features, however it requires the additional communications between the machine tool agents and the job agents. The machine tool agents generate proposals for each request based on the minimal completion time of the remaining machining features and send them to the coordination agents.

**Step 5: Selection of appropriate proposals by coordination agent**

The coordination agents scan all received proposals from the machine tool agents every RTIP, and assign the appropriate machine tool agents to the job agents. At present, we consider only the flow time of the job agents, and our goal is to minimize the average flow time of all the job agents. The flow time considered here includes the machining time, the transportation time, the re-fixturing time and the tool changing time. The constraints of the model are that only one machine tool agent is selected for each job agent and only one job agent has been assigned to each machine tool agent. The followings summarize the formulas representing the optimization problems considered here.

Parameters:

$$\mathrm{MP} = \{\mathrm{mp_r}: (\mathrm{mt_j}, \mathrm{fx_f}, \mathrm{ct_t}) | \mathrm{r} = 1, \dots, \mathrm{R}\}, \mathrm{R} = |\mathrm{MP}|, \tag{13}$$

$$\mathrm{MT} = \{\mathrm{mt_j} | \mathrm{j} = 1,2, \dots \mathrm{n}\}, \mathrm{n} = |\mathrm{MT}|, \tag{14}$$

$$\mathrm{FI} = \{\mathrm{ft_f} | \mathrm{f} = 1,2, \dots \mathrm{F}\}, \mathrm{F} = |\mathrm{FI}|, \tag{15}$$

$$\mathrm{CT} = \{\mathrm{ct_t} | \mathrm{t} = 1,2, \dots \mathrm{T}\}, \mathrm{T} = |\mathrm{CT}| \tag{16}$$

where,
$\mathrm{mp_r}$: ID of machining process, $\mathrm{mt_j}$: ID of machine tools, $\mathrm{ft_f}$: ID of fixtures, $\mathrm{ct_t}$: ID of cutting tools.
$\mathrm{FT}^{j}_{(i,r)}$: Estimation of completion time of job agent $i$ (i = 1,2,..m) according to the machining process $\mathrm{mp_r}$ (r = 1,2,..R) with machine tool agent $\mathrm{mt_j}$ (j = 1,2,..n).

Design variables:

$$x^{j}_{(i,r)} = \begin{cases} 1: \text{if the machine tool agent } \mathrm{mt_j} \text{ is selected for job agent } i \text{ according to the} \\ \quad \text{machining process } \mathrm{mp_r} \\ 0: \text{otherwise.} \end{cases}$$

Mathematical Model:

$$\text{Minimize} \quad Z = \sum_{j=1}^{n} \sum_{i=1}^{m} \sum_{r=1}^{R} x_{(i,r)}^{j} \, FT_{(i,r)}^{j} \tag{17}$$

$$\sum_{j=1}^{n} \sum_{r=1}^{R} x_{(i,r)}^{j} + \text{Dummy}_i = 1, \quad i = 1,2,..m \tag{18}$$

$$\sum_{i=1}^{m} \sum_{r=1}^{R} x_{(i,r)}^{j} + \text{Dummy}_j = 1, \quad j = 1,2,..n \tag{19}$$

$$x_{(i,r)}^{j} = 0,1 \tag{20}$$

We add dummy variables to equations (18) and (19) to change the constraints of sets of equations. Equation (17) is the objective function that is the total of the estimated flow time of all the job agents. Equation (18) is a constraint that only one machine tool agent is selected for each job agent. Equation (19) is a constraint that only one job agent has been assigned to each machine tool agent. The model described in equations (17)-(20) is an assignment problem and can be solved as a linear programming model. We can release the equation (20) from the model and apply linear techniques and the optimal solution will be integer. We can use other objective functions such as minimizing the manufacturing costs and minimizing the average of tardiness of all jobs with the above model.

After solving the above model, the coordination agents inform both the job agents and the machine tool agents that the machining features sent from the job agents shall be machined by the selected machine tools. This means that the coordination agents dynamically generate the process plans and the production schedules of the job agents and the machine tool agents. The job agents and the machine tool agents selected here carry out the requested machining processes in the next step. Therefore, the statuses of these agents are changed, and the status data are stored in the status boards. All the agents monitor the status data if necessary.

**Step 6: Preparation for next operation**

When the machine tool agents complete the machining operations of the job agents, the job agents modify their process plan networks. That is, the job agents delete the corresponding nodes representing the group of the machining features which was completed by the machine tool agents. New nodes of the process plan networks are generated to specify the groups of the machining features to be machined in the next step. The procedures presented in Steps 2 to 6 are repeated until the job agents do not have any remaining machining features.

### 3.2.4 Synchronization

The synchronization of negotiation between different agents is important issue for developing the multi agent architecture. The Petri nets (Proth & Xie 1996) are used, in the case study, for synchronizing the messages and the negotiation protocols between the different agents. This Petri nets control both the sequence and the timing of the interaction and the messages between the agents. Each Petri net represents one agent or interacting agents. Fig. 8 shows an example of the interaction between the agents for generating and sending the requests to the request board of the machine tool agents and generating the proposals by the machine tool agents. These Petri nets are linked with each other with global transition (transitions, $t_2$, $t_4$, $t_8$, $t_{14}$, $t_{17}$ in Fig. 8).

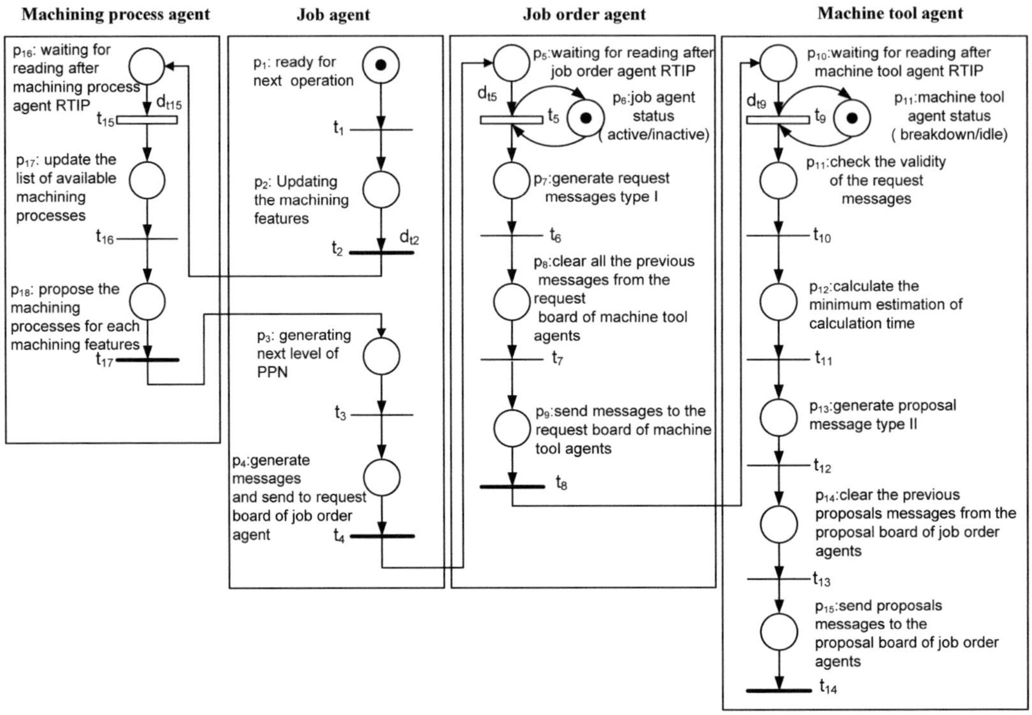


Fig. 8. Synchronizing agents for generating requests and proposals

### 3.2.4 Simulation Software and Experimental Results

A prototype of the agent based integrated process planning and scheduling system and the graphical presentation system have been developed for the case studies. The system developed here is able to simulate the distributed decision makings of the agents, the negotiation processes among the agents, and also the manufacturing processes in the FMS. The coordination agent use ILOG CPLEX optimization engine for solving the integer programming model of the coordination and for assigning the job agents to machine tool agents. Some case studies have been carried out to verify the applicability and the effectiveness of the proposed system to the integrated process planning and scheduling problems in the FMSs. The FMS considered here includes 7 machine tools and 4 job types. Fig. 9 shows the geometries of the job agents and their manufacturing features including cylinder and box type shape for the case studies. The detailed information of the machining features and the machining resources of the case studies are brought in the previous paper (Tehrani et al., 2007). The RTIP in the simulation is set to be 2 sec. for the machine tool agents, 3 sec for coordination agents and 4 sec. for the job agents.

#### 3.2.4.1 Efficiency of the proposed architecture

Two case studies have been done to evaluate the impact of introducing the coordination agents in multi agent systems. We compare the results with the dispatching rules which the job agents applying SPT dispatching rules for selecting the machine tools for their manufacturing operations without assisting from the coordination agents.

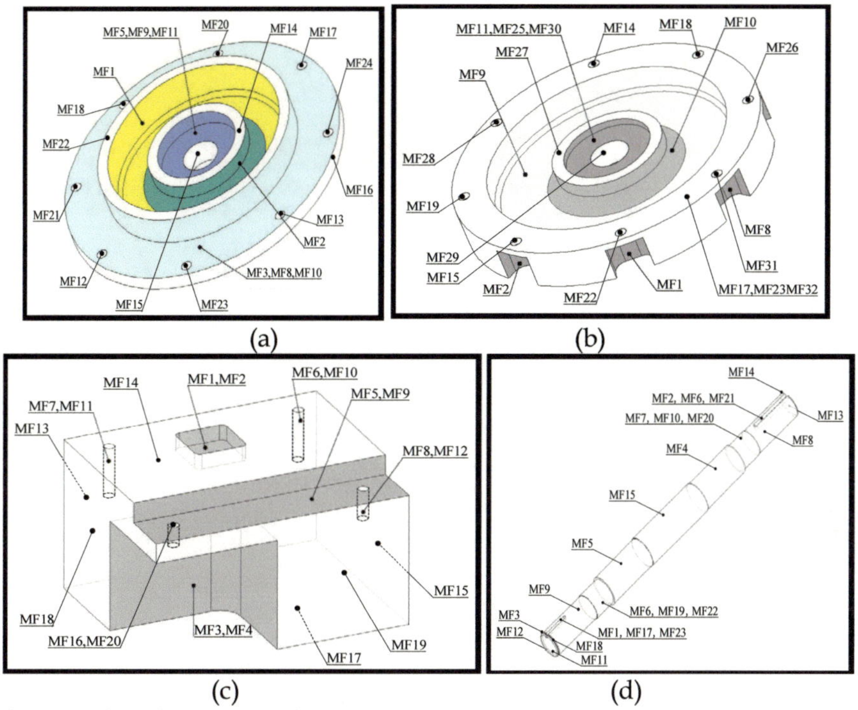


Fig. 9. Jobs considered in case studies.

Fig. 10 summarizes the comparison of the proposed architecture and the previous method from the view points of the average flow time of all the job agents and the calculation time for coordination. In the Fig. 10 the vertical axis gives the flow time of the individual job agents and the horizontal axis shows the individual job agents and their types.

It is understood, from Fig. 10(a) and (b), that the multi-agent systems with the coordination agents generate more suitable process plans and schedules from the viewpoint of the average flow time of the all the job agents. As you can see, the average flow time has been improved 10.9% and 10.39% for the cases (a) and (b) of Fig. 10, respectively. It is because that the mathematical programming methods applied here are suitable to reduce the average flow time of the job agents of the job shop process planning and scheduling problems. The calculation time for coordination is enough short and the proposed method is suitable for the real time application, when we have enormous number of job agents and machine tool agents.

### 3.2.4.2 Robustness of the proposed architecture

An additional experiment is also carried out to assess the robustness of the proposed architecture against the malfunction of the machine tools. The original process plans and schedules are shown for 10 job agents in the Gantt chart of Fig. 11 (a). In the experiment, the machine tool "MT14" is broken down at simulation time 4811 sec. and the recovery time is assumed to be 5000 sec. As you can see in the Gantt chart of Fig. 11 (b), the proposed architecture can dynamically generate alternative process and schedule to cope with the malfunctions of the machine tools. The job agents can be dynamically allocated to another manufacturing route in the process plan networks and new process plans for jobs 7,6,4,3 and job 2 has been generated dynamically.

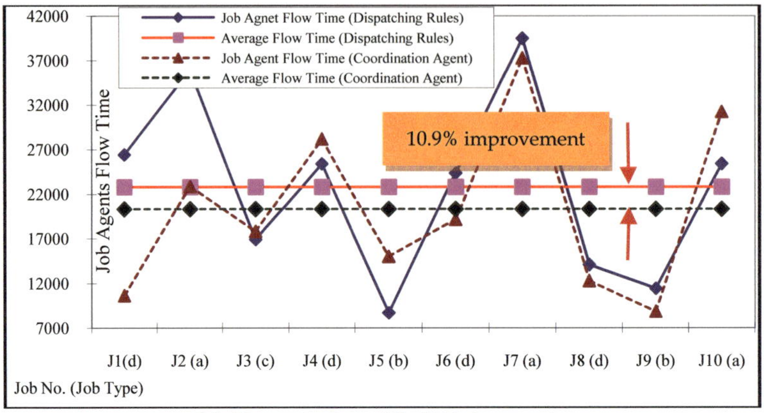


(a) Case study with 10 job agents

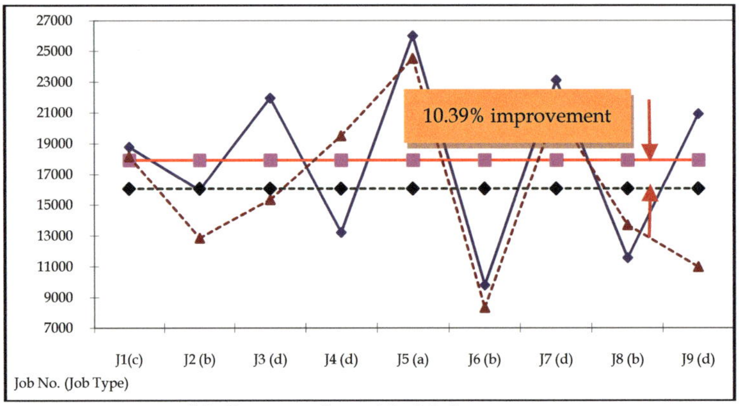


(b) Case study with 9 job agents.

Fig. 10. Case study and comparison with previous result.

In the other experiments, the following unforeseen changes have been considered in the job specifications.

1. Change the roughness of the machining features
   - Job 03, MF16 at simulation time 3000
   - Job 10, MF18 at simulation time 10000
2. Add a new machining feature to the job
   - Job 02, MF21 at simulation time 7000
   - Job 04, MF24 at simulation time 5000
   - Job 05, MF25 at simulation time 2900
3. Change the size of machining feature
   - Job 10, MF16 at simulation time 10000
   - Job 03, MF21 at simulation time 6500

The results are shown the Gantt chart of Fig. 11 (c). As shown in Gantt chart Fig. 11 (c), the proposed architecture can dynamically generate updated process plans and schedules to cope with the changes of job specifications.

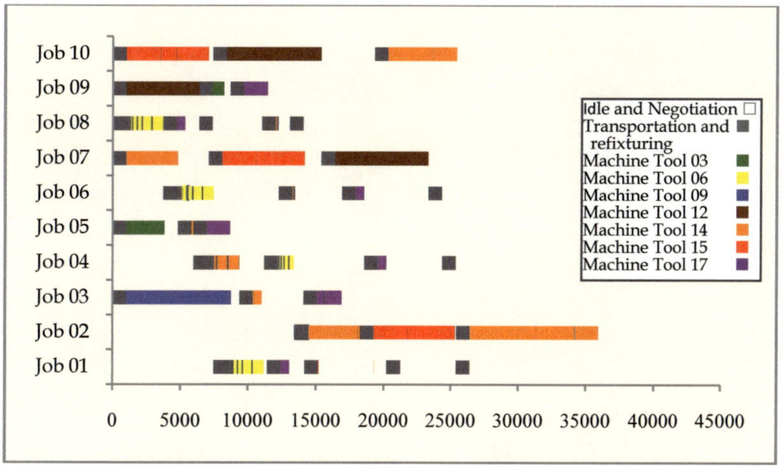


(a) Original schedule without unforeseen changes

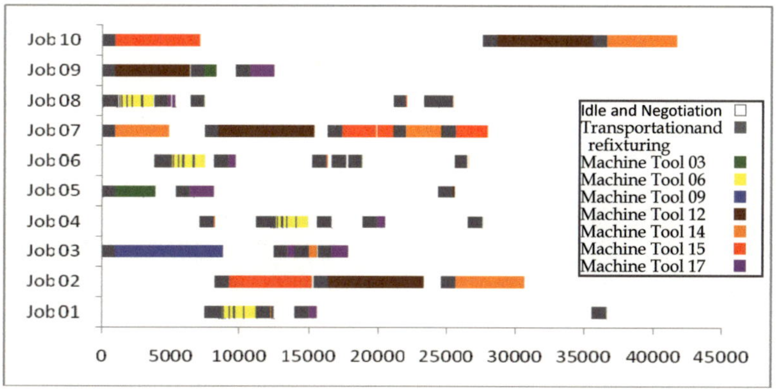


(b) Modified schedule for malfunction of machine tool "MT14"

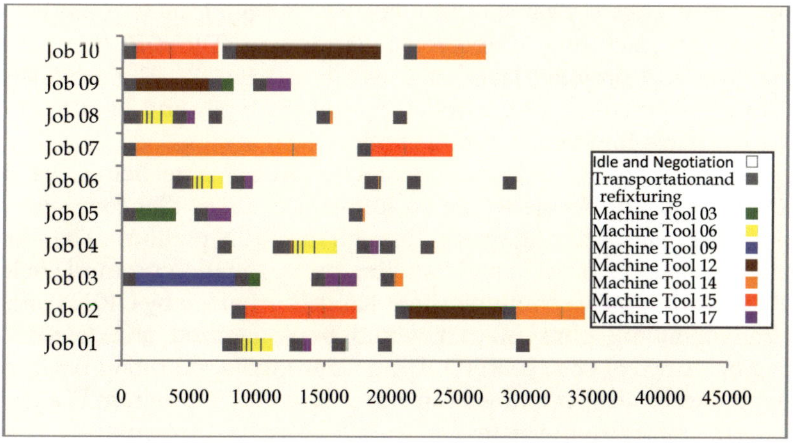


(c) Modified schedule for job specification changes

Fig. 11. Gantt chart for case study of robustness

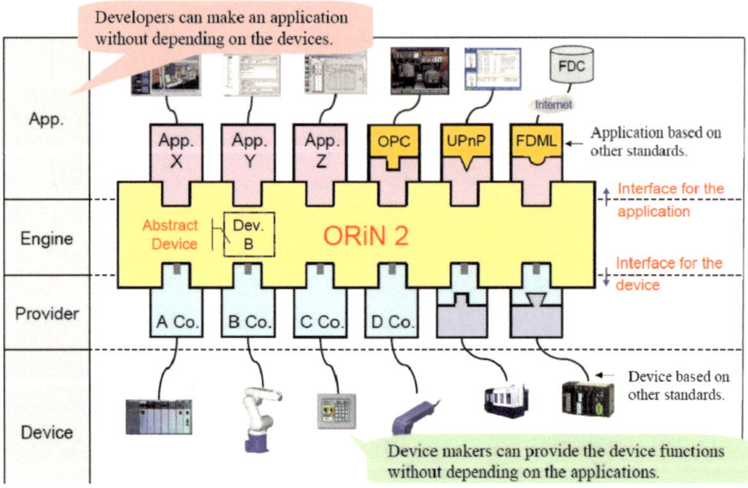


Fig. 12. Two layers of ORIN architecture

## 4. Realizing the agent manufacturing system

In spite of the promising perspective of these emergent distributed and intelligent approaches, until now the industrial applications of control systems developed in the context of reconfigurable manufacturing systems are extremely rare and the implemented functionalities are normally restrict, being very slow the adoption of these concepts by industry (Marik & McFarlane 2005).

We have collaboration with DENSO Wave Co. for realizing the agent manufacturing system through the ORIN architecture. ORIN 2.0 (Open Robot Interface for Network) provides integrated interface to access to the devices on the network (Hibino et al., 2006). You can easily access the data inside the devices from application software by using ORIN regardless of the manufacturers, devices or specifications of communication protocols. ORIN is a Distributed Real Manufacturing Simulation Environment (DRMSE) that consists of two layers; engine layer and provider layer as shown in the Fig. 12. The provider layer has a function to absorb a difference of controller equipment types and emulators. The engine layer provides interfaces for manufacturing applications.

ORIN proposes a hardware and software architecture for realizing the agent based manufacturing system. The agents would be software modules that communicate with the real hardware in the manufacturing system through the ORIN platform. The communication between agents for making decision and handling the negotiation protocol could been done and synchronized through the communication channels provided by ORIN platform. The job agents and corresponding physical part would be recognized and traced through the manufacturing by using bar code or RFID. The machine tools and robots could be connected directly through their controller and we can also define and re-program PLCs and different controller of the manufacturing systems.

In our research, we have successfully integrated our agent based simulation program with ORIN architecture. A barcode reader (DENSO AT10Q-SM) and a bar code generator (DENSO QRdraw Ad) have been connected to the agents through the ORIN architecture. The job agent receives the information from kanban by barcode reader. The bar code generator has been applied for generating the kanban cards including the job agent information, the disturbances and the job specification changes. The job agents and the machine tool agents can communicate and exchange data real timely through the ORIN architecture with the corresponding hardware in the manufacturing system.

## 5. Conclusion

Manufacturing companies at the beginning of 21th century have to face a dynamic environment where economical, technological and customer trends change rapidly, requiring the increase of flexibility and agility to react to unexpected disturbances, maintaining the productivity and quality parameters. The traditional manufacturing control systems are adapted on a case-by- case basis, requiring an expensive and huge time-consuming effort to develop, maintain or re-configure. The missing re- configurability is derived from the lack of agility to support emergency (change and unexpected disturbances). The challenge is to develop innovative, agile and reconfigurable architectures for distributed manufacturing control systems, using emergent paradigms and technologies. Multi-agent systems and HMSs are two promising paradigms to build this new class of distributed and intelligent manufacturing control systems. In this chapter, the manufacturing control systems, especially using artificial intelligence techniques to develop it, namely multi-agent systems and HMSs, was reviewed. Two case studies have been discussed in detail and their contributions, results and benefits of applying agent and holonic manufacturing control have been reviewed.

In first case study, a new real-time scheduling methods for the HMS are proposed to select a suitable combination of the CNC machine tool (CMT) holons and the job holons which carry out the machining process. A distributed decision-making procedure is proposed to select a suitable combination of the CMT holons and the job holons for the next machining processes, based on the utility values for the candidates. Some case studies of the real-time scheduling have been carried out to verify the effectiveness of the proposed methods. It was shown, through case studies, that the proposed methods are effective to improve the objective functions of the individual holons. In the second case study, a multi-agent system was proposed for the integrated process planning and scheduling systems for the FMSs. A systematic procedure was proposed to generate suitable process plans of the jobs and suitable schedules of the machine tools. The proposed method is able to solve the process planning and scheduling problems concurrently and dynamically, with use of the mathematical optimization methods and search algorithms of the process plan networks. Some case studies have been carried out to verify the applicability of the proposed method to the integrated process planning and scheduling problems in the FMSs including 7 machine tools and 10 jobs. It was shown, through the case studies, that the proposed multi-agent architecture is capable to generate appropriate process plans and schedules. It was also shown that the proposed architecture generates alternative process plans dynamically, to cope with the malfunctions of the machine tools and unforeseen job specification changes.

In the future research, we are trying to expand the architecture for other objective functions and multi objective integrated process planning and scheduling. We also are trying to develop general agents according to DCOM technology and defining interfaces for them that make agents possible to connect directly to ORIN to communicate with manufacturing hardware, real timely.

## 6. References

CMV, (1998). *Visionary Manufacturing Challenges for 2020, Committee on Visionary Manufacturing*. National Academic Press, Washington, DC, USA.

Baker, A. (1998). A survey of factory control algorithms which can be implemented in a multi-agent heterarchy: dispatching, scheduling and pull. Journal of Manufacturing Systems, Vol. 17, No. 4, pp. 297–320.

Brussel, H.V., Wyns, J., Valckenaers, P. & Bongaerts, L. (1998). Reference architecture for holonic manufacturing systems: PROSA. *Computers in Industry*, Vol. 37, No. 3, pp. 255–274.

Colombo, A., Schoop, R. & Neubert, R. (2006). An agent-based intelligent control platform for industrial holonic manufacturing systems. IEEE Transactions on Industrial Electronics, Vol. 53, No. 1, pp. 322–337.

Diltis, D., Boyd, N. & Whorms, H. (1991). The evolution of control architectures for automated manufacturing systems. *Journal of Manufacturing Systems*, Vol. 10, No. 1, pp. 63–79.

Hibino, H., Inukai, T. & Fukuda, Y. (2006). Efficient Manufacturing system implementation based on combination between real and virtual factory, *International Journal of Production Research*, Vol. 44, No. 18-19, pp. 3897-3915.

Koestler, A. (1969). *The Ghost in the Machine*. Arkana Books, London.

Leitao, P. (2009). Agent-based distributed manufacturing control: A state-of-the-art-survey. *Engineering Applications of Artificial Intelligence*, Vol. 22, pp. 979-991.

Leitao, P. & Restivo, F. (2006). ADACOR: a holonic architecture for agile and adaptive manufacturing control. *Computers in Industry*, Vol. 57, No. 2, pp. 121–130.

Marik, V. & McFarlane, D. (2005). Industrial adoption of agent-based technologies. *IEEE Intelligent Systems*, Vol. 20, No. 1, pp. 27–35.

Proth, M. & Xie, J., X. (1996). *Petri Net a Tool for Designing and Management of Manufacturing System*, John Willey and Sons.

Russel, S. & Norvig, P. (1995) Artificial Intelligence, A Modern Approach. Prentice- Hall, New Jersey.

Sepehri, M., M & Tehrani, H. (2005). Dynamic scheduling architecture for AGVs and machines in holonic manufacturing system with Petri nets, *International Journal of Industrial Engineering-Theory Applications And Practice*, Vol. 12, No. 2, pp. 132-142.

Shen, W. M., Wang, L. & Hao, Q. (2006). Agent-Based Distributed Manufacturing Process Planning and Scheduling: A State-of-the-Art Survey, *IEEE Transaction on System, Man, and Cybernetics-Part C: Application and Reviews*, Vol. 36, No. 4, pp. 563-577.

Tehrani, H., Sugimura, N., Tanimizu Y. & Iwamura, K. (2007). A Search Algorithm for Generating Alternative Process Plans in Flexible Manufacturing System, *Journal of Advanced Mechanical Design, System, and Manufacturing*, Vol. 1, No. 5, pp. 706-716.

Wang, L. H., Shen, W. M. & Hao, Q. (2006). An Overview of Distributed Process Planning and Its Integration with Scheduling. *International Journal of Computer Applications in Technology*, Vol. 26, No. 1-2, pp. 3-14.

Winkler, M. & Mey, M. (1994). Holonic manufacturing systems. European Production Engineering.

Wooldridge, M. (2002). *An Introduction to Multi-Agent Systems*. Wiley, New York.

Wooldridge, M. & Jennings, N. (1995). Intelligent agents: theory and practice. *The Knowledge Engineering Review*, Vol. 10, No. 2, pp. 115–152.

Wyns, J. (1999). Reference Architecture for Holonic Manufacturing Systems–The Key to Support Evolution and Reconfiguration, PhD Dissertation, *Department of Mechanical Engineering, Katholieke Universiteit Leuven*.

6

# Materials handling in flexible manufacturing systems

Dr. Tauseef Aized
*Professor, Department of Mechanical, Mechatrnics and Manufacturing Engineering, KSK Campus, University of Engineering and Technology, Lahore, Pakistan*

## 1. Introduction

Material handling can be defined as an integrated system involving such activities as moving, handling, storing and controlling of materials by means of gravity, manual effort or power activated machinery. Moving materials utilize time and space. Any movement of materials requires that the size, shape, weight and condition of the material, as well as the path and frequency of the move be analyzed. Storing materials provide a buffer between operations. It facilitates the efficient use of people and machines and provides an efficient organization of materials. The considerations for material system design include the size, weight, condition and stack ability of materials; the required throughput; and building constraints such as floor loading, floor condition, column spacing etc. The protection of materials include both packaging and protecting against damage and theft of material as well as the use of safeguards on the information system to include protection against the material being mishandled, misplaced, misappropriated and processed in a wrong sequence. Controlling material includes both physical control as well as status of material control. Physical control is the orientation of sequence and space between material movements. Status control is the real time awareness of the location, amount, destination, origin, ownership and schedule of material. Maintaining the correct degree of control is a challenge because the right amount of control depends upon the culture of the organization and the people who manage and perform material handling functions.

Material handling is an important area of concern in flexible manufacturing systems because more than 80 % of time that material spends on a shop floor is spent either in waiting or in transportation, although both these activities are non-value added activities. Efficient material handling is needed for less congestion, timely delivery and reduced idle time of machines due to non-availability or accumulation of materials at workstations. Safe handling of materials is important in a plant as it reduces wastage, breakage, loss and scrapes etc.

## 2. Principles of material handlings

The material handling principles provide fundamentals of material handling practices and provide guidance to material handling system designers. The following is a brief description of material handling principles.

### 2.1 Planning principle

All material handling should be the result of a deliberate plan where the needs, performance objectives and functional specification of the proposed methods are completely defined at the outset. In its simplest form a material handing plan defines the material (what) and the moves (when and where); together they define the method (how and who).

### 2.2 Standardization principle

Standardize handling methods and equipments wherever possible. Material handling methods, equipment, controls and software should be standardized within the limits of achieving overall performance objectives and without sacrificing needed flexibility, modularity and throughout anticipation of changing future requirements.

### 2.3 Ergonomic principle

Human capabilities and limitations must be recognized and respected in the design of material handling tasks and equipment to ensure safe and effective operations. Equipments should be selected that eliminates repetitive and strenuous manual labor and which effectively interacts with human operators and users.

### 2.4 Flexibility principle

Use methods and equipments that can perform a variety of tasks under varying operating conditions.

### 2.5 Simplification

Simplify material handling by eliminating, reducing or combining unnecessary movements and equipments.

### 2.6 Gravity

Utilize gravity to move material wherever possible.

### 2.7 Layout

Prepare an operation sequence and equipment layout for all viable system solutions and then select the best possible configuration.

### 2.8 Cost

Compare the economic justification of alternate solutions with equipment and methods on the basis of economic effectiveness as measured by expenses per unit handled.

### 2.9 Maintenance
Prepare a plan for preventive maintenance and scheduled repairs on all material handling equipments.

### 2.10 Unit load principle
A unit load is one that can be stored or moved as a single entity at one time, such as a pallet, container or tote, regardless of the number of individual items that make up the load. Unit loads shall be appropriately sized and configured in a way which achieves the material flow and inventory objectives at each stage in the supply chain.

### 2.11 Space utilization principle
Effective and efficient use must be made of all available space. In work areas, cluttered and unorganized spaces and blocked aisles should be eliminated. When transporting loads within a facility, the use of overhead space should be considered as an option.

### 2.12 System principle
Material movement and storage activities should be fully integrated to form a coordinated, operational system which spans receiving, inspection, storage, production, assembly, packaging, unitizing, order selection, shipping, transportation and the handling of returns. Systems integration should encompass the entire supply chain including reverse logistics. It should include suppliers, manufacturers, distributors and customers.

### 2.13 Automation principle
Material handling operations should be mechanized and/or automated where feasible to improve operational efficiency, increase responsiveness, and improve consistency and predictability.

### 2.14 Environmental principle
Environmental impact and energy consumption should be considered as criteria when designing or selecting alternative equipment and material handling systems.

### 2.15 Life cycle cost principle
A thorough economic analysis should account for the entire life cycle of all material handling equipment and resulting systems. Life cycle costs include capital investment, installation, setup and equipment programming, training, system testing and acceptance, operating (labor, utilities, etc.), maintenance and repair, reuse value, and ultimate disposal

## 3. Material Transport Equipment

International Materials Management Society has classified equipment as (1) conveyor, (2) cranes, elevators, and hoists, (3) positioning, weighing, and control equipment, (4) industrial vehicles, (5) motor vehicles, (6) railroad cars, (7) marine carriers, (8) aircraft, and (9) containers and supports. The following provides the details of material transport equipments.

### 3.1 Conveyor Systems

A Conveyor is used when a material is moved very frequently between specific points and the path between points is fixed. Conveyors combined with modern identification and recognition systems like bar code technologies have played a significant role in the transportation and sorting of a large variety of products in modern warehouses. Some of the common types of conveyors are:

- Roller conveyor
- Skate- wheel conveyor
- Belt conveyor
- In- floor towline conveyor
- Overhead trolley conveyor
- Cart-on-track conveyor

#### 3.1.1 Roller Conveyor

In roller conveyors, the pathway consists of a series of rollers that are perpendicular to the direction of travel. Loads must possess a flat bottom to span several rollers which can be either powered or non-powered. Powered rollers rotate to drive the loads forward in roller conveyor. The following figure shows a roller conveyor.

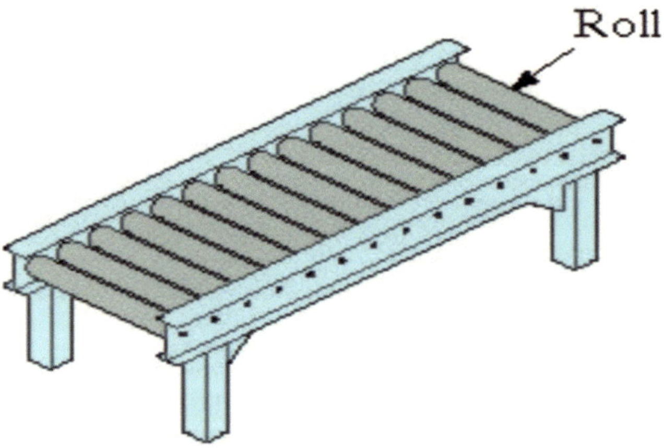


Fig. 1. Roller conveyor

#### 3.1.2 Skate-Wheel Conveyor

Skate-wheel conveyors are similar in operation to roller conveyor but use skate wheels instead of rollers and are generally lighter weight and non-powered. Sometimes, these are built as portable units that can be used for loading and unloading truck trailers in shipping and receiving. Figure 2 shows a skate-wheel roller.

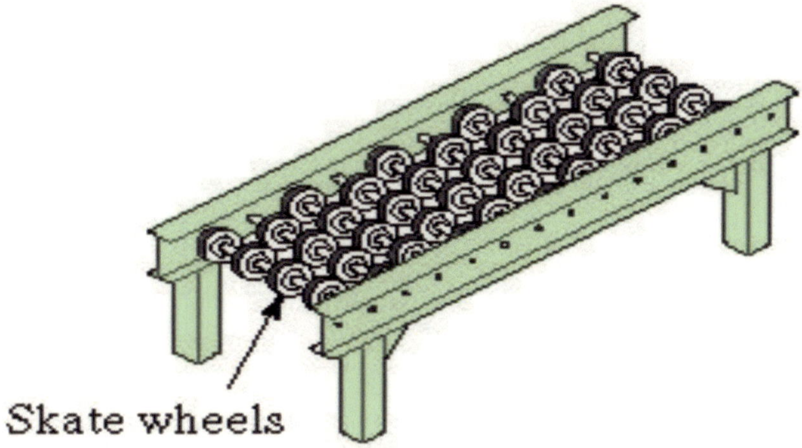


Fig. 2. Skate-wheel conveyor

### 3.1.3 Belt Conveyor

A belt conveyor is a continuous loop with forward path to move loads in which the belt is made of reinforced elastomeric support slider or rollers used to support forward loop. There are two common forms:

- Flat belt (shown)
- V-shaped for bulk materials

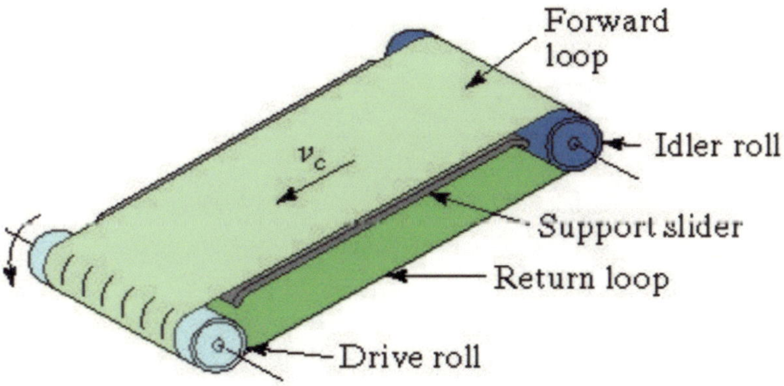


Fig. 3. Belt conveyor

### 3.1.4 In-Floor Tow-Line Conveyor

These are four-wheel carts powered by moving chains or cables in trenches in the floor. Carts use steel pins (or grippers) to project below floor level and engage the chain (or pulley) for towing. This allows carts to be disengaged from towline for loading and unloading purpose as is shown in Figure 4.

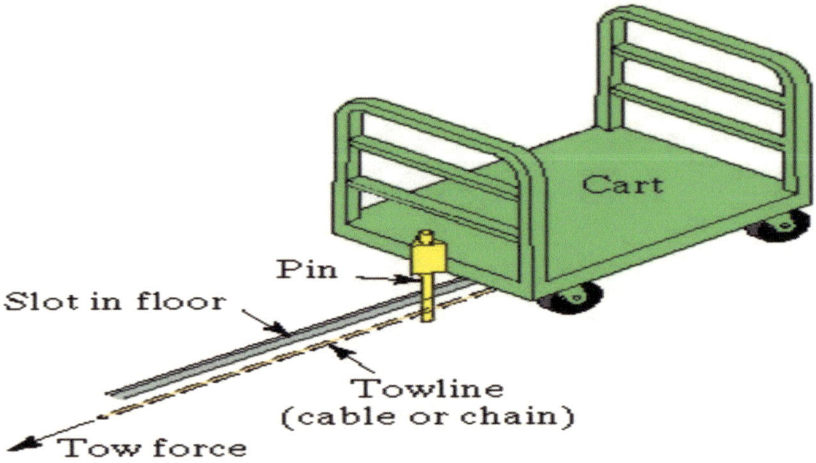


Fig. 4. In-floor two-line conveyor.

### 3.1.5 Overhead Trolley Conveyor

A trolley is a wheeled carriage running on an overhead track from which loads can be suspended. Trolleys are connected and moved by a chain or cable that forms a complete loop and are often used to move parts and assemblies between major production areas. Figure 5 shows an overhead trolley conveyor.

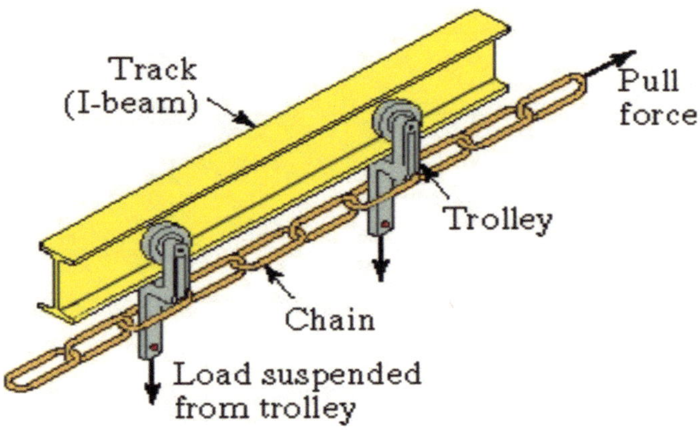


Fig. 5. Over-head trolley conveyor

### 3.1.6 Cart-On-Track Conveyor

Carts ride on a track above floor level and are driven by a spinning tube. The forward motion of cart is controlled by a drive wheel whose angle can be changed from zero (idle) to 45 degrees (forward). It is shown in the following figure.

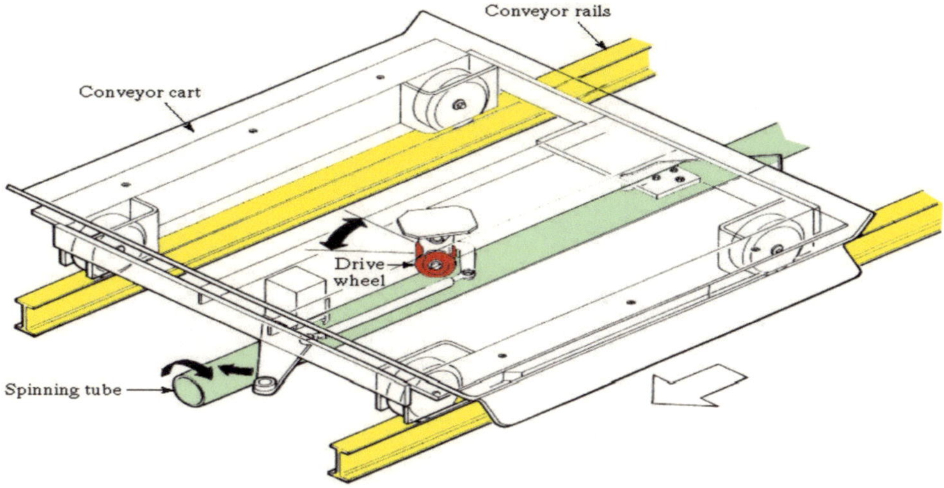


Fig. 6. Cart-on-track coveyor.

### 3.2 Cranes and Hoists

Cranes are normally used for transferring materials with some considerable size and weight and for intermittent flow of material. In general, loads handled by cranes are more varied with respect to their shape and weight than those handled by a conveyor. Hoists are frequently attached to cranes for vertical translation that is, lifting and lowering of loads. They can be operated manually, electrically, or pneumatically. Cranes usually include hoists so that the crane-and-hoist combination provides

- Horizontal transport
- Vertical lifting and lowering

This class of material handling equipments can typically lift & move a material up to 100 tons. A hoist consists of one or more fixed pulley & one or more rotatable pulley & a hook to attach load with it. The number of pulleys in hoist determines its mechanical advantage which is the ratio of load lifted & deriving force. Hoist with mechanical advantage of four are shown below:

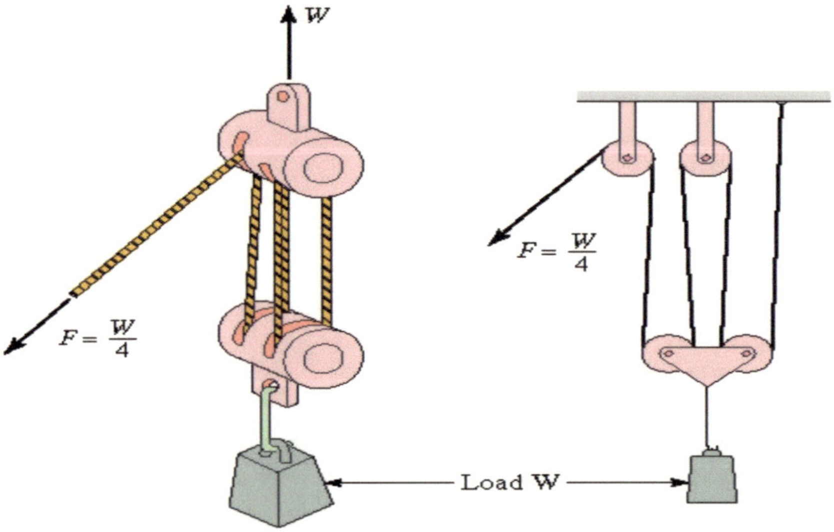


Fig. 7. (a) Sketch of the hoist (b) diagram to illustrate mechanical advantage

There are different types of cranes that are used in industrial applications. Some of these are discussed below.

### 3.2.1 Bridge Crane

A bridge crane consist of one or two horizontal girder or beam suspended between fixed rail on either end which are connected to the structure of building. The hoist trolley can be moved along the length of bridge & bridge can be moved the length of rail in building. These two capabilities provide motion along X-axis & Y-axis whereas hoist can provide motion in the z-axis. Their application includes heavy machinery fabrication. They have ability to carry load up to 100 tons.

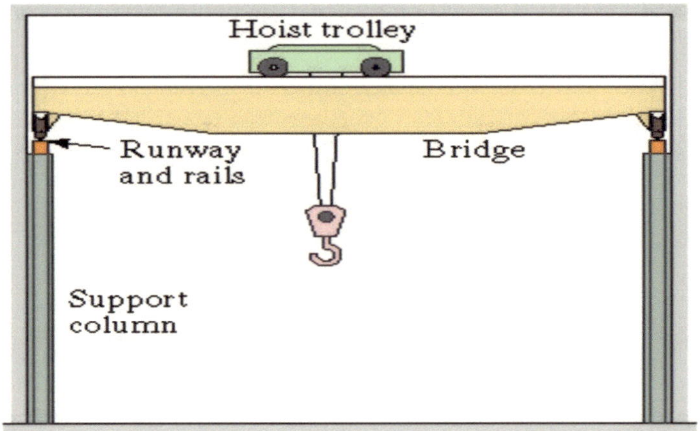


Fig. 8. Bridge crane

### 3.2.2 Half-gantry crane

Half gantry crane is distinguished from bridge crane by the presence of one or two vertical supporting elements which support horizontal girder. Gantry cranes may be half or double.Half gantry has one supporting vertical element whereas double gantry crane has two vertical supporting legs.

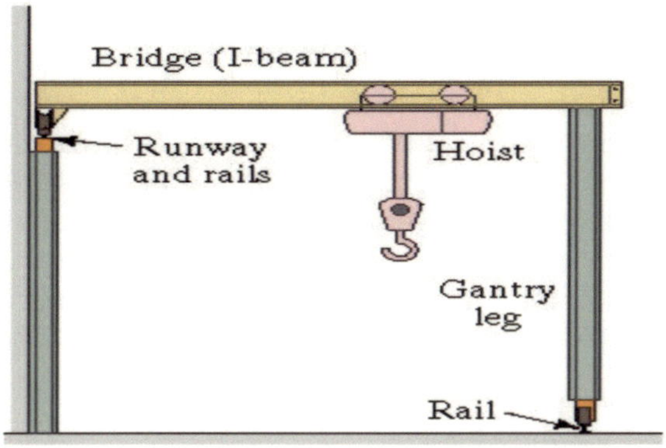


Fig. 9. Half gantry crane

### 3.2.3 Jib Crane

Jib cranes consist of a rotating arm with a hoist that runs along its length. The arm usually revolves on an axis which can be a fixed, ground-mounted post, or can be a wall or ceiling-mounted pin.

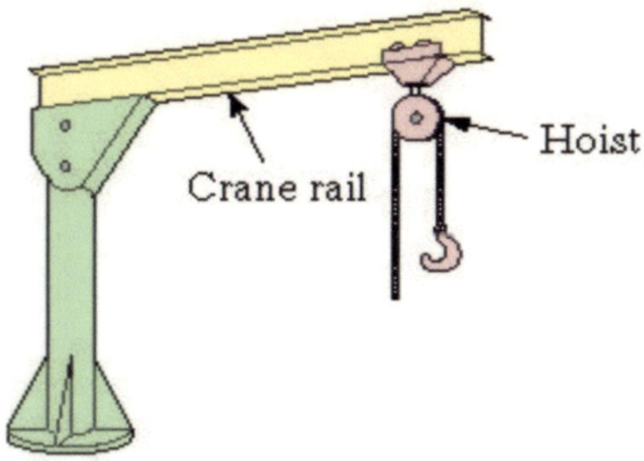


Fig. 10. Jib Crane

Wall-bracket mounted jib cranes are usually the least expensive jib cranes, but they require the most headroom and exert more force on their mounting wall. Cantilever jib cranes place the arm at the top, allowing for maximum lift when used in situations with limited headroom. They also exert less force on the wall on which they're mounted. Tie rod jib cranes make use of a tie rod between the arm and the mounting area. More inexpensive jib cranes feature manually operated chain hoists, while sophisticated cranes use an electric chain hoist. Jib cranes are used when the desired lifting area resides within a (semi-)circular arc.

#### 3.2.4 Stacker Crane

It is similar to a bridge crane. The major difference is that, instead of using a hoist, the stacker crane uses a mast with forks or a platform to handle unit loads. Stacker cranes are generally used for storing and retrieving unit loads in storage racks, especially in high-rise applications.

## 4. Automated Retrieval and storage equipments

Storage equipments can be in the form of racks, shelves, bins and drawers. Among these, storage rack is probably the most common form of storage equipment. There are numerous variants and configurations of storage racks, which include single-deep, double-deep rack, cantilever rack etc. and configurations that are designed to facilitate specific storage and retrieval operations drive-through, flow-through etc. More sophisticated retrieval and storage system combine the use of storage equipment, storing and retrieval machines and control that are manifested in a modern automated storage/ retrieval system.

## 5. Automated Guided Vehicles

An Automated Guided Vehicle System (AGVS) is a material handling system that uses independently operated, self-propelled vehicles guided along defined pathways in the facility floor. It is an automated material handling system which moves along predefined and preprogrammed path along an aisle from one station to another. The main parts of an AGV include structure, drive system, steering mechanism, power source (battery) and onboard computer for control.

### 5.1 Types of AGV

The following are common types of AGVs.

#### 5.1.1 Driverless Automated Guided Train

These are the first type of AGVS to be introduced around 1954.Its typical application is moving heavy payloads over long distances in warehouses and factories without intermediate stops along the route

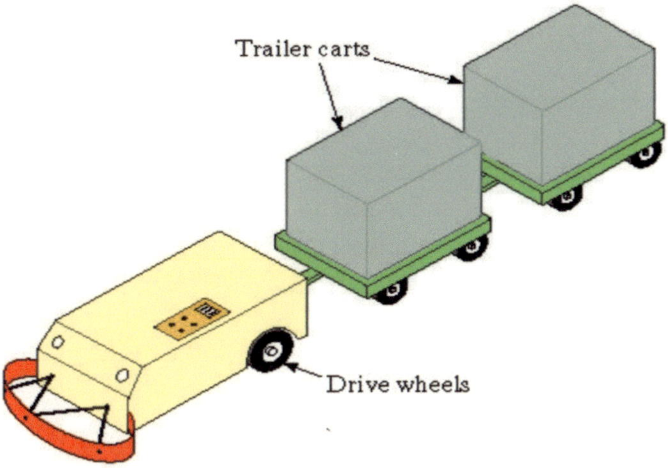


Fig. 11. Driverless automated guided vehicle

### 5.1.2 AGV Pallet Truck

These are used to move palletized loads along predetermined routes. Vehicle is backed into loaded pallet by worker; pallet is then elevated from floor. Worker drives pallet truck to AGV guide path and programs destination.

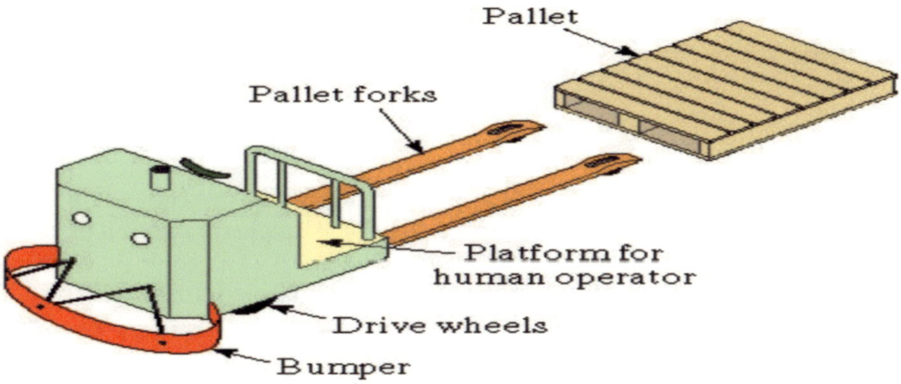


Fig. 12. AGV pallet truck

### 5.1.3 Unit Load Carrier

These are used to move unit loads from station to station and are often equipped for automatic loading/unloading of pallets using roller conveyors, moving belts, or mechanized lift platforms.

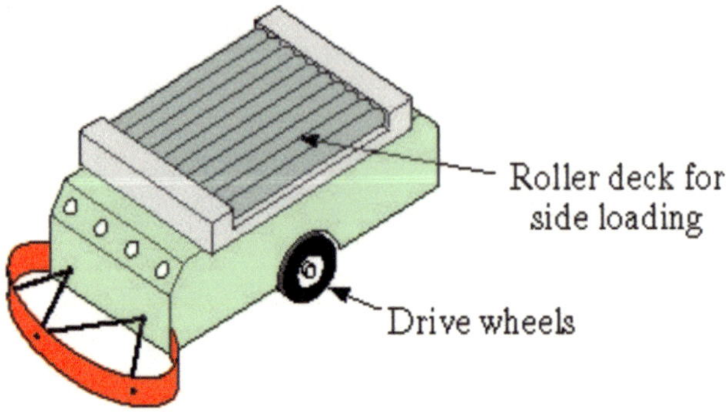


Fig. 13. Unit load carrier

### 5.1.4 Light load AGV

It can be applied for smaller loads. These are typically used in electronics assembly and office environments as mail and snack carriers.

### 5.1.5 Assembly AGV

These are used as assembly platforms, for example car chassis, engines etc., by carrying products and transport them through assembly stations.

### 5.1.6 Forklift AGV

It has the ability to pick up and drop off palletized loads both at floor level and on stands. Generally, these fork lift AGVs have sensors on forks for pallet interfacing.

### 5.1.7 Rail-Guided Vehicles

These are self-propelled vehicles that ride on a fixed-rail system. These vehicles operate independently and are driven by electric motors that pick up power from an electrified rail. Fixed rail system may be:

i. Overhead monorail - suspended overhead from the ceiling
ii. On-floor - parallel fixed rails, tracks generally protrude up from the floor

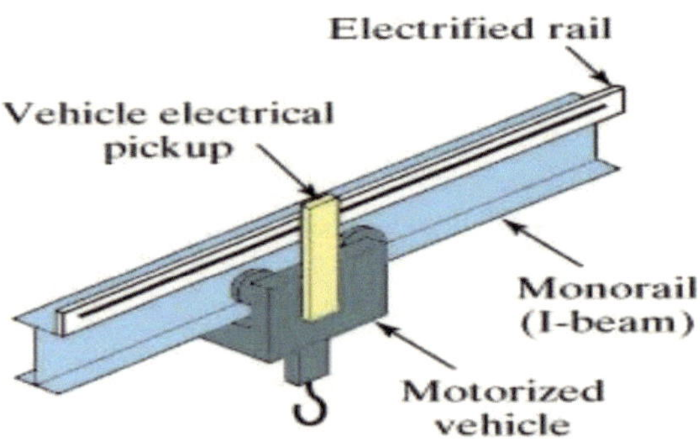


Fig. 14. Rail guided vehicle

### 5.2 AGVS System Management

AGVS is a complex system and a number of parameters need to be considered which include:
Guide-path layout
Number of AGVs required
Operational and transportation control

#### 5.2.1 Guide-path layout

The guide-path layout defines the possible vehicle movement path. Links and nodes that represent the action points such as pick-up and drop-off points, maintenance areas and intersections represent the path. The guide-path can be divided into four types:

1. Unidirectional single lane guide-path
2. Bi-directional single lane guide-path
3. Multiple lanes
4. Mixed guide-path.

Generally bidirectional single lane is considered the most cost effective and widely used layout.

#### 5.2.2 Number of AGVs required

It is important to estimate the optimum number of AGVs required for a system as too many AGVs will congest the traffic while too few means larger idle time for workstations in a system. Generally, the number of AGVs required is the sum of the total loaded and empty travel time and waiting time of the AGVs divided by the time an AGV is available.

#### 5.2.3 Operational and Transportation Control

The operation and transportation consists of vehicle dispatching, vehicle routing and traffic control issues. Once a demand arises for an AGV, a choice needs to be made regarding the vehicle to be dispatched among the pool of vehicles available. In an event when several workstations need servicing, a choice is to be made as to which workstation is to be serviced. The selection criteria can be applied for assigning the vehicles or workstations based on one or a combination of the following:
A random vehicle
Longest idle vehicle
Nearest vehicle
Farthest vehicle
Least utilized vehicle
Random workstation
Nearest workstation
Farthest workstation
Maximum queue size
Minimum remaining queue size
First come fist served
Unit load arrival time, due time or priority.
In order to dispatch an AGV to any workstation, it is necessary to find the shortest feasible path from the existing position. While selecting the shortest path it is necessary to consider

only those paths which are free and not occupied by vehicles. It may also be necessary to consider the future positions of the vehicles in the route in addition to their current occupied positions. In identifying the traffic control systems for AGVs movement, the approaches that can be used are forward sensing control, zone sensing control and combinatorial control. In forward sensing control, an AGV is equipped with obstruction detecting sensors that can identify another AGV in front of it and slow down or stop. This helps in improving the AGV utilization due to closer allowable distance between vehicles. However, this approach may not be able to detect the obstacles at intersections and around corners. This is generally useful for long and straight path which is divided into zones. Once an AGV enters a zone, it becomes unavailable for other AGVs which may introduce system inefficiency. The main advantages derived from the use of AGVs in manufacturing environment are:
Dispatching, tracking and monitoring under real time control which help in planned delivery.
Better resource utilization as AGVs can be economically justified.
Increased control over material flow and movement
Reduced product damage and routing flexibility
Increased throughput because of dependable on-time delivery.

## 6. Industrial Robots

Industrial robots are very useful material handling devices in an automated environment. An industrial robot is a reprogrammable multifunctional manipulator designed to move materials, parts, tools, or other devices by means of variable programmed motions and to perform a variety of other tasks. It is also defined as a machine formed by a mechanism including several degrees of freedom often having the appearance of one or several arms ending in a wrist capable of holding a job, tool and inspection device. It is automatically controlled, reprogrammable, multipurpose manipulative machine with several reprogrammable axes which is either fixed in place or mobile for use in industrial automation applications.

### 6.1 Robot components
The following are basic components of an industrial robot.

#### 6.1.1 Manipulator
It is a mechanical unit that provides motions similar to those of human arm and hand. The end of wrist can reach a point in space having a specific set of coordinates in specific orientation.

#### 6.1.2 End effector
It is attached with the end of wrist in a robot. It is a special purpose tooling which enables the robot to perform a particular job. Depending on the type of work, end effector may be equipped with any of the following:

a) Grippers, hooks, vacuum cups, and adhesive fingers for material handling
b) Spray guns for painting
c) Attachments for different kinds of welding processes.

### 6.1.3 Control system

It is a brain of a robot which gives commands for the movements of the robot. It stores the data to initiate and terminate movements of the manipulator. It interfaces with the computers and other equipments such as manufacturing cells or assembly operations.

### 6.1.4 Power supply

It supplies the power to the controller and manipulator. Each motion of manipulator is controlled and regulated by actuators that use an electrical, pneumatic or hydraulic power.

## 6.2 Robot Types

Robots are generally classified as Cartesian or rectilinear, cylindrical, polar or spherical jointed arms. They are also classified, from material handling point of view, as under:

### 6.2.1 Pick and place robot

It is also called fixed sequence robot and is programmed for a specific operation. Its movements are from point to point and cycle is repeated. These robots are simple and inexpensive and are used to pick and place materials.

### 6.2.2 Playback robot

This robot learns the work and motions from operator who leads the playback robot and its end effector through the desired path. The robot memorizes and records the path and sequence of motions and can repeat them continuously without any further action or guidance by the operator.

### 6.2.3 Numerically controlled robot

It is a programmable type of robot and works same as the numerical control machines. The robot is servo controlled by digital data and its sequence of movements can be changed with relative ease.

### 6.2.4 Intelligent robot

It is capable of performing some of the functions and tasks carried out by humans and is equipped with a variety of sensors with usual and tactile capabilities. It can perform tasks such as moving among a variety of machines on a shop floor avoiding collisions. It can recognize, select and properly grip the correct work piece.

## 6.3 Robot applications in Material handling

The major applications in material handling include:

1. Industrial robots are used to load/ unload materials during operations.
2. These are used to transfer the material from one conveyor to another.
3. These are used in palletizing and de-palletizing in such a way that parts/ materials are taken from conveyor and are loaded on to a pallet in a desired pattern and sequence and vice-versa.
4. These are very effective in automated assembly where repetitive work is required.

5. Intelligent robots can be used to automatically pick the right work piece without interference of operator and hence improves quality and pace of work.

## 7. References

M.P. Groover. "Automation, Production systems and computer integrated manufacturing" Second edition. Pearson-Prentice Hall, 2008.

K. Sareen and C. Grewal."CAD/CAM: Theory and concepts" S. Chand & Co. 2009.

C. R. Alavala. " CAD/CAM: Concepts and applications" Prentice-Hall, 2008.

P. N. Rao. " CAD/CAM: Principles and applications" McGraw-Hill, 2004.

C. R. Asfahl. "Robots and manufacturing automation" Second edition, John-Wiley and sons.1992.

M. P. Groover and E. W. Zimmers. Jr. " CAD/CAM: Computer added design and manufacturing" Pearson-Prentice Hall, 2009.

G. Chryssolouris, "Manufacturing systems: Theory and Practice" Springer-Verlag,1992.

# 7

# Scheduling methods for hybrid flow shops with setup times

Larysa Burtseva[a], Victor Yaurima[b] and Rainier Romero Parra[c]
[a]*Autonomous University of Baja California, Mexicali,*
[b]*CESUES Superior Studies Center, San Luis Rio Colorado, Sonora,*
[c]*Polytechnic University of Baja California, Mexicali,*
*Mexico*

## 1. Introduction

Many real manufacturing systems process a large number of product variants in the same flow. These products may differ in some optional components; consequently, the processing time on a machine differs from one product to the next, and the need to prepare one or more machines before beginning or after the finishing of jobs is frequently presented. The preparation activities are: machine adjustment and feeders preparation to process a next job, dismantling after a previous job, machine calibrating, inspection of accessories or tools, cleaning of the machines and adjacent areas, etc. In the scheduling theory, the time required to shift from one job to another on a given machine is defined as additional production cost or setup time. The scheduling problems, which consider the setup times, have a high computational complexity. Pinedo (2008) presents a proof of the NP-hardness of the single machine case with setup consideration. They are more complex when the resource model has the parallel machine environment.

The time that a job spends on a machine includes three phases: setup, processing, and removal. In the majority of investigations dedicated to production planning and scheduling it is assumed that the setup/removal times are negligible or nonseparable, therefore they are included in the job processing time, and hence are ignored. The nonseparable setup time assumption simplifies the analysis, and these problems can be formulated and solved as standard scheduling problem. However, an explicit treatment of the setup times in most applications is required and represents a special interest, because machine setup time is a significant factor for production scheduling in many cases. It may easily consume more than 20% of available machine capacity if it is not well handled (Pinedo, 2008).

Numerous examples of scheduling problems which consider separable setup times are given in the literature, including electronics manufacturing, automobile assembly plant, the packaging industry, textile industry, steel manufacturing, airplane engine plant, label sticker manufacturing company, semiconductor industry, maritime container terminal, ceramic tile manufacturing sector, as well as in electronics industry in sections for inserting components on printed circuit boards (PCB), where this kind of problems is frequent.

The purpose of this chapter is to present a class of deterministic scheduling problems in a multi-stage parallel machine environment called hybrid flow shop with setup times and appropriate methods for its resolution. The chapter includes a description of model with necessary definitions and notations; concepts of product family and batch, which are important elements of setup time analysis as well as a classification of setup times and problems that each category produces. The last section is focused on problems with sequence-depended setup times in hybrid flow shops. A review of investigated cases is explained, including the application of genetic algorithms for this kind of scheduling problems: structure of a genetic algorithm and description of several crossover operators appropriated to use based on previous investigations of authors. This section includes an algorithm and an example of a complex problem solution. A conclusion is presented at the chapter end.

## 2. Hybrid flow shop with setup times

In the scheduling theory, a multi-stage production process with the property that all products have to pass through a number of stages in the same order is classified as a flow shop. In a simple flow shop, each stage consists of a single machine, which handles at most one operation at a time. It is more realistic to assume that, at every stage, a number of machines that operate in parallel are available. This model is known as a hybrid flow shop (HFS). Some stages may have only one machine, but for the model to be qualified as a HFS, at least one stage must have multiple machines in parallel. These machines can be identical, or have different capacities. Each job is processed by at most one machine at each stage. The flow of products in the plant is unidirectional; each product is processed at only one machine in each stage.

The HFS models are common in the industry, which have the same technological route for all products as a sequence of stages, and any stages have a group of machines to realize the same operation. Various process industries, such as chemical, textile, metallurgical, semiconductors, printed circuit board, pharmaceutical, oil, food, and automobile manufacture, can be modeled as a HFS. In such industries, at some stages the facilities are duplicated in parallel to increase the overall capacities or to balance the capacities of the stages, or either to eliminate or to reduce the impact of bottleneck stages on the shop floor capacities.

Among scheduling problems which consider separable setup times in parallel machine environment, there is a class of problems of a high computational complexity, where setup from one product to another occurs on a machine; and machine parameters, which have to be changed during a setup, differ according to the production sequence. It leads to sequence-dependent setup times and consequently to sequence-dependent setup costs.

A HFS with setup times has the following characteristics:

- There are $k$ stages of processing in a linear order: 1, 2, ..., $k$.
- Each of the $n$ jobs visits the stages in this order, though all jobs do not need to visit all stages. Stages may be skipped for a particular job, but the process flow for each job is the same.

- Each stage has a predetermined number of parallel machines. However, the number of machines varies from stage to stage.
- The processing time for every job on every machine that it visits is known in advance and is constant.
- A job represents the processing of an item or a set of identical items (a container, a pallet, a box, a lot or a part) called batch.
- The jobs can belong to different job families. Jobs from the same family may have different processing times, but they can be processed on a machine after another without requiring any adjustment of machine in between.
- Every job is to be processed on one machine at a time without preemption and a machine processes no more than the job at a time. When an operation is started on a machine, it must be finished without interruption.
- Typically, buffers are located between stages to store intermediate products.
- The problem consists of assigning the jobs to machines at each stage and sequencing the jobs assigned to the same machine so that some optimality criteria are minimized.

The following index are used to describing the problems: $j$ for job, $j = 1,\ldots, n$, $i$ for stage, $i = 1, 2, \ldots,k$; $m_i$ for number of machines at the stage $i$; $l$ for machine index, $l = 1, 2, \ldots, m_i$.

The three-field notation $\alpha|\beta|\gamma$ is used to describe all details of considered HFS problem variant. The $\alpha$ field denotes the shop configuration, including the shop type and machine environment per stage. The $\alpha$ field discomposes into four parameter, i.e. $\alpha_1$, $\alpha_2$, $\alpha_3$, and $\alpha_4$, positioned as $\alpha_1\alpha_2$, $(\alpha_3\alpha_4^{(1)}, \alpha_3\alpha_4^{(2)}, \ldots, \alpha_3\alpha_4^{(\alpha_2)})$. Here, parameter $\alpha_1$ indicates the considered shop, and parameter $\alpha_2$ indicates the number of stages. For the HFS notation, FH is in the $\alpha_1$ position, and the value of parameter $\alpha_2$ has to be major that one. For each stage, parameters $\alpha_3$ and $\alpha_4$ indicate the machine set environments. More specifically, $\alpha_3$ indicates information about the type of the machines while $\alpha_4$ indicates the number of machines in the stage.

The possible machine set environments on the stage $i$ of a HFS are:

1. *Single machine* (1): a special case; any stages (not all) in a HFS can have only one machine.
2. *Identical machines in parallel* ($Pm_i$): job $j$ may be processed on any of $m_i$ machines;
3. *Uniformed machines in parallel* ($Qm_i$): the $m_i$ machines in the set have different speeds; a job $j$ may be processed on anyone machine of set, however its processing time is proportional of the machine speed.
4. *Unrelated machines in parallel* ($Rm_i$): a set of $m_i$ different machines in parallel. The time that a job spends on a machine depends on the job and the machine.

When there are several consecutive stages with the same machine set environments, the parameters $\alpha_3$ and $\alpha_4$ can be grouped as $((\alpha_3\alpha_4^{(i)})_{i=s}{}^{k})$, where $s$ and $k$ are the index of the first and the last consecutive stage, respectively. For example, the notation $FH4$, $(1,(P2^{(i)})_{i=2}{}^{3},R3^{(4)})$ refers to a HFS configuration with four stages where there are one machine at the first stage, two identical machines in parallel at second and third stages and three unrelated parallel machines in the fourth stage.

The $\beta$ field provides the shop properties; also other conditions and details of the processing characteristics, which may enumerate multiple entries, also may be empty if they are not.

The following model properties are frequently associated with a setup time HFS scheduling problem:

*batch*(*b*) Batch processing. A machine is able to process up to *b* jobs continuously without any setup.

*brkdown* Machine breakdown implies that a machine may not be continuously available.

*fmls* Job families. The *n* jobs belong to *F* different job families. Jobs from the same family may have different processing times, but they can be processed on a machine after another without requiring any setup in between.

$M_{jk}$ Machine eligibility restrictions. Processing of job *j* is restricted to the set *Mj* of machines at stage *k*.

$r_j$ Release dates. The job *j* cannot start processing before release data $r_j$.

*R* Removal time. Machines become free only after the setup of the job has been removed.

$S_{si}$ Sequence-independent setup times. The setup time of machine depends only on the job to process and does not depend on the previous job.

$S_{sd}$ Sequence-dependent setup times. The setup time of machine required to process next job depends on the previous job.

$w_j$ The priority factor denoting the weight or importance of job *j* relative to the other jobs of system.

The $\gamma$ field establishes the objective to be minimized. The more common objective functions to minimize in a HFS scheduling problem are: $C_{\max}$ as maximum completion time; $F_{\max}$ as maximum flow time; $L_{\max}$ as maximum lateness; $T_{\max}$ as maximum tardiness; $E_{\max}$ maximum earliness, among others. The most used objective function to be minimized in a HFS scheduling problem is the completion time when the last job to leave the system, referred to a makespan or $C_{\max}$.

A HFS standard scheduling problem with *k* stages and a number of the identical parallel machines in each stage is denoted as $FHk$, $((PM^{(i)})_{i=1}{}^{k}) \mid \mid C_{max}$. In this case, the formula defines a HFS with *k* stages, $|M^{(i)}|$ identical machines in parallel on stage *i*, *i* = 1, ..., *k*; there are not any special parameter $\beta$, and the objective is the makespan minimizing.

Figure 1 illustrates the physical relationship between machines and stages, which corresponds to the notation $FH3$, $(1, P3^{(2)}, R2^{(3)}) \mid M_{j3}, S_{sd} \mid T_{\max}$, referring to tri-stage HFS. The stage 1 has one machine, stage 2 has three identical machines in parallel, and stage 3 has two parallel unrelated machines; $M_{j3}$ and $S_{sd}$ indicate that there are machine eligibility restrictions at stage 3 and setup times depended on the sequence of jobs. The objective is the maximum tardiness minimizing. Moreover, the figure shows that there are unlimited buffers between stages to storage unfinished products, so called Work In Process (WIP).

A production system, to be classified as a HFS has to be flexible. It is important to know the differences between a flexible production system and a traditional one; what exactly means the concept of flexibility and what justifies the use of specific production planning models for flexible production systems. Automated manufacturing systems display flexibility in multiple and intertwined ways, pertaining to the equipment, processes, products,

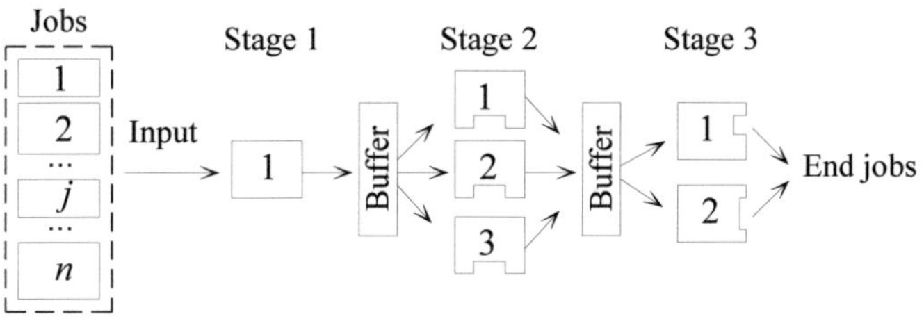


Fig. 1. Resource model for a tri-stage HFS.

production volumes, etc. Among the more important concepts, are the following (Crama, 1997), (Vairaktarakis, 2004):

1. *machine flexibility,* the ability of the machines to perform various types of operations without requiring a prohibitive effort in switching from one operation to another;
2. *material handling flexibility,* the ability of the material handling system to move different parts efficiently for proper positioning and processing through the manufacturing facility;
3. *operation flexibility*, the ability to realize it in different ways;
4. *processing flexibility,* that means that jobs may skip stages or there is a set of part types that the system can produce without major setups;
5. *routing flexibility,* the ability of a manufacturing system to produce a part by alternate routes through the system.

A planning production model with sets machines in parallel has to comply with one of these concepts to be classified as a flexible flow shop tacking in account that the flow of products in the plant is unidirectional. The hybridizing occurs when any products require special manufacture conditions, e.g., different qualities and capacities of machines at the same stage, assignment any jobs on certain machines, and another special conditions.

Meanwhile, the HFS has been studied since the 70$^{th}$, the researcher put much attention to this model and some new designs were discovered on the recent years. This fact probably implicates confusions in the terminology and notations. Actually, there is not in the literature a conventional classification of this kind of flow shops. A variety of known models should be interpreted as a HFS or its special case.

There are:

*Flexible flow shop* (FFS); a HFS in the parallel identical machine environment when the machines in each set are identical (*processing flexibility* within a production stage which is derived from the ability to process a job on any parallel machine at stage). Some authors, as e.g., Pinedo (2008), Jungwattanakit et al., (2009) do not use the notion HFS, and describe the more complex configurations as a FFS with not identical parallel machines at least on one stage. Moreover, a variety of authors do not differ between terms of FFS and HFS referring this model as a flexible (hybrid) flow shop (Allahverdi et al., 2008); or use the HFS term in parallel identical machine environment (Naderi et al., 2009).

*Flexible flow line* (FFL) and *Flow shop with multiple processors* (FSMP or MPFS) are equivalent to a FFS (Lin & Liao, 2003). Zandieh at al. (2006) considering that the HFS is known commonly as a flexible flow line, because the flow of jobs in that system is unidirectional.

*Hybrid flexible flow shop or Flexible hybrid flow line* (HFFL); this model is equivalent to a HFS where jobs might skip stages (*processing flexibility* across production stages) (Ruiz & Vazquez-Rodriguez, 2010), (Allahverdi et al., 2008)

*Parallel HFS* (PHFS) system represents a HFS decomposed into smaller HFS sub-designs operated in parallel. More specific, a PHFS is composed of a number of independent sub-designs each of which is a HFS of the unidirectional routing (*routing flexibility*) (Vairaktarakis, 2004).

The HFS scheduling problems which consider setup times are among the most difficult classes of scheduling problems. It is known, that a one-machine sequence-dependent setup scheduling problem is equivalent to a traveling-salesman problem which is NP-hard, even for a small system, the complexity of this problem is beyond the reach of existing theories (Pinedo, 2008). A HFS restricted to two processing stages, even in the simplest case when one stage contains two identical machines and the second only a single machine, is already NP-hard, according to Gupta (1988). Moreover, the special case where there is a single machine per stage, known as the flow shop, and the simplest case where there is a single stage with several machines, known as the parallel machine environment, are also NP-hard (Glover & Laguna., 1997). The total number of possible solutions for a HFS to be $n!(\Pi_{i=1}{}^{k}\, m_i)^n$ while the number of possible solutions in a regular flow shop scheduling problem is $n!$ The complicity of a HFS scheduling problem with setup time condition depends essentially on setup time nature.

## 3. Batch processing

A technical similarity between products of a plant often reflects an obvious grouping of them into product groups. Products can be sorted out into groups according to their design attributes, which include part shape, size, surface texture, material type, raw material estate, or according to their manufacturing attributes. The technical similarities of the products within a group permit reduce essentially the setups number on a machine, when a setup from one product to another occurs and hence manufacturing time would be decreased and consequently machine usage time would be improved.

This idea is adapted by the Group Technology (GT) (Andrés et al., 2005). The GT concept is based on the simplification and standardization process. It was dedicated originally to single machine environment to reduce setup times. This concept was further extended to the production planning in productive systems which have some available resources in each of the stages of production and not negligible setups known as the HFS problem with setup times dependent on the sequence (Li, 1997).

From the GT surge the concepts of product *family* and *batch*. The jobs are supposed to be partitioned into $F$ families, $F \geq 1$. A batch is a set of jobs of the same family. Batching occurs only if setup costs or times are not negligible and several jobs of the same product type have to be produced. When the processing is realized in batches (lots, pallets, containers, boxes), the operations processed simultaneously start together and complete together, with just a single setup in the beginning. Their processing time depends only on the family of the batch.

When one batch is completed, the resources have to be adjusted for the next batch. The time needed for the setup depends on the families of both adjacent batches. A batch is called feasible if it can be processed without any tool switches. While families are supposed to be given in advance, batch formation is a part of the decision making process. To *batch-sizes* calculating has to decide how many units must be produced consecutively. In (Liu & Chang, 2000) is indicated that the processing in large batches may increase machine utilization and reduce the total setup time. However, large batch processing increases the flow time. There is a tradeoff between flow time and machine utilization by selecting batch size and scheduling. According to the GT, no family can be split, only a single batch can be formed for each family.

Batch setup models are further partitioned into *batch availability* and *job availability* models. According to the batch availability model, all the jobs of the same batch become available for processing and leave the machine together. Two rules that define the processing time of a batch are distinguished (Lushchakova & Strusevich, 2010):

- In the case of sequential batch processing, also known as "sum-batch", the processing time of a batch on machine is equal to the total processing times of its jobs;
- In the case of simultaneous batch processing, also known as "max-batch", the processing time of a batch on machine is equal to the largest processing time of its jobs.

In the job availability model, each job's start and completion times are independent on other jobs in its batch.

The term of *family* denotes initial job partitioning, while the term of *batch* is used to denote a part of the solution. Many publications use the term batch to denote the initial job partitioning and they use different names like sub-batch, lot, sub-lot, etc., to denote a set of jobs of the same family processed consecutively on the same machine. In the literature, a job availability model is considered, if not stated otherwise.

Li (1997) gives an example of scheduling problem from an airplane engine plant, Pratt and Whitney Inc. (PWI). The blade line, one of the production lines at PWI, characterized as a two-stages HFS, produces various types of blades used in airplane engines. Each stage of the blade line at PWI has a different number of machines. The types of blades that have similar processing requirements are grouped into families. A major setup is required if a machine at any stage switches from one family of blades to the other. A minor setup is required if a machine switches from one type of blade to another type in the same family. Since setup times are not insignificant and unit processing times for all types of blades are very short, the plant processes each type of blade in batches (lots).

The *batch setup time* (cost) can be machine dependent or sequence (of families) dependent. It is sequence-dependent if its duration (cost) depends on the families of both the current and the immediately preceding batches, and is sequence-independent if its duration (cost) depends solely on the family of the current batch to be processed.

A HFS scheduling problems with setup times which consider job processing in batches can be *sequence-dependent* as well as *sequence-independent*. Most studies assume that either no setup has to be performed or that setup times are sequence-independent and there is only a single unit of each product type. In this case, a job's setup time may be added to its process time. However, if setup times are sequence-dependent or if several jobs of the same product type have to be produced, setups have to be considered explicitly.

In a non-batch processing environment, a setup time (cost) is incurred prior to the processing of each job. The corresponding model can also be viewed as a batch setup time (cost) model in which each family consists of a single job.

## 4. Classification of HFS with setup times

The setup times, defined as the time required to shift from one job to another on a given machine, are considered as separable and non-separable from the processing operation.

The *non-separable* setup times are either included in the processing times or are negligible, and hence are ignored. There exist some situations in which the nonseparable setup and removal operations must be modeled and closely coordinated. Such situations are common in automatic production systems which involve intermediate material handling devices, like an automatic guided vehicles and robots, loading and unloading (Crama, 1997), (Kim et al., 1997), (Pinedo, 2008).

When these operations are *separable*, i.e. they are not a part of processing operation, the structure of the breakdown time when a job belongs to a machine is as follows (Cheng et al., 2000):

1) Setup time that is independent on the job sequence. This operation consists of activities such as fetching the required details, and fixtures, and setting them up on the machine.
2) Setup time that is dependent on the job to be processed. The carrying out of this operation includes the time required to put the job in the jigs and fixtures and to adjust the tools.
3) Processing time of the job being processed.
4) Removal time that is independent on the job that has been processed. This operation includes activities such as dismounting the jigs, the fixtures and/or tools, inspecting/sharpening of the tools, and cleaning the machine and the adjacent area.
5) Removal time that is dependent on the job that just has been processed. This operation includes activities such as disengaging the tools from the job, and releasing the job from the jigs and fixtures.

Three phases of job processing can be grouped as following: the separable setup, the processing, and the separable removal times represented by items (1, 2), (3), and (4, 5), respectively. When separable setup/removal times are not negligible in the scheduling problem, they should be explicitly considered.

The setup times which are separable from the processing times, could be *anticipatory* (*detached*) or *non-anticipatory* (*attached*). A setup is anticipatory if it can be started before the corresponding job or batch becomes available on the machine. In such a situation, the idle time of a machine can be used to complete the setup of a job on a specific machine. Otherwise, a setup is non-anticipatory and the setup operations can start only when the job arrives at a machine as the setup is attached to the job. Further, setup times of a job at a specific machine could be dependent on the job immediately preceding that job or be independent on it. If it is not stated explicitly that setups are non-anticipatory.

As follows, a classification of HFS scheduling problems with setup times, derivate from the classification presented in (Cheng et al., 2000), is described. These problems generally fall into the following four board categories depending on practical situations (Figure 2):

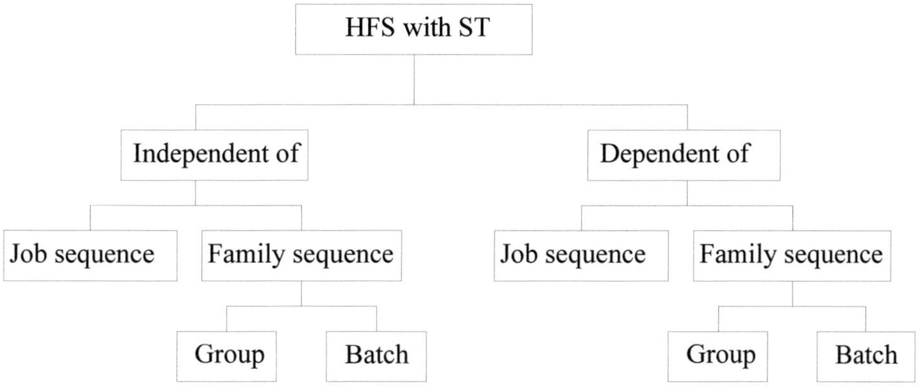


Fig. 2. Classification of HFS with setup times.

- HFS with sequence-independent job setup times.
- HFS with sequence-dependent job setup times.
- HFS with sequence-independent family setup times:
  - HFS with sequence-independent group setup times;
  - HFS with sequence-independent batch setup times.
- HFS with sequence-dependent family setup times:
  - HFS with sequence-dependent group setup times;
  - HFS with sequence-dependent batch setup times.

*HFS with sequence-independent job setup times.* The setup times are separable from the processing times and sequence-independent, i.e., depend only on the job to process and do not depend on the job sequence on this machine. Such setup times could be either detached or attached to the processing times. However, if this setup time is attached to a job, the idle time of a machine cannot be used, and hence the setup time have to be considered as a part of the processing time and the problem can be formulated and solved as a standard HFS problem. For this reason, only *detached* sequence-independent job setup times represent a special interest. Further, removal times could be either zero or positive. The removal times, if they are present, have to be included in makespan definition.

Three papers with different setup/removal restrictions are mentioned as references. In (Kim et al., 1997) the $C_{max}$ minimizing problem for FFS with two stages, independent setup times and negligible removal times is considered. A scheduling rule similar to the Johnson's rule is suggested to minimize makespan. Another work, (Low, 2005), addresses to a HFS with $J$ stages and unrelated parallel machines at each stage, independent setup and dependent removal times. The objective is to minimize total flow time in the system. A simulated annealing-based heuristic is proposed to solve the addressed problem in a reasonable running time. In (Harjunkoski & Grossmann, 2002) is considered a HFS model where setup times are included, but they are only dependent on the machine and not on the job. The objective is to minimize job assignment costs and one-time machine-initialization costs.

*HFS with sequence-dependent job setup times* (Li, 1997), (Naderi et al., 2009).This situation occurs when the part of the setup of job *i* can be used for processing the next job *j*. It implicates that the removal time of the job *i* will depend on the job *j* to be processed next. On the other hand, the setup time of job *j* also depends on the job *i* being processed currently, because the setup part of the previous job *i* can be used for the next job *j*. The net effect of these two factors is that the setup time of job *j* depends on the immediately preceding job *i*. Sequence-dependent setup times are of the anticipatory (detached) type because their nature. Therefore, the setup information cannot move with the job; and sequence-dependent setup times cannot be of the attached type as information about the currently processed job *i*; and the next job *j* requires to determine the needed setup time to be processed.

*HFS with sequence-independent/dependent family setup times.* In the above two categories the setups are associated with individual jobs. However, in many real-world situations, the processing of jobs is realized taking in account the job family. When jobs belonging to the same family scheduled contiguously, they only need a common setup operation, and so called *family setup times* (FST) are involved. The job partitioning into families implicates two next situations:

- the setup operation arises only when a machine shifts from processing a job in one family to processing a job in another family;
- a job containing several identical items may be split into multiple sublots and the setup operation arises only when a machine shifts from processing the sublot of one job to processing a sublot of another job.

  In general, these FST scheduling problems require two interrelated decisions:
- the size and number of the sublots of each family where the items of a single sublot are processed together;
- the scheduling of each sublot through the HFS where each sublot requires setup on each machine. According to the GT assumption, a family does not split into sublots, and the jobs of the same family are processed together. It refers to so called HFS with *group setup time* (GST) problem. However, if the families are split, it requires a solution of an interrelated *batching* problem to find the optimal size of each sublot. It refers to a so called *HFS with batch setup times* (BST) problems.

Since the BST problems require the solution of two interrelated problems (that of batching and scheduling), these problems are relatively harder to solve than their corresponding GST problems requiring only the solution of a scheduling problem (Monma & Potts, 1989).

The setup times in the Sequence-Independent Family Setup Times problem could be either attached or detached. Since each sublot in case of BST (or family in case of GST problem) consists of multiple jobs, the sequence-independent setup time cannot be added to the processing time of any one of these jobs, as the first job in the sequence of the sublot or batch is not known until the scheduling problem is solved. However, for the sequence-dependent BST and GST problems, the sequence-dependent sublot or batch setup times are only of the detached type.

Batch scheduling problem with setup times arises frequently in process industries, parts manufacturing environments and cellular assembly systems (such as chemical, pharmaceutical, food processing, metal processing, printing industries and semiconductor testing facilities). Detailed surveys of the recent publications about HFS with setup times might be consulted in (Ribas et al., 2010), (Ruiz, 2010), (Allajverdi, et al., 2008), (Zandieh et al., 2006).

## 5. HFS with sequence-dependent setup times

### 5.1 Investigated problems

In recent years, many researchers put attention to the HFS problem with sequence-dependent setup times in consequence of its complexity, the variety of models as realistic as theoretical, and used tools to the algorithm creation. On Table 1 are summarized publications dedicated to the investigation of the HFS with sequence-dependent setup times. The first column indicates the year of publication, the second is the bibliographical reference, the third describes the problem; and the final column shows the resolution method, type of approach as well as other case details.

| Year | Author | Problem | Comments |
|---|---|---|---|
| 1991 | Guinet | $FHm,((PM^{(k)})_{k=1}^{m}) \mid S_{sd} \mid \{C_{\max},\overline{T}\}$ | ad-hoc heuristics, textile industry |
| 1993 | Adler et al. | $FMm,((RM^{(k)})_{k=1}^{m}) \mid S_{ad} \mid \overline{T}^{w}$ | DR, packging industry |
| | Voss | $FHm,((PM^{(1)},1^{(2)}) \mid S_{sd} \mid C_{\max}$ | TS and heuristics |
| 1995 | Aghezzaf et al. | $FHm,((RM^{(k)})_{k=1}^{m}) \mid S_{sd} \mid \{C_{\max},F_{\max},\overline{F}\}$ | MPF, heuristics, carpet manufacturing |
| 1997 | Li | $FH2,((1^{(1)},PM^{(2)})) \mid batch,S_{sd},split \mid C_{\max}$ | heuristics, major and minor setups, airplane engine plant |
| 2000 | Liu & Chang | $FHm,((PM^{(k)})_{k=1}^{m})) \mid S_{sd},block \mid \overline{E}^{w}+\overline{T}^{w}$ | MPF based heuristics |
| 2003 | Kurz & Askin | $FHm,((PM^{(k)})_{k=1}^{m})) \mid S_{sd} \mid C_{\max}$ | heuristics |
| | Lin & Liao | $FHm,((PM^{(k)})_{k=1}^{2})) \mid S_{sd}^{(1)},M_{j}^{(2)} \mid wT_{\max}$ | heuristic, label sticker manufacturing |
| 2004 | Kurz & Askin | $FHm,((PM^{(k)})_{k=1}^{m})) \mid S_{sd} \mid C_{\max}$ | MPF, MPR-GA |

| | | | |
|---|---|---|---|
| 2005 | Andres et al. | $FH3,((PM^{(k)})_{k=1}^{3}))\mid S_{sd}\mid other$ | MPF, GT |
| | Pearn et al. | $FHm,((PM^{(k)})_{k=1}^{m}))\mid S_{sd}\mid \overline{T}$ | MPF, heuristics, packaging industry |
| | Tang et al. | $FHm,((PM^{(k)})_{k=1}^{m}))\mid S_{sd}\mid C_{\max}$ | NN |
| 2006 | Ruiz & Maroto | $FHm,((RM^{(k)})_{k=1}^{m}))\mid S_{sd},M_j\mid C_{\max}$ | MPR-GA |
| | Zandieh | $FHm,((PM^{(k)})_{k=1}^{m}))\mid S_{sd}\mid C_{\max}$ | Artificial Immune System |
| 2007 | Chen et al. | $FH3,((RM^{(k)})_{k=1}^{3}))\mid S_{sd},block,prec\mid C_{\max}$ | MPF, lower bounds, TS, container terminal |
| | Voss & Witt | $FHm,((PM^{(k)})_{k=1}^{m})\mid S_{sd}\mid \overline{T}^{w}$ | MPF, DR, heuristics. Multi-project RCPSP, steel manufacturing |
| 2008 | Jungwattanakit et al. | $FHm,((RM^{(k)})_{k=1}^{m}))\mid S_{sd},r_j\mid \alpha C_{\max}+(1-\alpha)\overline{U}$ | MPF, heuristics, GA, SA, TS |
| | Ruiz et al. | $FHm,((RM^{(k)})_{k=1}^{m}))\mid skip,rm,lag,S_{sd},M_j,prec\mid C_{\max}$ | MPF, heuristics |
| 2009 | Jungwattanakit et al. | $FHm,((RM^{(k)})_{k=1}^{m}))\mid S_{sd},r_j\mid \alpha C_{\max}+(1-\alpha)\overline{U}$ | MPF, heuristics, DR, GA |
| | Naderi et al., a | $FHm,((PM^{(k)})_{k=1}^{m}))\mid S_{sd},transport\mid \{\overline{F},\overline{T}\}$ | SA |
| | Yaurima et al. | $FHm,((RM^{(k)})_{k=1}^{m}))\mid S_{sd},M_j,buffer\mid C_{\max}$ | MPR-GA, electronic indusry |
| | Naderi et al., b | $FHm,((PM^{(k)})_{k=1}^{m}))\mid S_{sd}\mid \{\overline{F},\overline{T}\}$ | SA |
| | Alfieri | $FHm,((PM^{(k)})_{k=1}^{m}))\mid S_{sd},reentry,batch\mid several$ | simulation, heuristics, TS |

Table 1. Investigated problems of HFS with sequence-dependent setup times.

The table shows that in recent years the publications are dedicated to more complex models of the problem, with unrelated parallel machine environment, release times, limited buffers, lags, and machine eligibility among others. A general framework to solve the problem includes: dispatching rules (DR), neural networks (NN), tabu search (TS), multiple permutation representation (MPR), local search (LS), simulated annealing (SA), genetic algorithms (GA). Mathematical programming formulation (MPF) is developed for many models.

The procedures to seek a solution of a HFS problem can be classified into two principal categories (Quadt & Kuhn, 2007): *optimal procedures* and *heuristics*. The literature does not report application of any optimal procedure, like Dynamic Programming or Branch & Bound methods, to solution of a HFS problem with sequence-dependent setup times. Heuristics do not necessarily find an optimal solution. However, they are usually faster than optimal procedures and some of them are used for realistic problem sizes. Heuristics may be split into *holistic* and *decomposition* approaches. Holistic approaches consider the complete scheduling problem in an integrated way. A simple holistic approach is to use dispatching rules to select the next job that has to be produced whenever a machine becomes idle. The use of such heuristics is very common in HFS, see, e.g., (Adler et al., 1993), (Voss & Witt, 2007), (Jungwattanakit et al., 2009).

Most holistic procedures are local search methods or metaheuristics. In HFS, a job consists of several operations, one for each production stage. Thus, a HFS schedule assigns machine for each operation as well as a production sequence for each machine. A move to a neighboring schedule may change the machine assignment of an operation or its position in the sequence. This neighborhood is very large. Hence, local search procedures and metaheuristics must find ways to limit the size of the neighborhood. This can be done by only allowing certain moves that appear promising. Examples of metaheuristic techniques application are given in (Tang et al., 2005), (Chen et al., 2007), (Naderi et al., 2009), (Yaurima et al., 2009), among others.

In contrast to holistic approaches, decomposition approaches divide the problem into segments that are considered consecutively, with respect to the production stages, the individual jobs, or the sub-problems to be solved (batching, loading, and sequencing), e.g., (Li, 1997), (Alfieri, 2009).

The HFS scheduling problems with sequence-dependent setup times are among the most difficult classes of scheduling problems. When a practical problem of large instance sizes does not require of a fast result obtain, a good approximate solutions are achieved through a genetic algorithm (GA).

### 5.2 GA approach

A GA is a well known search technique used to find solutions to optimization problems. It was proposed by Holland (1975). All GA act according to the scheme represented on Figure 3. Candidate solutions are encoded by chromosomes (also called genomes or individuals). The set of initial individuals forms the population. Fitness values are defined over the individuals and measure the quality of the represented solution. The genomes are evolved through the genetic operators generation by generation to find optimal or near-optimal solutions. Three genetic operators are repeatedly applied: selection, crossover, and mutation. The selection picks chromosomes to mate and produce offspring. The crossover combines two selected chromosomes to create next generation of chromosomes. The mutation randomly reorganizes the structure of genes in a chromosome, so that a new combination of genes may appear in the next generation. The individuals evolve until some stopping criterion is met.

The first paper was by Ruiz and Maroto (2006) where application of GA techniques to HFS problem with sequence-dependent setup times had been realized. There were considered the makespan minimization criterion on a $m$-stage problem with unrelated parallel machines, sequence-dependent setup times and machine eligibility. The proposed GA was

superior to

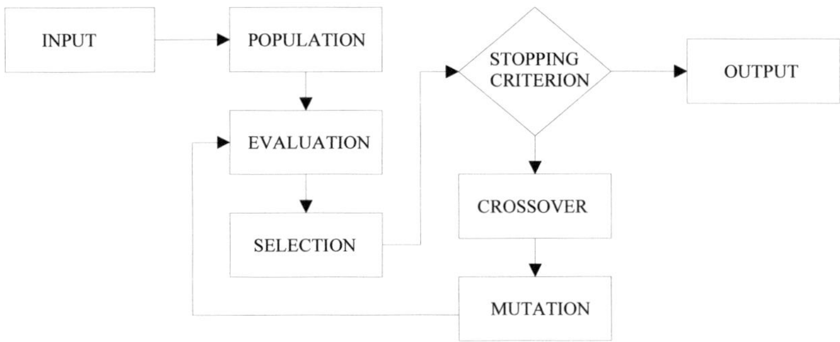


Fig. 3. GA scheme

a wide range of heuristics and other metaheuristics, among them, ACO based heuristics, TS procedures, other GAs, SA and deterministic procedures. A similar problem, with unrelated parallel machines at each stage, and setup times, was approached in (Jungwattanakit, et al. 2008) using GAs and later in (Jungwattanakit, et al., 2009) applying several heuristics including DR, GA, tailored heuristics, TS and SA. In these two papers, the authors study a linear combination of the makespan and a number of tardy jobs as an objective. Recently, in (Yaurima, et al., 2009) is proposed a GA for a complex HFS problem considering makespan minimization on an *m*-stage problem with sequence-dependent setup times, unrelated parallel machines, machine eligibility and limited buffers.

### 5.3 Crossover operators

The crossover operator is an important factor for a good performance of the GA. The following crossover operators should be considered to solve the examined problem according to previous investigations of authors.

1. OBX - *Order Based Crossover* (Gen, 1997), (Figure 4).

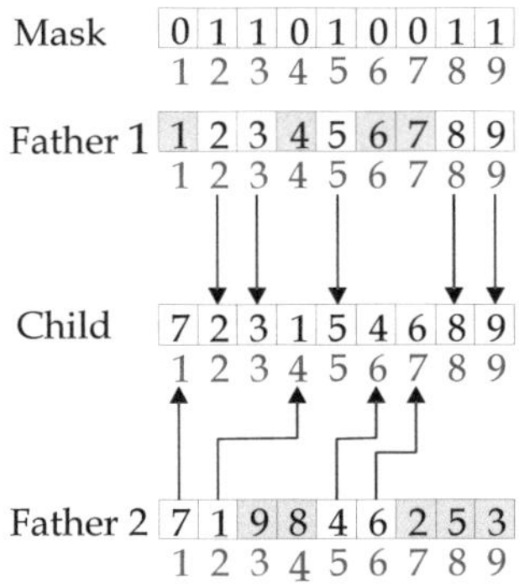


Fig. 4. OBX Crossover

This operator is based on a binary mask. The values of the mask equal to one indicate that the corresponding sequence of elements from parent 1 to child, is copied. The remaining elements from parent 2 are copied. The mask values are generated randomly and uniformly in all crossover operations.

2. PPX - *Precedence Preservative Crossover* (Bierwirth, et al., 1996), (Figure 5).

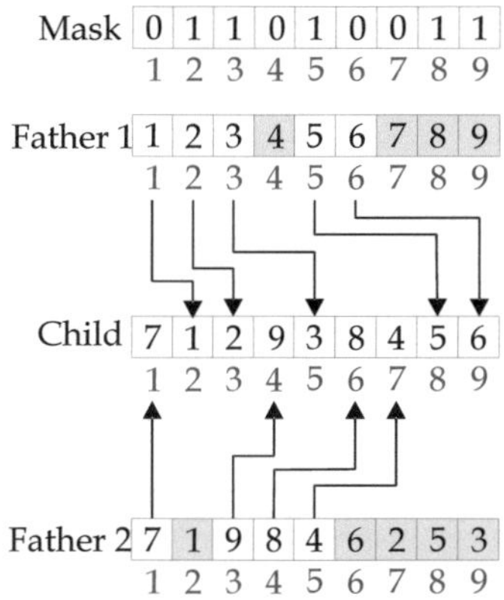


Fig. 5. PPX Crossover

This operator is based on a binary mask. The values of the mask equal to 1 indicate that corresponding elements from parent 1 are copied to child and values equal to 0 indicates that the elements are copied from parent 2, according to each value of the mask one at a time alternately.

3. OSX - *One Segment Crossover* (Guinet & Salomon, 1996), (Figure 6).

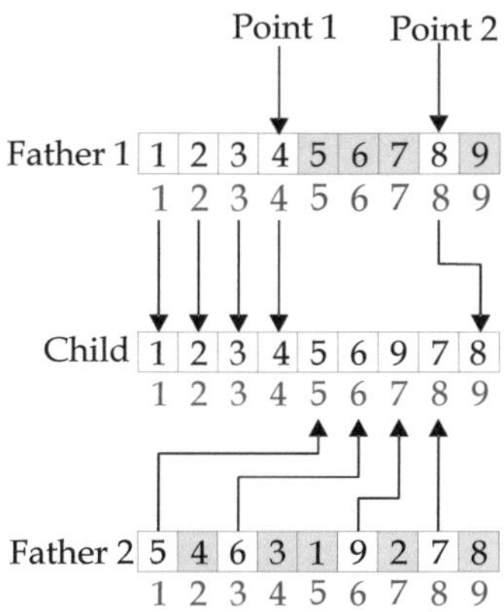


Fig. 6. OSX Crossover

Two points are randomly chosen. The elements from parent 1 since position 1 to the first place are copied. Elements from parent 2 since first point to the second point are copied. Finally, the items from parent 1 since the second point to last position are copied, considering not copied elements.

4. TP - *Two Point* (Michalewicz, 1996), (Figure 7).

Two points are randomly chosen. The elements from parent 1 since first position to the first point and since second point to the last position are copied. The elements from parent 2 since first point to the second point are copied.

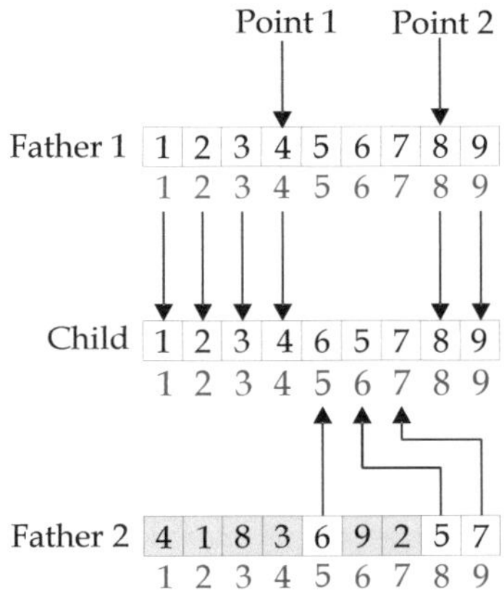


Fig. 7. TP Crossover

5. SB2OX - *Similar Block 2-Point Order Crossover* (Ruiz & Maroto, 2006), (Figure 8).

The common blocks in both parents (at least two consecutive identical jobs) are copied to the children, then two random cut points are defined and the section between these two points directly copied to children. The missing elements of each offspring are copied in the relative order of the other parent.

6. ST2PX - *Setup Time Two Point Crossover* (Yaurima, et al., 2009), (Figure 9).

In this crossover operator the sequence-dependent setup time is considered. Two points randomly in the sequence are chosen. The elements since first position to the first point and since second point to the last position, are copied from parent 1. The elements since first point to the second point are copied from parent 2 according to the minimum setup time of one machine randomly chosen from the first stage.

### 5.4 A problem of makespan minimizing in a HFS with multiple constrains

A complex problem of makespan minimizing in a HFS with sequence-dependent setup times, unrelated machines, availability constraints and limited buffers is presented. The real case of the television production environment is considered (Yaurima, et al., 2009).

Different television models are distinguished by their set of PCBs. The monthly production plan is developed based on current requirements, machines availability and resource constrains. It is updated daily depending on the final section requirements. It is examined the auto-Insertion section, where various PCB types are manufactured with automated machines for 70 television models, 45 machines and production units of different brands are dealt with.

The auto-insertion section is represented by a HFS with six stages (operations) common for all PCB types. However, some PCBs do not require all six operations. Each stage consists of several insertion machines in parallel, and they are dedicated to the certain types of component processing. At each instant of time, each machine works on at most one PCB, and each PCB is processed by at most one machine. The PCBs are moving along the assembly line, from one machine to another until it became a complete unit.

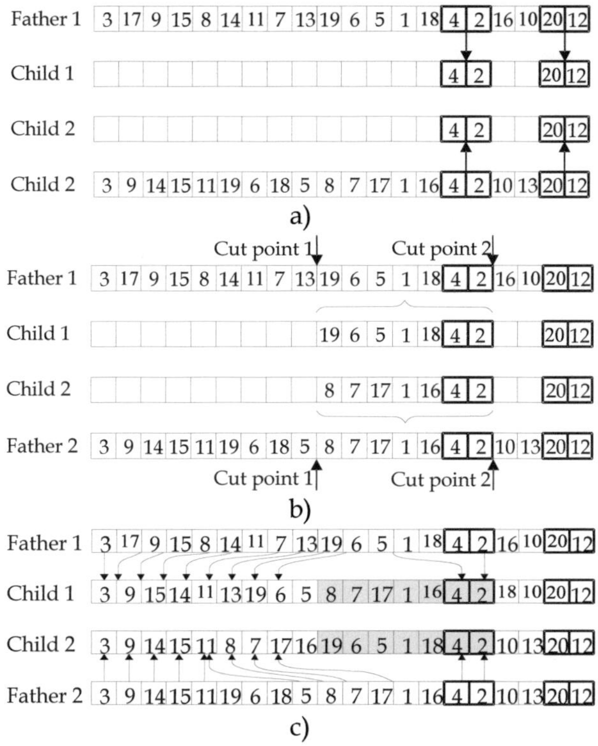


Fig. 8. SB2OX Crossover:

a) the common jobs in both parents are copied over to the offspring;

b) jobs before a randomly chosen cut point are inherited from the direct parent;

c) the missing elements in the offspring are copied in the relative order of the other parent.

The flow is determined by technological constraints. Machines of different brands with identical functionality but with different speeds or capabilities are included in the stage. The processing time depends on the machine brand. It is considered scheduling in the presence of machine eligibility restrictions when not all machines can process all PCBs, and machine availability restrictions when the use of machines depends on their current state: active or in maintenance service. Adjustment of the machine and the preparation of its feeder are required when the board type is changed. The feeders have different capacities (number of slots). For example, machines could have 60 slots or 80 slots. The time needed for adjustment essentially depends on the board type previously processed in the machine. It cannot be neglected in the television PCB production environment. Hence, a sequence-dependent setup time is needed. Each machine has a limited capacity buffer for storing WIP. If the storage is filled to full capacity, the production on this machine is blocked.

The problem is modeled as a HFS with the following constraints: (1) From two to six successive stages with the common flow pattern for all PCB types; (2) Stages with unrelated machines; (3) Machine eligibility/availability; (4) Sequence-dependent setup time; (5) Limited buffers. The goal is to find a schedule that minimizes the total production time.

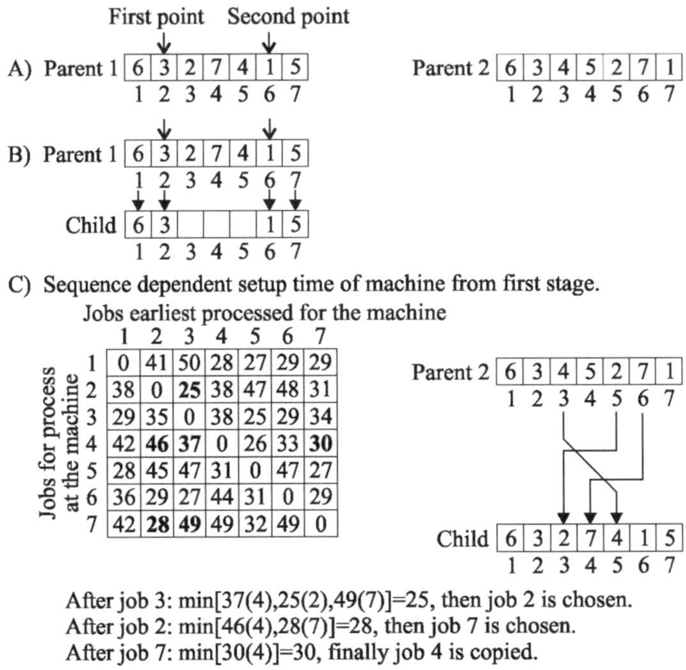


Fig. 9. ST2PX Crossover

The problem is denoted as $FHm,((RM^{(i)})_{i=1}^{(m)} \mid S_{sd}, M_j, Block \mid C_{\max}$. The next is the problem statement: Let a set $N$ of $n$ jobs, $N = \{1,2,...,n\}$ given at time 0 has to be processed in a set $M$ of $m$ consecutive production stages, $M = \{1,2,...,m\}$, without preemption. The objective to minimizing is total completion time known as makespan. On stage $i \in M$, a set $M_i = \{1,2,...,m_i\}$ of unrelated parallel machines is given, where $|M_i| \geq 1$.

Each job has to be processed by exactly one machine at each stage. Let $p_{i,l,j}$ be the processing time of job $j \in N$, on machine $l \in M_i$, at stage $i$. A machine based sequence-dependent setup time is considered. Let $S_{i,l,j,k}$ be the setup time on machine $l$, at stage $i$, when processing job $k \in N$, after processing job $j$. A set of eligible machines that can process job $j$ at stage $i$, is denoted as $E_{i,j}$, $1 \leq |E_{ij}| \leq m_i$. For each machine $l \in M_i$ a limited buffer for jobs is given. A maximal storage capacity in front of each machine $l$ is $b_{i,l}$, $1 \leq |b_{i,l}| \leq n$.

Many authors separate sequencing and assignment decisions in the HFS problems. To solve this problem, a way proposed by Ruiz and Maroto (2006) is used, where the assignment of jobs to machines in each stage is done by a evaluation function. In the HFS with no setup times and no availability constraint assignment of the job to the first available machine would result in the earliest completion time of the job. In the HFS with unrelated parallel machines it is demonstrated that if the first available machine is very slow for a given job, assigning the job to this machine can result in a later completion time compared with assignment to other machines. With the consideration of the setup times this problem becomes worse. To solve it, in our algorithm, a job is assigned to the machine that can finish the job at the earliest time at a given stage, taking into consideration different processing speeds, setup times, machine availability, and buffer size.

The calculation of the total completion time $C_{max}$ is as follow: Let $\pi$ be a job permutation or sequence; $\pi_{(j)}$ be the job at the $j$th position in the sequence, $j \in N$. Each job has to be processed at each stage, so $m$ tasks per job are considered. Let $L_{i_l}$ be the last job assigned to machine $l$ at stage $i$, $l \in M_i$. Let $S_{i_l, L_{i_l}, \pi_{(j)}}$ be the setup time of machine $l$ at stage $i$ when processing of job $\pi_{(j)}$ after having processed the previous work assigned to this machine $l(L_{i_l})$. Let $C_{i,\pi_{(j)}}$ be the completion time of job $\pi_{(j)}$ at stage $i$, $i \in M$, then

$$C_{i,\pi_{(j)}} = \min_{l=1}^{m_j}\{\max\{C_{i,L_{i_l}} + S_{i_l,L_{i_l},\pi_{(j)}} ; C_{i-1,\pi_{(j)}}\} + p_{i_l,\pi_{(j)}}\}$$

The makespan is calculated as follows:

$$C_{max} = \max_{j=1}^{n}\{C_{m,\pi_{(j)}}\}$$

The GA, was tuned up by the following parameters elected in the parameter calibration step: crossover ST2PX; mutation Swap; crossover probability 0.8; mutation probability 0.1; population size 200. The execution steps of this algorithm ($GA_{SBC}$) are presented below.

**Algorithm. GA*SBC*.**

```
Input: The population of P_size individuals.
Output: An individual of length n.
generate_population
regeneration = 1
while not stopping_criterion do
        for i=0 to P_size
                evaluate_objective_function(i)
        keep_the_best_individual_found()
        if actual_best_makespan >= previous_best_makespan
                iterations_without_improvement = iterations_without_improvement +1
                if iterations_without_improvement = 25
                        if regeneration = 10
                                stopping_criterion = true
                        else
                                sort_the_population_in_ascending_order_of_Cmax()
                                regenerate_population()
                                regeneration = regeneration+1
                                iterations_without_improvement = 0
        select_individuals_by_the_binary_tournament_selection
        crossover ST2PX with probability 0.8
        mutation SWAP with probability 0.1
```

### 5.5 Example

The following example illustrates this algorithm execution. Let is considered an instance with parameters $n$ = 7, $m$ = 3, $m_1 = m_2$ = 2, and $m_3$=1. Let Table 2 sets up eligibility, and Table 3 processing times. The number -1 means that the machine $l$ is not eligible or not available for the job $j$. Table 4 shows sequence-dependent setup times of job $k$ if job $j$ precedes to job $k$. Table 5 shows the limited buffer sizes.

| Job $j$ | Stage $i$ | | |
|---|---|---|---|
| | 1 | 2 | 3 |
| 1 | {1} | {1} | {1} |
| 2 | {2} | {1,2} | {1} |
| 3 | {1,2} | {1,2} | {1} |
| 4 | {1,2} | {2} | {1} |
| 5 | {1,2} | {1,2} | {1} |
| 6 | {1} | {2} | {1} |
| 7 | {2} | {2} | {1} |

Table 2. A set of eligible machines at stage $i$ that can process job $j$

| | Stage $i$ | 1 | 1 | 2 | 2 | 3 |
|---|---|---|---|---|---|---|
| | Machine $l$ | 1 | 2 | 1 | 2 | 1 |
| Job $j$ | 1 | 54 | -1 | 69 | -1 | 60 |
| | 2 | -1 | 76 | 75 | 67 | 55 |
| | 3 | 58 | 93 | 51 | 82 | 75 |
| | 4 | 59 | 95 | -1 | 52 | 88 |
| | 5 | 75 | 62 | 58 | 73 | 93 |
| | 6 | 50 | -1 | -1 | 52 | 61 |
| | 7 | -1 | 57 | -1 | 66 | 93 |

Table 3.The processing time $p_{i,l,j}$ of job $j$, on machine $l$, at stage $i$.

| | Job $k$ | 1 | 2 | 3 | 4 | 5 | 6 | 7 |
|---|---|---|---|---|---|---|---|---|
| Job $j$ | 1 | 0 | 41 | 50 | 28 | 27 | 29 | 29 |
| | 2 | 38 | 0 | 25 | 38 | 47 | 48 | 31 |
| | 3 | 29 | 35 | 0 | 38 | 25 | 29 | 34 |
| | 4 | 42 | 26 | 37 | 0 | 26 | 33 | 30 |
| | 5 | 28 | 45 | 47 | 31 | 0 | 47 | 27 |
| | 6 | 36 | 29 | 27 | 44 | 31 | 0 | 29 |
| | 7 | 42 | 28 | 49 | 49 | 32 | 49 | 0 |

Table 4. Sequence-dependent setup times for the first machine

| Stage $i$ | 1 | 1 | 2 | 2 | 3 |
|---|---|---|---|---|---|
| Machine $l$ | 1 | 2 | 1 | 2 | 1 |
| Buffer $b_{i,l}$ | 2 | 2 | 3 | 2 | 3 |

Table 5. Limited buffers

Let a population with 10 individuals is generated (Figure 10). Figure 11 presents the fitness value of each individual. The best solution is represented by the individual 2 with makespan 817. The population is ordered and regenerated: 20% best individuals are kept, 40% are replaced by simple Insert mutation of the best individual, and reminding worst 40% are replaced by randomly generated individuals. Figure 12 shows the regeneration result.

Figure 13 shows result of the binary selection. The ST2PX crossover is applied with probability 0.8 (Figure 14). Let is assumed that the first point is at position 2, and the second point is at position 6 (Fig. 14A). Elements from position 1 to position 2 of parent 1 are copied to the child. Elements from position 6 to position 7 (last position) are copied from parent 1 (Fig. 14B). The remaining positions of the child are filled with best elements from parent 2, taking into account the sequence-dependent setup times (Fig. 14C). Three jobs (4, 2 and 7) can be processed at position 3 after processing job 3 at position 2. Hence, three setup times (37, 25, 49) are compared, and job 2 with minimal setup time 25 is chosen. Two setup times (46 and 28) are compared for position 4, and job 7 is chosen. The last job (4) is copied to position 5. Finally, the SWAP mutation is applied with probability 0.1 (Fig. 15). Fig. 16 shows the Gantt chart of the final result.

Job sequences

| Individuals | 1 | 2 | 3 | 4 | 5 | 6 | 7 |
|---|---|---|---|---|---|---|---|
| 1 | 7 | 1 | 4 | 2 | 5 | 3 | 6 |
| 2 | 6 | 3 | 2 | 7 | 4 | 1 | 5 |
| 3 | 3 | 2 | 7 | 4 | 5 | 6 | 1 |
| 4 | 5 | 6 | 1 | 3 | 2 | 7 | 4 |
| 5 | 3 | 7 | 2 | 1 | 5 | 4 | 6 |
| 6 | 4 | 7 | 3 | 5 | 1 | 2 | 6 |
| 7 | 7 | 3 | 1 | 4 | 6 | 5 | 2 |
| 8 | 1 | 2 | 3 | 5 | 6 | 7 | 4 |
| 9 | 6 | 3 | 4 | 5 | 2 | 7 | 1 |
| 10 | 4 | 2 | 5 | 7 | 3 | 1 | 6 |

Fig. 10. Initial population

Job sequences

| Individuals | 1 | 2 | 3 | 4 | 5 | 6 | 7 | $C_{max}$ |
|---|---|---|---|---|---|---|---|---|
| 1 | 7 | 1 | 4 | 2 | 5 | 3 | 6 | 865 |
| 2 | 6 | 3 | 2 | 7 | 4 | 1 | 5 | 817 |
| 3 | 3 | 2 | 7 | 4 | 5 | 6 | 1 | 846 |
| 4 | 5 | 6 | 1 | 3 | 2 | 7 | 4 | 873 |
| 5 | 3 | 7 | 2 | 1 | 5 | 4 | 6 | 856 |
| 6 | 4 | 7 | 3 | 5 | 1 | 2 | 6 | 847 |
| 7 | 7 | 3 | 1 | 4 | 6 | 5 | 2 | 869 |
| 8 | 1 | 2 | 3 | 5 | 6 | 7 | 4 | 881 |
| 9 | 6 | 3 | 4 | 5 | 2 | 7 | 1 | 833 |
| 10 | 4 | 2 | 5 | 7 | 3 | 1 | 6 | 878 |

Fig. 11. Fitness value of each individual

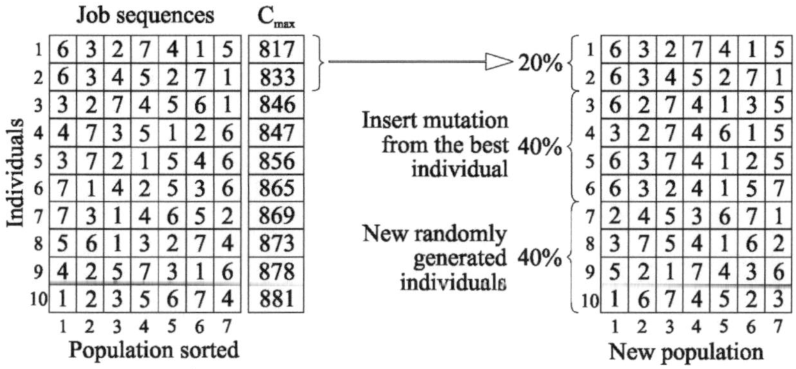


Fig. 12. Regeneration procedure

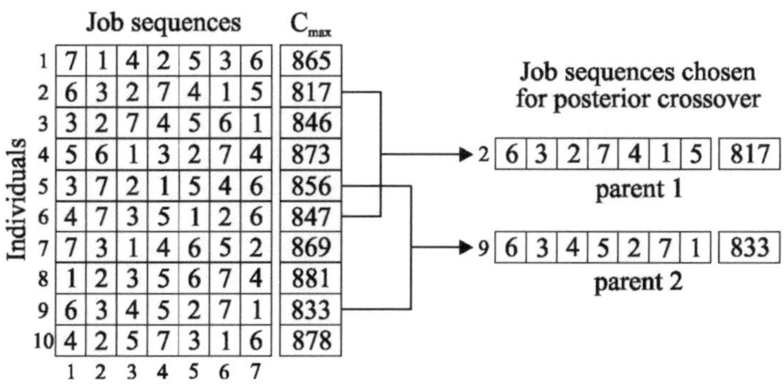


Fig. 13. Binary selection

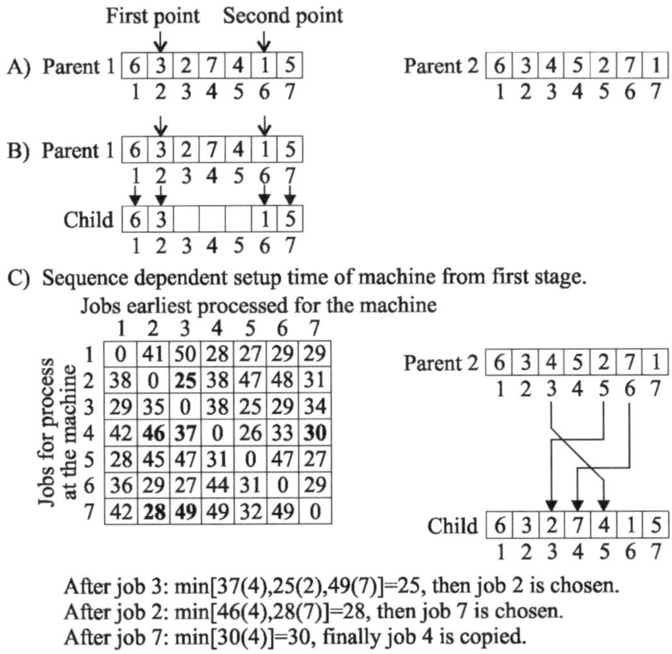


Fig. 14. ST2PX crossover application

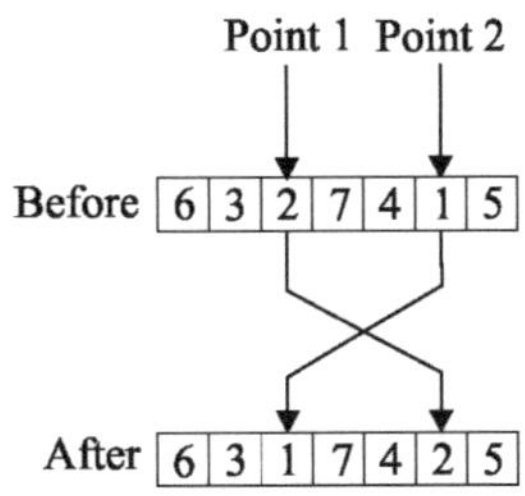


Fig. 15. SWAP mutation

## 6. Conclusion

There are several applications of the HFS scheduling problems which consider setup times in industry, and the variety of models as realistic as theoretical is practically innumerable; then this field of study will attract always the researcher attention. The hardest situation involving setup times is HFS problem with sequence-dependent setup times. It is among the most difficult classes of scheduling problems. Due the complexity, artificial intelligence and metaheuristic techniques should be used for practical problems with multistage parallel machine environment and large instance sizes, in particularity, evolutionary algorithms.

Actually, the authors are exploring a mixed model which consist of a HFS combined with a number of assemble lines. There are considered setup times of machines. The problem involves splitting of lots. Its solution consumes all topics exposed in this chapter.

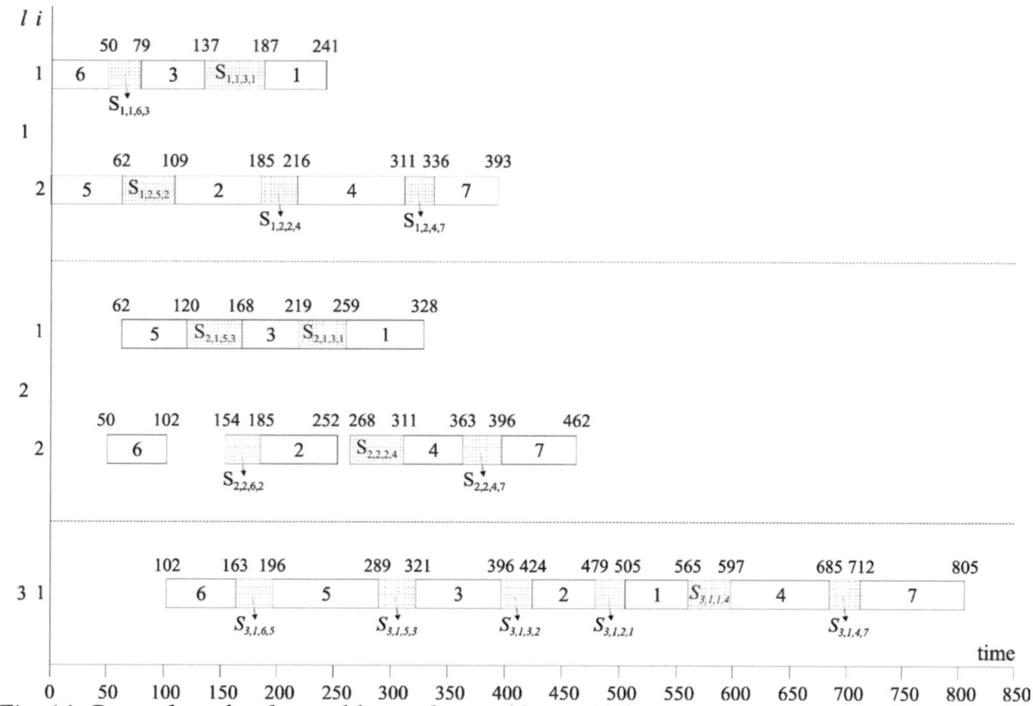

Fig. 16. Gantt chart for the problem solution ($C_{max}$ = 805)

## 7. References

Adler, L.; Fraiman, N.; Kobacker, E.; Pinedo, M.; Plotnicoff, J.C. & Wu, T.P. (1993). Bpss: a scheduling support system for the packaging industry. *Operations Research,* Vol. 41, No. 4, (July-August 1993) 641–648, ISSN 0030-364X

Aghezzaf, E.-H.; Artiba, A.; Moursli, O. & Tahon, C. (1995). Hybrid flowshop problems, a decomposition based heuristic approach, *Proceedings of the International Conference on Industrial Engineering and Production Management,* IEPM'95, FUCAM-INRIA. pp. 43–56.

Agnetis, A.; Pacifici, A.; Rossi, F.; Lucertini, M.; Nicoletti, S.; Nicolo, F.; Oriolo, G.; Pacciarelli, D. & Pesaro, E. (1997). Scheduling of flexible flow lines in an automobile assembly plant. *European Journal of Operational Research.* Vol. 97, No. 2, (March 1997) 348–362, ISSN 0377-2217

Alfieri, A. (2009). Workload simulation and optimisation in multi-criteria hybrid flowshop scheduling: a case study. *International Journal of Production Research.* Vol. 47, No. 18, (January 2009) 5129– 5145, ISSN 0020-7543.

Allahverdi, A.; Ng, C. T.; Cheng, T. C. E.& Kovalyov, M. Y. (2008) A survey of scheduling problems with setup times or costs. *European Journal of Operational Research,* 187, No. 3, (June 2008) 985–1032, ISSN 0377-2217

Andres, C.; Albarracin, JM.; Tormo, G.; Vicens, E. & Garcia-Sabater, JP. (2005). Group technology in a hybrid flowshop environment: a case study. *European Journal of Operational Research.* vol. 167. No. 1, (November 2005) 272–81, ISSN 0377-2217

Arthanary, L. & Ramaswamy, K. (1971). An extension of two machine sequencing problem. *Journal of the Operational Research Society of India*,Vol. 8. No. 4. 10–22.

Bierwirth, C.; Mattfeld, D. & Kopfer, H. (1996). On permutation representations for scheduling problems. Parallel Problem Solving from Nature - PPSN IV, Vol. 1141. pp. 310-318. LNCS, Springer. ISBN 978-3-540-61723-5, Berlin

Botta-Genoulaz, V. (2000). Hybrid flow shop scheduling with precedence constraints and time lags to minimize maximum lateness. *International Journal of Production Economics*. Vol. 64. No. 1-3, (March 2000) 101–111, ISSN 0925-5273

Brah, S. A. & Hunsucker, J. L. (1991). Branch and bound algorithm for the flow shop with multiple processors. *European Journal of Operational Research*. Vol. 51. No. 1, (March 1991) 88-99, ISSN 0377-2217

Chen, B. (1995). Analysis of classes of heuristics for scheduling a two-stage flow-shop with parallel machines at one stage. *Journal of the Operational Research Society*. Vol. 46. No. 2. (February 1995) 234–244, ISSN 0160-5682

Chen, L.; Bostel, N.; Dejax, P.; Cai, J.C.; Xi, L.F. (2007). A tabu search algorithm for the integrated scheduling problem of container handling systems in a maritime terminal, *European Journal of Operational Research*. Vol. 181, No. 1, (August 2007) 40–58, ISSN 0377-2217

Cheng, T. C. E.; Gupta, J. N. D. & Wang, G. (2000). A review of flowshop scheduling research with setup times. *Production and Operations Management Society*. Vol. 9, No. 3, (September 2000) 262-282, ISSN 1059-1478

Crama, Y. (1997). Combinatorial Optimization Models for Production Scheduling in Automated Manufacturing Systems. *European Journal of Operational Research*. Vol. 99, No. 1 ( May 1997) 136–153, ISSN 0377-2217

Gen, M. & Cheng, R. (1997). Genetic algorithms & engineering optimization. *John Wiley & Sons*, ISBN 0-471-31531-1, NY

Glover, F. & Laguna, M. (1997). Tabu search. *Kluwer Academic Publishers*, ISBN:079239965X Boston

Guinet, A. (1991). Textile production systems: a succession of non-identical parallel processor shops. *Journal of the Operational Research Society*. Vol. 42, No. 8, (August 1991) 655–671, ISSN 0160-5682

Guinet, A., & Solomon, M. M. (1996). Scheduling hybrid flowshops to minimize maximum tardiness or maximum completion time. *International Journal of Production Research*. Vol. 34, No. 6, (June 1996) 1643-1654, ISSN 0020-7543

Gupta, J.N.D. (1988), Two-stage, hybrid flowshop scheduling problem. *Journal of the Operational Research Society*. Vol. 39, No. 4, (April 1988) 359-364, ISSN 0160-5682

Gupta, J. N. D. & Tunc, E. A. (1994). Scheduling a two-stage hybrid flowshop with separable setup and removal times. *European Journal of Operational Research*. Vol. 77, No.3, (September 1994) 415–428, ISSN 0377-2217

Gupta, J.N.D.; Strusevich, V.A. & Zwaneveld, C. (1997). Two-stage no-wait scheduling models with set-up and removal times. *Computers & Operations Research*. Vol. 24. No. 11, (November 1997) 1025–1031, ISSN 0305-0548

Harjunkoski, I. & Grossmann, I.E. (2002). Decomposition techniques for multistage scheduling problems using mixed integer and constraint programming methods. *Computers & Chemical Engineering*. Vol. 26, No. 11, (November 2002) 1533–1552, ISSN 0098-1354

Holland, J. H. (1992). Adaptation in Natural and Artificial Systems. *Ann MIT Press*, 0-262-08213-6, Boston

Jungwattanakit, J.; Reodecha, M.; Chaovalitwongse, P.; & Werner, F. (2008). Algorithms for flexible flow shop problems with unrelated parallel machines, setup times, and dual criteria. *International Journal of Advanced Manufacturing Technology*. Vol. 37, No. 3–4 (May 2008) 354–370, ISSN 0268-3768

Jungwattanakit, J.; Reodecha, M.; Chaovalitwongse, P.; & Werner, F. (2009). A comparison of scheduling algorithms for flexible flow shop problems with unrelated parallel machines, setup times, and dual criteria. *Computers & Operations Research*. Vol. 36, No. 2, (Febrary 2009) 358–378, ISSN 0305-0548

Kim, J. S.; Kang, S. H. & Lee, S. M. (1997). Transfer batch scheduling for a two-stage flowshop with identical parallel machines at each stage. *Omega-International Journal of Management Science*. Vol. 25, No. 5, ( October 1997) 547–555, ISSN 0305-0483

Kurz, M.E.; Askin, R.G. (2003). Comparing scheduling rules for flexible flow lines. *International Journal of Production Economics*. Vol. 85, No. 3, (November 2003) 371–388, ISSN 0925-5273

Kurz, M.E.; Askin, R.G. (2004). Scheduling flexible flow lines with sequence-dependent setup times. *European Journal of Operational Research*. Vol. 159, No. 1, (November 2004) 66–82, ISSN 0377-2217

Leon, V.J. & Ramamoorthy, B. (1997). An adaptable problem-space-based search method for flexible flow line scheduling. *IIE Transactions*. Vol. 29, No. 2, (February 1997) 115–125, ISSN 0740-817X

Li, S. (1997). A hybrid two-stage flowshop with part family, batch production, major and minor setups. *European Journal of Operations Research*. Vol. 102. No. 1, (Octubre 1997) 142–156, ISSN 0377-2217

Lin, H.-T. & Liao, C.-J. (2003). A case study in a two-stage hybrid flow shop with setup time and dedicated machines. *International Journal of Production Economics*. Vol. 86, No. 2, (November 2003) 133–143, ISSN 0925-5273

Liu, Ch.-Y. and Chang, Sh.-Ch. (2000). Scheduling Flexible Flow Shops with Sequence-Dependent Setup Effects. *IEEE Transactions on Robotics and Automation*. Vol. 16, No. 4, (August 2000) 408-419, ISSN 1042-296X

Low, Ch. (2005). Simulated annealing heuristic for flow shop scheduling problems with unrelated parallel machines. *Computers & Operations Research*. Vol. 32, No 8, (August 2005) 2013–2025, ISSN 0305-0548

Lushchakova, I. N. & Strusevich, V. A. (2010). Strusevich. Scheduling incompatible tasks on two machines. *European Journal of Operational Research*. Vol. 200. No. 2, (January 2010) 334–346, ISSN 0377-2217

Michalewicz, Z. (1996). Genetic *Algorithms + Data Structures = Evolution Programs*. Springer-Verlag, ISBN 3-540-606776-9, Berlin

Monma, C. L. & Potts, C. N. (1989). On the Complexity of Scheduling with Batch Setups. *Operations Research*. Vol. 37, No. 5. (September 1989) 798–804, ISSN 0030-364X

Moursli, O. & Pochet, Y. (2000). A branch-and-bound algorithm for the hybrid flowshop. *International Journal of Production Economics*. Vol. 64. (Mach 2000)113–125, ISSN 0925-5273

Naderi, B.; Zandieh, M.; Khaleghi, A.; Ghoshe Balagh & Roshanaei, V. (2009). An improved simulated annealing for hybrid flowshops with sequence-dependent setup and transportation times to minimize total completion time and total tardiness. *Expert Systems with Applications.* Vol. 36. No. 6, (August 2009) 9625–9633, ISSN 0957-4174

Naderi, B.; Zandieh, M.; Roshanaei, V. (2009). Scheduling hybrid flowshops with sequence dependent setup times to minimize makespan and maximum tardiness, *International Journal of Advanced Manufacturing Technology.* Vol. 41, No. 11–12, (April 2009). 1186–1198, ISSN 0268-3768

Pearn, W. L.; Chung, S. H.; Yang, M. H. & Chen, C. Y.(2005). The integrated circuit packaging scheduling problem (icpsp): A case study. *International Journal of Industrial Engineering-Theory Applications and Practice.* Vol. 12. No. 3. 296–307, ISSN 1943-670X

Pinedo, M. L. (2008). Scheduling: Theory Algorithms, and Systems, Springer Science+Business Media, ISBN: 978-0-387-78935-4, NY.

Quadt, D. & Kuhn, D. (2007). A taxonomy of flexible flow line scheduling procedures. *European Journal of Operational Research.* Vol. 178, No. 3 (May 2007) 686–698, ISSN 0377-2217

Ribas, I.; Leisten, R. J. & Framiñan, M. (2010). Review and classification of hybrid flow shop scheduling problems from a production system and a solutions procedure perspective. *Computers & Operations Research.* Vol. 37, No. 8, (August 2010) 1439–1454, ISSN 0305-0548

Ruiz, R. & Maroto, C. (2006). A genetic algorithm for hybrid flowshops with sequence dependent setup times and machine eligibility. *European Journal of Operational Research.* Vol. 169, No. 3 (March 2006) 781–800, ISSN 0377-2217

Ruiz, R.; Sivrikaya, F.; Şerifoğlu, T. U. (2008). Modeling realistic hybrid flexible flowshop scheduling problems. *Computers & Operations Research*, Vol. 35, No. 4, (April 2008) 1151–1175, ISSN 0305-0548

Ruiz, R. & Vazquez-Rodriguez, J. A. (2010). The hybrid flow shop scheduling problem. *European Journal of Operational Research.* Vol. 205. No. 1, (August 2010) 1–18, ISSN 0377-2217

Tang, L.X.; Liu, W.X. & Liu, J.Y. (2005). A neural network model and algorithm for the hybrid flow shop scheduling problem in a dynamic environment. *Journal of Intelligent Manufacturing.* Vol. 16, No. 3, (June 2005) 361–370, ISSN: 0956-5515

Vairaktarakis, G. (2004). Flexible Hybrid Flowshop, *In: Handbook of Scheduling: Algorithms, Models, and Performance Analysis.* J. Y.-T. Leung, 5-1 – 5-33, ISBN: 1-58488-397-9, Boca Raton, Florida.

Voss, S. & Witt, A. (2007). Hybrid flow shop scheduling as a multi-mode multi-project scheduling problem with batching requirements: A real-world application. *International Journal of Production Economics.* Vol. 105, No. 2, (Febrary 2007) 445–458, ISSN 0925-5273

Yaurima, V.; Burtseva, L. & Tchernykh, A. (2009). Hybrid Flowshop with Unrelated Machines, Sequence Dependent Setup Time, Availability Constraints and Limited Buffers. *Computers & Industrial Engineering*, Vol. 56, No. 4, (May 2009) 1452-1463, ISSN 0360-8352

Zandieh, M.; Ghomi, S.M.T. Fatemi & Husseini, S.M. Moattar. (2006) An immune algorithm approach to hybrid flow shops scheduling with sequence-dependent setup times, *Applied Mathematics and Computation.* Vol. 180, No. 1, ( September 2006) 111–127, ISSN 0096-3003

8

# ACO-based Multi-objective Scheduling of Identical Parallel Batch Processing Machines in Semiconductor Manufacturing

Li Li, Pan Gu, Fei Qiao, Ying Wu and Qidi Wu
*Tongji University*
*China*

## 1. Introduction

The batch processing machines (BPMs) have the ability to process more than one job together (called a batch). So the scheduling problem of the BPMs concerns not only the priorities of the jobs obtaining the processing service of a BPM, but the number of the jobs processed together on them. According to diverse classified criteria (such as the number of the BPMs and the job families), the scheduling problem of the BPMs can be further divided into several styles, e.g., a single BPM scheduling problem (SBPM), identical parallel BPMs scheduling problem (PBPM) , non-identical PBPM, the BPMs scheduling problem with compatible job families and the BPMs scheduling problem with incompatible job families.

In this paper, we address the BPMs scheduling problem in a semiconductor wafer fabrication facility (fab), in where there are many BPMs, such as diffusion machines, oxidation machines and dry strip machines. The jobs processed on those machines cannot be batched together unless they use the same recipe of those BPMs. As a result, the scheduling problem of those BPMs is abstracted as identical PBPM with incompatible job families. In a fab, because most of upstream and downstream machines of the BPMs are non-BPMs, jobs must be batched or split regularly during their fabrication processes. Therefore, a good scheduling solution of those BPMs is essential to efficiently utilize their capacity and satisfy the requirements of their downstream machines to balance the fab-wide workload and achieve better fab-wide operational performance.

In recent years, there have been many studies of the BPMs scheduling problem. Mathirajan and Sivakumar (Mathirajan & Sivakumar, 2006) have reviewed 98 articles published between 1986 and 2004 on this topic, and the research has considerably evolved since 2004. For example, to minimize the makespan or average flow time of the jobs, Chien and Chen (Chien & Chen, 2007) developed a genetic algorithm (GA) for batch sequencing combined with a novel timetabling algorithm to handle waiting time constraints, frequency-based setups, limited machine availability and a rolling horizon-based scheduling mechanism for scheduling of furnaces for semiconductor fabrication. Chou et al. (Chou et al., 2006) presented a hybrid GA for SBPM with arbitrary job release times. To meet due date requirements from customers, Gupta and Sivakumar (Gupta & Sivakumar, 2007) presented a dynamic scheduling method for SBPM with a look-ahead batching strategy to control the delivery performance between earliness and tardiness measures. Erramilli and Mason (Erramilli & Mason, 2006) proposed a mixed integer program and a simulated annealing (SA)-based heuristic method to solve SBPM to minimize the total weighted tardiness (TWT). To be applicable to a real production environment, some research work has also considered upcoming jobs. Solomon et al.(Solomon et al., 2002) presented a dispatching policy for the BPMs that incorporated knowledge of future arrivals, the status of critical machines in subsequent processing, and setup times into batch processing scheduling. Mönch et al. (Mönch et al., 2006) proposed a simple heuristic method based on the Apparent Tardiness Cost (ATC) Dispatching Rule to minimize the TWT on PBPM with incompatible job families and unequal job ready times, in which inductive decision trees and neural networks from machine learning were used to estimate the look-ahead parameter. Liu et al. (Liu et al., 2007) proved that SBPM of minimizing the total tardiness was NP-hard even if the machine capacity was only two jobs. Accordingly, most studies have used heuristic rules (e.g.,(Gupta & Sivakumar, 2007; Solomon et al., 2002; Mönch et al., 2006)) or metaheuristic searching methods (e.g., (Chien & Chen, 2007; Chou et al., 2006; Erramilli & Mason, 2006)). Although heuristic rules can reach a solution quickly, they are myopic algorithms that pursue local optimization without considering global optimization. Consequently, the metaheuristic searching methods (such as GA and SA) have been gradually adopted to obtain global optima.

Ant Colony Optimization (ACO), inspired by the foraging behavior of real ant colonies, is a population-based approach developed by Dorigo in 1992 (Dorigo M, 1992). ACO has been successfully applied to several NP-hard combinatorial optimization problems, such as the Traveling Salesman Problem (TSP), Quadratic Assignment Problem (QAP), Vehicle Routing Problem (VRP), Job-Shop Scheduling Problem (JSP), Flow-Shop Scheduling Problem (FSP), etc.(Dorigo M. & Stützle T., 2004). However, few researchers have applied ACO to solve the BPMs scheduling problem. Only Srinivasa Raghavan and Venkataramana (Srinivasa & Venkataramana, 2006) used an ACO algorithm to solve a static scheduling problem of minimizing the TWT of PBPM with incompatible job families.

In this paper, firstly, we build an identical PBPM model concerned the practical considerations of incompatible jobs, dynamic job arrivals, sequence-dependent setup times and the qual-run requirements of advanced process control (APC). Then, we propose an ACO-based solution to simultaneously minimize the TWT and makespan of the jobs. Finally, the effectiveness of the proposed method is demonstrated by a variety of simulation experiments. The simulation results show that the proposed method produces smaller TWT and makespan than the common Apparent Tardiness Cost-Batched Apparent Tardiness Cost (ATC-BATC) rule, Max Batch Size rule (MBS), a GA and an Ant System algorithm (AS).

The rest of this paper is organized in four sections. In Section 2, we present the problem description and assumptions. Then we outline the ACO-based solution in Section 3. In section 4, we show computational experiments and results. Finally, we give conclusions and future research topics in section 5.

## 2. Problem Description and Assumptions

### 2.1 Problem description

With the 3-field notation, the PBPM scheduling problem in this paper can be denoted as

$$M|A_{ij}, Q_i, Batch, incompatible|\ min(\sum_i \sum_j w_{ij}Tij\ +\ max_{i,j}(F_{ij})) \tag{1}$$

where $M$ is the number of the BPMs in PBPM; $A_{ij}$ is the arrival time of job $j$ of family $i$ ; $Q_i$ is the qual-run time of family $i$ ; $w_{ij}$ and $Tij$ are the weight, the tardiness and the completion time of job $j$ of family $i$ , respectively.
There are two way to solve a PBPM scheduling problem. One is to distribute the jobs to PBPM first, then batch the jobs and determine the priorities of the batches on each BPM. The other is to batch the jobs first, then distribute the batches to PBPM and sequence the batches on each BPM. Balasubramanian et al.(Balasubramanian et al., 2004) have shown by extensive simulations that the second way achieved better solutions with less computation time. Therefore, we have adopted the second style.
There are two main constraints when forming the batches. First, only jobs belonging to the same family can be processed together. Second, the number of the jobs in a batch cannot exceed the capacity of the PBPM (i.e., maximum batch size constraint). Another important consideration is the trade-off between the waiting time for forming a full batch and the waste of the PBPM capacity.
Distributing and sequencing the batches are the same as in other problems of scheduling parallel machines. The issues to consider are the hot lots, workload balance and the utilization of the PBPM. It is also worthwhile to consider the trade-off between the setup times of scheduling and the qual-run requirements of APC. In real semiconductor manufacturing environments, APC could achieve the best quality result by frequent changeovers between jobs from different families, possibly avoiding the need for qual-runs. However, frequent changeovers cost setup time and cause capacity loss. Instead of achieving the best quality, APC determines a parameter range which represents acceptable quality for every job family. Based on this range, APC provides a threshold value $n_i$ for each job family $i$ . If a machine has been processing no less than $n_i$ jobs (or batches for BPMs) from family $j(j \neq i)$ ,then before the next time it processes jobs from family $i$ ,a qual-run is required on that machine. In a qual-run, no real job is processed. A blank wafer is processed to obtain the status of the machine so that the operator can properly set the machine parameters to achieve high quality results. Before the result of the qual-run is available, jobs cannot be processed on that machine. Therefore, a trade-off between the time lost for setups and the time lost due to qual-runs is required. However, most related research has not considered the qual-run requirements of APC. We have found only the studies of Cai et al. (Cai et al., 2007) and Patel (Patel N-S, 2004) that incorporated the constraint of process control into scheduling decisions.

### 2.2 Problem assumptions

The assumptions involved in the PBPM scheduling problem include:
(i) The machines in the PBPM are identical;
(ii) The PBPM scheduling problem is considered with a schedule horizon (e.g., one shift, one day or several days), within which the scheduling plan of the jobs from multiple families is decided;
(iii) The processing time of a batch on one machine is independent of the number of the jobs in the batch;
(iv) Once processing begins on a batch, no job can be removed from or added to the machine until it finishes.
(v) There are sequence-dependent random setup times for changeovers between jobs from different families, and no setup times between jobs from the same family.

## 3. ACO-Based Solution

### 3.1 Build a search space

Before we use an ACO algorithm to find a solution, the first task is to build a search space for the ACO algorithm. In this paper, the search space is composed of nodes which are combinations of the batches and the BPMs in the PBPM.

For $N_i$ jobs, there are $C^1_{N_i} + C^2_{N_i} + ... + C^B_{N_i}$ ( $C$ is the combination operator) batching styles, subject to the constraint of maximum batch size. In the case of many jobs (especially with a number of dynamic arrival jobs), this kind of batching style will result in lower computation efficiency. In this paper, we form the batches using the time window concept (denoted by $\Delta t$ ) proposed by Mönch et al. (Mönch et al., 2005).

$$\Delta t = dt \times Avg(P_{ij}) \tag{2}$$

where $P_{ij}$ is the processing time of job $j$ of family $i$ ; $Avg(P_{ij})$ is the average processing time of the jobs; $dt$ is a distribution parameter of $\Delta t$ .

At each batching decision point $t$ ( $t$ is set as the earliest ready time of the jobs to be batched), the jobs of family $i$ with arrival (ready) time less than the upper boundary of the time window interval $t + \Delta t$ is denoted as $M(j, t, \Delta t) = \{ij | A_{ij} \leq t + \Delta t\}$ .Then, we batch the jobs in $M(j, t, \Delta t)$ subject to the maximum batch size constraint. We repeat the above process until all jobs have been assigned to a batch. The ready time of each batch equals the latest arrival time of the jobs in the batch. Finally, the search space (denoted as $S$ )is built with nodes composed of the batches and the BPMs in the PBPM.

Table 1 shows a simple example of building a search space. We assume that there are 2 machines in the PBPM and their batch size is 2 jobs. There are 2 families of jobs whose processing times are set to 10 min and 15 min, respectively. For each family, there are 3 jobs to be scheduled. Here $dt$ is set to 1. Then the time window $\Delta t$ can be computed as $\Delta t = 1 \times (10 + 10 + 10 + 10 + 15 + 15 + 15)/6 = 12.5\ min$ .The batches formed are shown in Table 2. These batches and machines constitute the search space $S = \{(l_{11}, m_1), (l_{12}, m_1), ((l_{11}, l_{12}), m_1), (l_{13}, m_1), (l_{21}, m_1), (l_{22}, m_1), (l_{23}, m_1), ((l_{22}, l_{23}), m_1), (l_{11}, m_2), (l_{12}, m_2), ((l_{11}, l_{12}), m_2), (l_{13}, m_2), (l_{21}, m_2), (l_{22}, m_2), (l_{23}, m_2), ((l_{22}, l_{23}), m_2)\}$ ,whose size is 16 nodes.

| $Job(l_{ij})$ | $l_{11}$ | $l_{12}$ | $l_{13}$ | $l_{21}$ | $l_{22}$ | $l_{23}$ |
|---|---|---|---|---|---|---|
| $A_{ij}(min)$ | 0 | 5 | 15 | 8 | 13 | 18 |

Table 1. An example of building a search space

| $Batches$ | $l_{11}$ | $l_{12}$ | $(l_{11}, l_{12})$ | $l_{13}$ | $l_{21}$ | $l_{22}$ | $l_{23}$ | $(l_{22}, l_{23})$ |
|---|---|---|---|---|---|---|---|---|
| $A_{Batch}(min)$ | 0 | 5 | 5 | 15 | 8 | 13 | 18 | 18 |

Table 2. The formed batches for the simple example

### 3.2 Find a solution with an ACO algorithm

The parameters used in the ACO algorithm to find a solution are defined in Table 3.

| *Parameter* | *Meaning* |
|---|---|
| $m$ | The index of the BPMs in PBPM |
| $B$ | The capacity of the BPMs in PBPM |
| $ij$ | The index of the jobs, which means job $j$ of family $i$ |
| $\Delta t$ | The time window for job batching |
| $dt$ | The distribution parameter of time window $\Delta t$ |
| $K$ | The number of the ants in the artificial ant colony |
| $k$ | The index of the artificial ants |
| $t_{max}$ | The maximum number of iterations |
| $t_1$ | The iteration index |
| $\delta$ | The minimum change of the minimum objective values in two consecutive iterations |
| $L^k_{tabu}$ | The tabu-list of ant $k$ |
| $L^k_{task}$ | The task-list of ant $k$ |
| $\tau_0$ | The initial pheromone on each arc |
| $WT_{ATC-BATC}$ | The TWT of the scheduling results obtained by ATC-BATC rule |
| $F_{ATC-BATC}$ | The makespan of the scheduling results obtained by ATC-BATC rule |
| $l$ | The task-list of ant $k$ |
| $c$ | A candidate node in $L^k_{task}$ |
| $\tau_{c_0 l}$ | The pheromone on the arc $(c_0, l)$ |
| $c_0$ | The last node selected by artificial ant $k$ using the same machine as $c$ |
| $q$ | A probability parameter $(0 \leq q \leq 1)$ |
| $\alpha$ | A parameter denoting the relative importance of the pheromone density and the heuristic factor |
| $\eta_{c_0 c}$ | The heuristic factor if $c$ is selected as the successor task of $c_0$ |
| $P_c$ | The processing time of $c$ |
| $U_{c_0 c}$ | The setup time for the changeover between $c_0$ and $c$ |
| $x_c$ | The qual-run parameter of $c$ |
| $A_c$ | The arrival time of $c$ |
| $Q_c$ | The qual-run time of $c$ |
| $F_{c_0}$ | The processing finish time of $c_0$ |
| $B_c$ | The batch size of $c$ |
| $W_c$ | The workload of the machine processing $c_0$ if $c$ is selected as the successor task of $c_0$ |
| $W_m$ | The workload of machine $m$ |
| $\gamma$ | A parameter to regulate the workload among the BPMs in PBPM |
| $min_m(F_{t_1-1})$ | The earliest finish time of the BPMs in PBPM in iteration $t_1 - 1$ |
| $max_m(F_{t_1-1})$ | The latest finish time of the BPMs in PBPM in iteration $t_1 - 1$ |
| $\xi$ | A parameter to reduce the pheromone trail on an arc used by an ant to make it less attractive to the following ants |
| $OV_{t_1}$ | The minimum objective value of the solutions in iteration $t_1$ |
| $OV_{t_1-1}$ | The minimum objective value of the solutions in iteration $t_1 - 1$ |
| $x, y$ | Two different nodes in the search space |

| | |
|---|---|
| $\tau_{xy}(t_1)$ | The pheromone on arc $(x,y)$ in iteration $t_1$ |
| $\tau_{xy}(t_1+1)$ | The pheromone on arc $(x,y)$ in iteration $t_1+1$ |
| $\Delta\tau_{xy}^{bs}$ | The new pheromone deposition related to the best-so-far solution |
| $\rho$ | The pheromone evaporation parameter |
| $T^{bs}$ | The best-so-far solution during the search process |

Table 3. The list of parameters used in the ACO algorithm

The detailed flowchart of the ACO algorithm is shown in Figure 1.

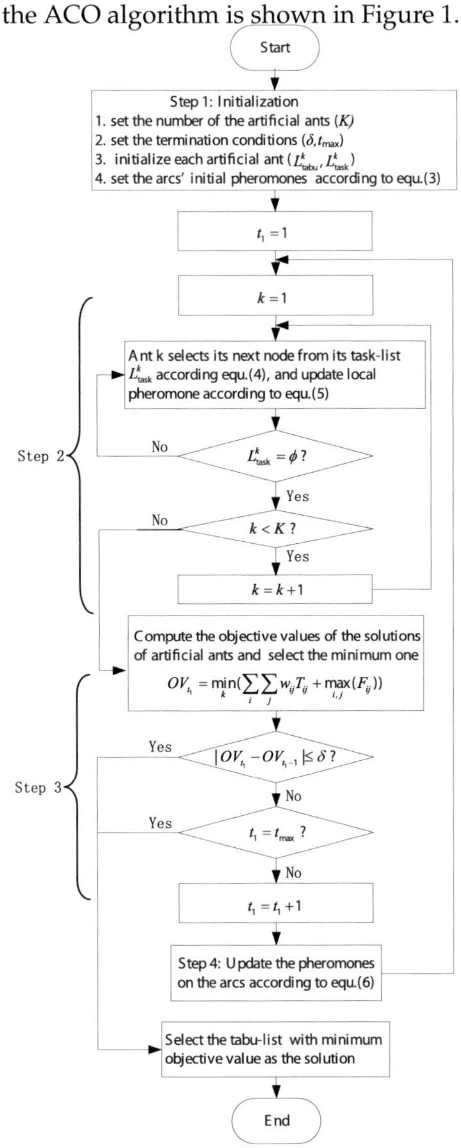


Fig. 1. The flowchart of the ACO algorithm

Step 1: Initialization. There are four main tasks in the initialization stage: to determine the number of the ants in the artificial ant colony, to set the termination conditions for the search, to initialize each artificial ant, and to set the initial pheromones on the arcs.
a) The number of ants in the artificial ant colony
Here we set the number of ants in the artificial ant colony to the number of nodes in the search space. For example, the number of artificial ants for solving the problem in Table 1 can be set to 16.
b) The termination conditions
Here we set two kinds of termination conditions. One is the maximum number of iterations (denoted by $t_{max}$ ). The other is the minimum change of the minimum objective values in two consecutive iterations (denoted by $\delta$ ).
c) Initialization of each artificial ant
First, we build a tabu-list and a task-list for each artificial ant (indexed with $k$ ), denoted by $L^k_{tabu}$ and $L^k_{task}$ , whose initial values are set to $\phi$ and $S$ ,respectively. Then, we distribute the start points (i.e., the nodes in the search space) randomly to each artificial ant. The node distributed to ant $k$ is added to $L^k_{tabu}$ , and deleted from $L^k_{task}$ . To guarantee that each job is processed only once, the nodes with the same job as the distributed node are also deleted from $L^k_{task}$ .Take the problem in Table 1 as an example, and assume that the node $(l_{11}, m_1)$ is assigned to ant 1 .Then the tabu-list and task-list of ant 1 become $\{(l_{11}, m_1)\}$ and $\{(l_{12}, m_1), (l_{13}, m_1), (l_{21}, m_1), (l_{22}, m_1), (l_{23}, m_1), ((l_{22}, l_{23}), m_1), (l_{12}, m_2), (l_{13}, m_2), (l_{21}, m_2), (l_{23}, m_2), ((l_{22}, l_{23}), m_2)\}$ ,respectively.
d) Initialization of the pheromone on each arc
The initial pheromone on each arc is set with the scheduling results obtained by the ATC-BATC rule, which also guarantees that the scheduling results achieved by the proposed ACO algorithm are no worse than those of the ATC-BATC rule.

$$\tau_0 = 1 / (K \times (WT_{ATC-BATC} + F_{ATC-BATC})) \tag{3}$$

Step 2: Each artificial ant searches for its solution. Artificial ant $k$ selects its next node $l$ from its task-list $L^k_{task}$ according to the so-called pseudorandom proportional rule, given by

$$l = \begin{cases} arg\ max_{c \in L^k_{task}} \{ \frac{\alpha\tau_{c_0c} + (1-\alpha)\eta_{c_0c}}{\Sigma_c \alpha\tau_{c_0c} + (1-\alpha)\eta_{c_0c}} \}, & q \leq q_0 \\ max_c(rand(0,1) \times \frac{\alpha\tau_{c_0c} + (1-\alpha)\eta_{c_0c}}{\Sigma_c \alpha\tau_{c_0c} + (1-\alpha)\eta_{c_0c}}, & otherwise \end{cases} \tag{4}$$

$$\eta_{c_0c} = (1 - \frac{P_c + U_{c_0c} + x_c Q_c + max((A_c - F_{c_0}), 0)}{max_c(P_c) + max_c(U_{c_0c}) + max_c(Q_c) + max((max_c(A_c) - F_{c_0}), 0)}) + \frac{B_c}{B} + \gamma \Delta W_c$$

$$\Delta W_c = \begin{cases} \frac{W_c}{max_m(W_m)}, & W_c \leq max_m(W_m) \\ \frac{W_c}{max_m(W_m)} - 1, & W_c > max_m(W_m) \end{cases}$$

$$\gamma = \frac{\Sigma_i \Sigma_j P_{ij} / M - min_m(F_{t_1 - 1})}{max_m(F_{t_1 - 1}) - \Sigma_i \Sigma_j P_{ij} / M}$$

$$x_c = \begin{cases} 1, & \text{if no less than } n_i \text{batches from family} j (j \neq c) \text{have been processed} \\ -1, & \text{if} (n_i - 1) \text{batches from family} j (j \neq c) \text{have been processed} \\ 0, & otherwise \end{cases}$$

Obviously, ant $k$ selects the node with the highest attractiveness indicated by the learned pheromone trails and the heuristic information with probability $q_0$, while with probability $1 - q_0$ it performs a biased exploration of the arcs. The heuristic factor $\eta_{c_0 c}$ simultaneously takes into consideration on $c$'s occupation time (including possible qual-run time, processing time, setup time and waiting time), the capacity utilization rate and the relative workload of the machine.

The selected node is added to $L^k_{tabu}$ and deleted from $L^k_{task}$.Meanwhile, the nodes with the same job as the selected node are also deleted from $L^k_{task}$.Then, the local pheromone trail is updated

$$\tau_{c_0 l} = (1 - \xi)\tau_{c_0 l} + \xi\tau_0 \tag{5}$$

The parameter $\xi$ allows artificial ants to increase their exploration of new arcs, and in practice avoids a stagnation behavior (i.e., the ants do not converge on a common path).

The process is repeated until $L^k_{task}$ is empty. Obviously, the tabu-list $L^k_{tabu}$ is the solution obtained by ant $k$'s search process.

Step 3: Determine whether the termination conditions are satisfied. First, we compute the objective values of the solutions obtained by the artificial ants. Then we select the minimum to compare with that of the last iteration. If the difference between these two consecutive minimum objective values is no more than a small positive value (denoted by $\delta$), we stop the search process. The tabu-list with the minimum objective value is taken as the solution. Otherwise, we determine whether $t_{max}$ has been reached. If the answer is yes, we select the tabu-list with the minimum objective value as the solution. Otherwise, we go to step 4.

Step 4: Pheromone updating. We update the pheromone values on the arcs with the best-so-far solution according to equation (6), and then repeat steps 2 and 3.

$$\tau_{xy}(t_1 + 1) = (1 - \rho)\tau_{xy}(t_1) + \rho\Delta\tau^{bs}_{xy}, \forall(x, y) \in T^{bs} \tag{6}$$

$$\Delta\tau^{bs}_{xy} = 1/min_k(\Sigma_i\Sigma_j w_{ij}T_{ij} + max_{i,j}(F_{ij})),\ 0 < \rho < 1$$

The pheromone trail update, both evaporation and new pheromone deposition, only applies to the arcs of the best-so-far solution, not to all the arcs. In this way, the computational complexity of the pheromone update at each iteration is reduced from $O(K^2)$ to $O(K)$.The deposited pheromone is discounted by a factor $\rho$, which results in the new pheromone trail being a weighted average between the old pheromone value and the newly deposited pheromone.

## 4. Computational experiments and results

### 4.1 The scheduling methods to be compared

We compared the performances of the proposed ACO algorithm with those of ATC-BATC, MBS, a GA (refer to (Balasubramanian et al., 2004)) and an Ant System algorithm (refer to (Li et al., 2008) ). As a common BPM scheduling rule, ATC-BATC has good performance on static BPM scheduling problems. To adapt it to dynamic jobs arrival, we made some small modifications to the rule. The decision flow of the modified ATC-BATC rule is as follows.

First, at each batching decision point $t$, compute the index of job $j$ of family $i$ which belongs to $M(j, t, \Delta t)$.

$$I_{ij}(t_0) = \frac{w_{ij}}{P_{ij}} exp(-\frac{(d_{ij} - P_{ij} - t_0 + (A_{ij} - t_0)^+)^+}{k_1\overline{P}}) \tag{7}$$

where $w_{ij}$ , $P_{ij}$ , $d_{ij}$ and $A_{ij}$ are the weight, the processing time, the due date and the arrival time of job $j$ of family $i$ ,respectively; $t_0$ is the scheduling decision point; $k_1$ is the look-ahead parameter, which usually ranges from 0.1 to 5; and $\overline{P}$ is the average processing time of the jobs. Then form batches by selecting jobs in a non-increasing order of $I_{ij}$ ,subject to the maximum batch size constraint. Repeat the above process until the batching is complete. Second, make a batch sequence in non-increasing order of $I_{bi}$ according to equation (8) and distribute the batches to BPMs of the PBPM in turn

$$I_{bi}(t_0) = \Sigma_{j=1}n_{bi}I_{ij}(t_0) \times min(\frac{n_{bi}}{B},1) \quad (8)$$

where $n_{bi}$ is the number of the jobs in a batch of family $i$ .

### 4.2 The problem cases for the simulations

We used the dry strip operations in a real wafer fab to demonstrate the proposed ACO algorithm. There are 3 identical machines for the dry strip operations in this wafer fab. Each machine has a capacity of 3 jobs and can process 4 different recipes. The threshold value for the qual-run requirements of each recipe is 3 jobs. The processing times for $recipe_1$ , $recipe_2$ , $recipe_3$ and $recipe_4$ are random variables from the uniform distributions $Uniform(90,100), Uniform(90,100), Uniform(70,80) and Uniform(90,100)$ , respectively. The setup times are from the distribution $Uniform(10,20)$ .The qual-run times for $recipe_1$ , $recipe_2$ , $recipe_3$ and $recipe_4$ conform to the distributions $Uniform(30,40), Uniform(30,40), Uniform(20,30)$ , and $Uniform(30,40)$ , respectively. The time unit is minutes.

### 4.3 Determining the distribution parameter $dt$ for time window $\Delta t$

Although Mönch et al.(Mönch et al., 2005) presented the concept of the time window, they did not specify how to determine its value. We ran a number of simulations on random problem cases (shown in Table 4) to determine the best value for the distribution parameter $dt$ of the time window $\Delta t$ . ATC-BATC was used for the simulations, with its look-ahead parameter $k_1$ set to 0.5 (according to extensive simulations). Values of $dt$ in the range from 0 to 2.0 at intervals of 0.25 were tested. The simulation results are shown in Table 5 and Figure 2. From the simulation results, we can reach the following conclusions.

| *Problemparameter* | *Valueused* | *Numberofvalues* |
|---|---|---|
| *Numberofjobs* | 20 | 1 |
| *Arrivaltimesofjobs* | $Uniform(-r\Sigma_i\Sigma_jP_{ij}/(BM), r\Sigma_i\Sigma_jP_{ij}/(BM))$<br>$r = 0.25, 0.50, 0.75, 1.0, 1.25, 1.5, 1.75, 2.0$ | 8 |
| *Duedatesofjobs* | $A_{ij} + P_{ij} + Uniform(0, Avg(P_{ij}))$ | 1 |
| *TimewindowΔt* | $dt \cdot Avg(P_{ij})$<br>$dt = 0, 0.25, 0.5, 0.75, 1, 1.25, 1.5, 1.75, 2.0$ | 9 |
| *Weightperjob* | $Uniform(0,1)$ | 1 |
| | *Totalparametercombinations* | 72 |
| | *Numberofproblemspercombination* | 5 |
| | *Totalproblems* | 360 |

Table 4. Problem cases for determining the time window $\Delta t$

| | $r=0.25$ | $r=0.5$ | $r=0.75$ | $r=1.0$ | $r=1.25$ | $r=1.5$ | $r=1.75$ | $r=2.0$ |
|---|---|---|---|---|---|---|---|---|
| $dt=0.25$ | 169.40 | 242.45 | 239.39 | 335.41 | 257.45 | 253.24 | 312.09 | 345.01 |
| $dt=0.50$ | 168.44 | 226.08 | 182.97 | 250.39 | 257.18 | 253.24 | 262.68 | 345.01 |
| $dt=0.75$ | 168.44 | 178.25 | 182.97 | 250.39 | 250.27 | 251.35 | 277.44 | 342.01 |
| $dt=1.00$ | 168.44 | 178.25 | 182.97 | 230.49 | 205.27 | 233.97 | 257.13 | 330.12 |
| $dt=1.25$ | 168.44 | 178.25 | 182.34 | 230.49 | 206.65 | 233.97 | 257.13 | 330.12 |
| $dt=1.50$ | 168.44 | 178.25 | 182.34 | 230.49 | 206.65 | 233.97 | 261.68 | 332.98 |
| $dt=1.75$ | 168.44 | 178.25 | 182.34 | 230.20 | 218.49 | 233.97 | 261.68 | 333.52 |
| $dt=2.00$ | 168.44 | 178.25 | 182.34 | 230.20 | 218.49 | 233.97 | 262.90 | 333.52 |

Table 5. The average objective values with variables $dt$ and $r$

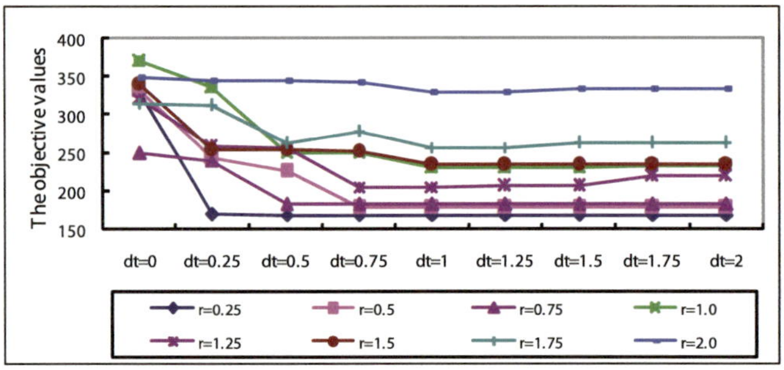


Fig. 2. The objective values with variables $dt$ and $r$

i) The parameter $dt$ interacts strongly with the arrival time distribution parameter $r$ ; the best choice is $dt = 1$ for most cases. ii) Larger $dt$ is not better than smaller $dt$ when $r$ is large. For example, when $r$ was set to 1.25, 1.75 or 2.0, the objective values with $dt = 1.5$ , $dt = 1.75$ or $dt = 2.0$ were more than with $dt = 1$ .

In addition, we simulated the same problem cases with the proposed ACO algorithm. The parameters $q_0$ , $\alpha$ , $\rho$ , $\delta$ , $\xi$ and $t_{max}$ were set to 0.5, 0.5, 0.1, 0.001, 0.1 and 100, respectively. The average improvements of ACO with different values of $r$ compared with ATC-BATC are shown in Figure 3. The improvement increased with $r$ , with an inflection in the curve at $r = 1$ . When $r$ was from 0.25 to 1 or from 1.25 to 2, the larger was $r$ , the better were the improvements by ACO. Furthermore, the improvements for $r$ above 1 were better than those for $r$ from 0.25 to 1. In the following simulations, we only consider $r$ values from 0.25 to 1.

### 4.4 Determining the values of the parameters in the ACO algorithm

a) Determine the probability parameter $q_0$ Tuning the parameter $q_0$ allows adjusting the degree of exploration and the choice of whether to concentrate the search around the best-so-far solution or to explore other solutions. We determined the probability parameter $q_0$ of the proposed ACO algorithm with the same problem cases shown in Table 4 with the time window $\Delta t$ 's distribution parameter $dt$ set to 1. The parameters $\alpha$ , $\rho$ , $\delta$ , $\xi$ and $t_{max}$ were set to 0.5, 0.1, 0.001, 0.1 and 100, respectively. The simulation results are shown in Figure 4. From these results, we can conclude that $q_0 = 0.2$ is the best selection for most cases.

b) Determine the pheromone importance parameter $\alpha$ We determined the pheromone importance parameter $\alpha$ with the same problem cases shown in Table 4 with $dt = 1$ and $q_0 = 0.2$ .

The parameters , $\rho$ , $\delta$ , $\xi$ and $t_{max}$ were set to 0.1, 0.001, 0.1 and 100, respectively. The simulation results are shown in Figure 5. In all cases, $\alpha = 0.7$ was the best choice.

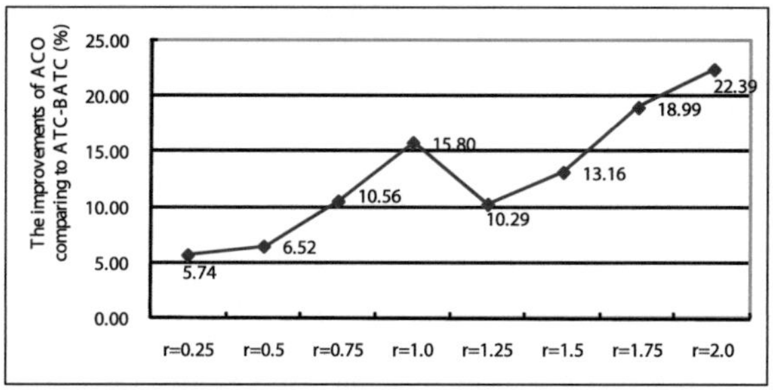


Fig. 3. Analysis of $r$ 's impacts on the ACO algorithm's performance

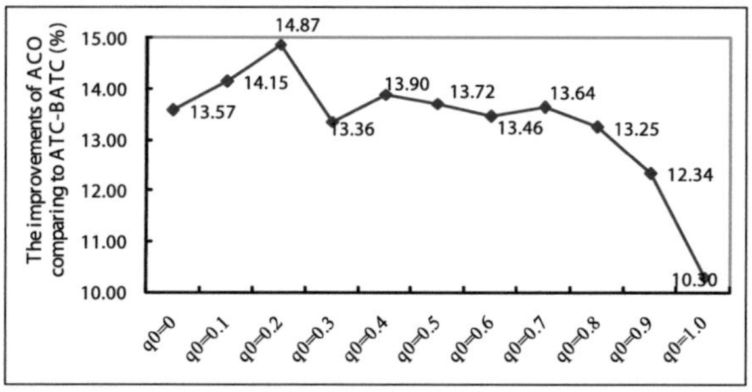


Fig. 4. The simulation results for determining the probability parameter $q_0$

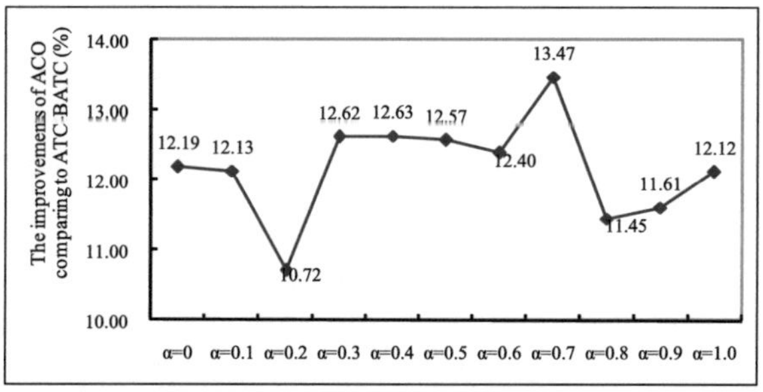


Fig. 5. The simulation results for determining the probability parameter $\alpha$

### 4.5 Comparison between ACO, ATC-BATC, MBS, GA and AS

With the above simulation results, the parameters of the ACO algorithm were set as Table 6. The parameters of the GA and the AS were set according to (Balasubramanian et al., 2004) and (Li et al., 2008), respectively.

The problem cases for comparing between ACO, ATC-BATC, MBS, GA and AS are shown in Table 7. In the simulations, we considered the impacts of the number and the arrival time

| *Parameter* | *Value* |
|---|---|
| $\alpha$ | 0.7 |
| $q_0$ | 0.2 |
| $\rho$ | 0.1 |
| $\delta$ | 0.001 |
| $\xi$ | 0.1 |
| $t_{max}$ | 100 |

Table 6. The parameters of the ACO algorithm

distribution of the jobs on the ACO algorithm's performance. The number of jobs was gradually increased by multiplying the number of machines and the number of recipes on each machine. The average improvements on the TWT and makespan of ACO are shown in Figure 6. From the simulation results, we can make the following conclusions.

| $Problemparameter$ | $Valueused$ | $Numberofvalues$ |
|---|---|---|
| $Numberofjobs$ | 20,32,44,56,68,80 | 6 |
| $Arrivaltimesofjobs$ | $Uniform(-r\Sigma_i\Sigma_j P_{ij}/(BM), r\Sigma_i\Sigma_j P_{ij}/(BM))$<br>$r = 0.25, 0.50, 0.75, 1.0$ | 4 |
| $Duedatesofjobs$ | $A_{ij} + P_{ij} + Uniform(0, Avg(P_{ij}))$ | 1 |
| $Timewindow\Delta t$ | $dt \cdot Avg(P_{ij}), dt = 1$<br>$dt = 0, 0.25, 0.5, 0.75, 1, 1.25, 1.5, 1.75, 2.0$ | 1 |
| $Weightperjob$ | $Uniform(0,1)$ | 1 |
| | $Totalparametercombinations$ | 24 |
| | $Numberofproblemspercombination$ | 10 |
| | $Totalproblems$ | 240 |

Table 7. The problem cases for comparing ACO and ATC-BATC

i) The value of the arrival time distribution parameter $r$ had an important impact on the ACO algorithm's average improvements on the TWT and makespan. Larger $r$ , i.e., the job arrivals were spread over a larger time range, resulted in better improvements on the TWT and makespan. In addition, the performance of MBS increasingly deteriorated with larger $r$ (shown in Figure 6(a)).

ii) Comparing to the heuristic rules (ATC-BATC and MBS), the number of jobs affected the ACO algorithm's average improvements on the average of the TWT and makespan. The more jobs, the better the average improvements, independent on $r$ 's value. However, comparing to the GA and AS, the impact of change in the number of jobs on the improvements of ACO on the average of the TWT and makespan fluctuated (shown in Figure 6(b)).

To further discuss the impacts of the batch size and the number of the recipes on the BPMs, and the number of the BPMs on the performance of the proposed method, it is assumed that the range of the number of the machines is from 3 to 5, the range of the batch size of a BPM is from 3 to 5, and the range of the number of the recipes of a BPM is from 4 to 6. Other conditions are the same as Table 7. The simulation results are shown in Figure 7. Obviously, the number and the capacity of the machines and the number of the recipes play an important role on the average improvements on the TWT and makespan of the ACO algorithm. Less machines, the bigger capacity and more recipes, the more average improvements on the TWT and makespan are. In addition, the performance of MBS increasingly improved with more recipes and bigger capacity.

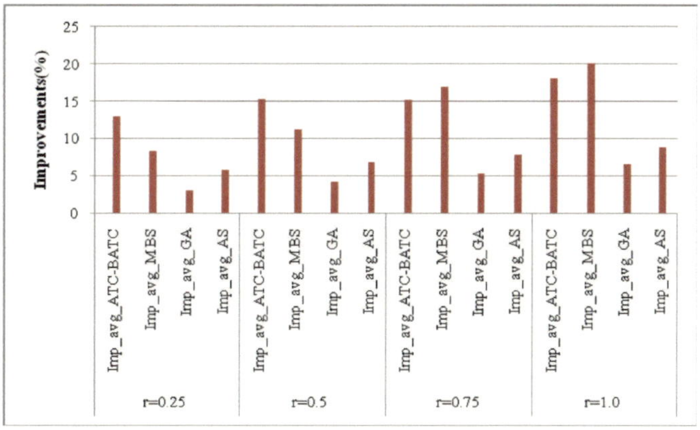


(a) The improvements by ACO with variable arrival time distribution of jobs * *Imp_avg_ATC_BATC* the average improvements on the TWT and makespan of ACO compared to ATC-BATC; the *Imp_avg_GA* : the average improvements on the TWT and makespan of ACO compared to GA; *Imp_avg_AS* : the average improvements on the TWT and makespan of ACO compared to AS

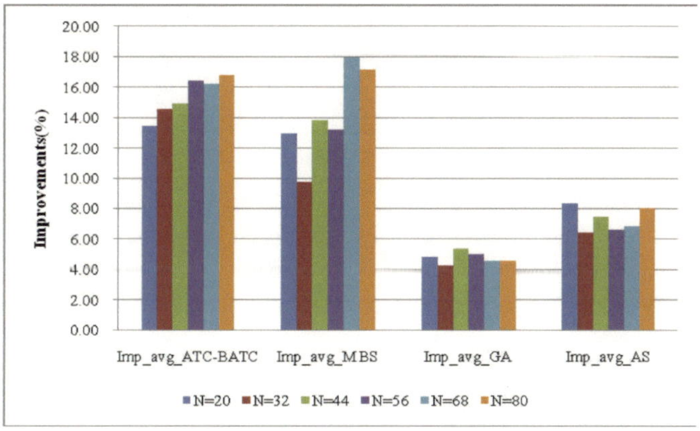


(b) The improvements by ACO with variable number of the jobs

Fig. 6. Simulation results for comparison between ACO and ATC-BATC, MBS, GA and AS

## 5. Conclusions

Batch processing machines play important roles in semiconductor wafer fabrication facilities. In this paper, we modeled the batch processing operations in a real wafer fab as an identical PBPM problem considering the practical complications of incompatible job families, dynamic job arrivals, sequence-dependent setup times and qual-run requirements of APC, and proposed an ACO algorithm to solve the problem with smaller TWT and makespan than

ATC-BATC, MBS, GA and AS. The main contributions of the paper are to create a method applicable in a production environment, to propose a better value for the time window $\Delta t$ from simulations, and to apply the ACO algorithm to obtain the solutions. Our next step is to integrate the ACO algorithm with the advanced planning and scheduling software of the real wafer fab.

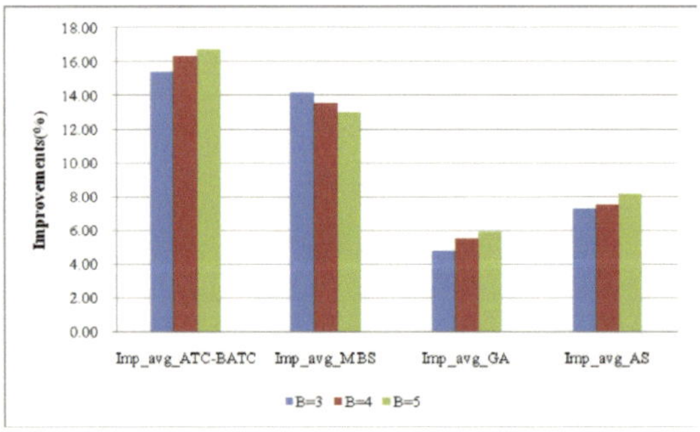


(a) The improvements by ACO with variable capacity of a BPM

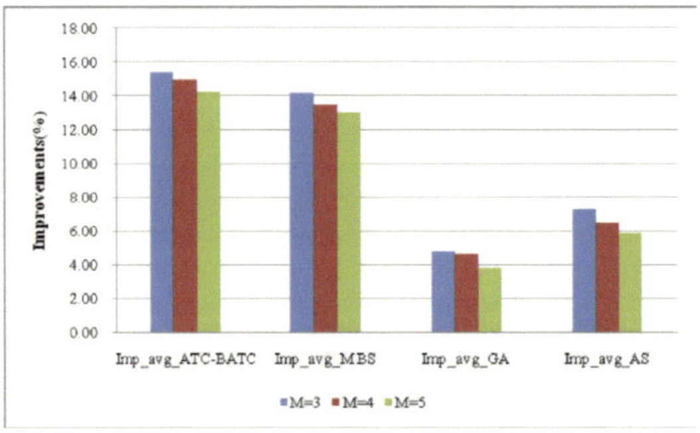


(b) The improvements by ACO with variable number of the BPMs

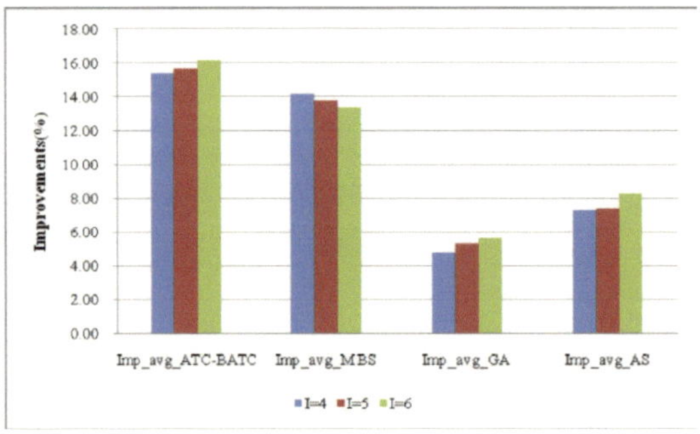


(c) The improvements by ACO with variable number of the recipes of a BPM

Fig. 7. The impacts of the batch size, the number of the recipes on the BPMs and the number of the BPMs

## 6. Acknowledgement

This work was supported in part by National Natural Science Foundation of China (No.50905129,No.70531020), Program for Young Excellent Talents in Tongji University (No.2006KJ006), the Grant from the Ph.D. Programs Foundation of Ministry of Education of China (No. 20070247007), and Shanghai Innovation Program (No. 09DZ1120600).

## 7. References

Mathirajan,M.& Sivakumar,A.I., (2006). A literature review, classification and simple meta-analysis on scheduling of batch processors in semiconductor. *International Journal of Advanced Manufacturing Technology*, Vol.29, No.9-10, July 2006,990-1001,ISSN 0268-3768

Chien C-F. & Chen C-H, (2007). A novel timetabling algorithm for a furnace process for semiconductor fabrication with constrained waiting and frequency-based setups. *OR Spectrum*, Vol.29, No.3, July 2007,391-419,ISSN 0171-6468

Chou F-D.; Chang P-C & Wang H-M (2006). A hybrid genetic algorithm to minimize makespan for the single batch machine dynamic scheduling problem. *International Journal of Advanced Manufacturing Technology*, Vol.31, No.3-4, Nov 2006,350-359,ISSN 0268-3768

Gupta A.K. & Sivakumar A.I. (2007). Controlling delivery performance in semiconductor manufacturing using look ahead batching. *International Journal of Production Research*, Vol.45, No.3, Feb 2007, 591-613(23), ISSN 0020-7543

Erramilli V. & Mason S.J. (2006). Multiple orders per job compatible batch scheduling. *IEEE Transactions on Electronics Packaging Manufacturing*, Vol.29, No.4, Oct 2006, 285-296, ISSN 1521-334X

Solomon M., Fowler J.W., Pfund M.et al.(2002). The inclusion of future arrivals and down-stream setups into wafer fabrication batch processing desicions. *Journal of Electionics Manufacturing*, Vol.11, No.2,2002,149-159, ISSN 0960-3131

Mönch L.; Zimmermann J. & Otto p (2006). Machine learning techniques for scheduling jobs with incompatible famlies and unequal ready times on parallel batch machines. *Engineering Applications of Artificil Intelligence*, Vol.19, No.3, Apr 2006, 235-245, ISSN 0952-1976

Liu L-L, Ng C-T & Cheng T-C-E (2007). Scheduling jobs with agreeable processing times and due dates on a single batch processing machine. *Theoretical Computer Science*, Vol.374, No.1-3, April 2007, 159-169, ISSN 0304-3975

Dorigo M. (1992) *Optimization, Learning and Natural Algorithms*, Ph.D. Thesis, Politecnico di Milano, Italy.

Dorigo M. & Stützle T (2004). *Ant colony optimization*, Cambridge,MIT Press.

Srinivasa Raghavan N.R. & Venkataramana M. (2006). Scheduling parallel batch processors with incompatible job families using ant colony optimization, *Proceeding of the 2006 IEEE International Conference on Automation Science and Engineering*, Oct 2006,pp.507-512, ISSN 1545-5955, Shanghai.

Balasubramanian H, Mönch L & Fowler J et al. (2004). Genetic algorithm based scheduling of parallel batch machines with incompatible job families to minimize total weighted tardiness. *International Journal of Production Research*, Vol.42, No.8,April 2004, 1621-1638, ISSN 1366-588X

Cai Y.W, Kutanoglu E, Hesenbein J et al.(2007). Single-machine scheduling problem with advanced process control constraints,*AEC/APC Symopsium XIX*, Indian Wells, Calif. USA, pp. 507-512.

Patel N-S (2004).Lot allocation and process control in semiconductor manufacturing - a dynamic game approach. *Proceeding of 43rd IEEE Conference on Decision and Control*,Vol.4, pp.4243-4248,,ISSN 0005-1179 ,Atlantis,Paradise Island, Nassau, Bahamas, Dec 2004

Mönch L, Balasubramanian H, Fowler J-W et al. (2005). Heuristic scheduling of jobs on parallel batch machines with incompatible job families and unequal ready times. *Computers and Operations Research* , Vol.32, No.11, Nov 2005, 2731-2750, ISSN 0305-0548

Li Li, Qiao F & Wu Q.D. (2008).ACO-Based Scheduling of Parallel Batch Processing Machines with Incompatible Job Families to Minimize Total Weighted Tardiness, *Ant Colony Optimization and Swarm Intelligence*, Vol.5217,219-226, ISBN 978-3-540-87526-0,Sept 2008,Springer, Berlin

# 9

# Axiomatic design of Agile manufacturing systems

Dominik T. Matt
*Free University of Bolzano*
*Italy*

## 1. Introduction

The trend of shifting abroad personnel-intensive assembly from Europe to foreign countries continues. Manufacturing systems widely differ in investment, demand and output. Since sales figures can hardly be forecasted, it is necessary to conceptualize highly flexible and adaptable systems which can be upgraded by more scale-economic solutions during product life cycle, even under extremely difficult forecasting conditions. Unlike flexible systems, agile ones are expected to be capable of actively varying their own structure. Due to the unpredictability of change, they are not limited to a pre-defined system range typical for so called flexible systems but are required to shift between different levels of systems ranges.

Modern manufacturing systems are increasingly required to be adaptable to changing market demands, which adds to their structural and operational complexity (Matt, 2005). Thus, one of the major challenges at the early design stages is to select an manufacturing system configuration that allows both - a high efficiency due to a complexity reduced (static) system design, and a enhanced adaptability to changing environmental requirements without negative impact on system complexity.

Organizational functional periodicity is a mechanism that enables the re-initialization of an organization in general and of a manufacturing system in particular. It is the result of converting the combinatorial complexity caused by the dynamics of socioeconomic systems into a periodic complexity problem of an organization.

Starting from the Axiomatic Design (AD) based complexity theory this chapter investigates on the basis of a long-term study performed in an industrial company the effects of organizational periodicity as a trigger for a regular organizational reset on the agility and the sustainable performance of a manufacturing system.

Besides the presentation of the AD based design template which helps system designers to design efficient and flexible manufacturing systems, the main findings of this research can be summarized as follows: organizational functional periodicity depends on environmentally triggered socio-economic changes. The analysis of the economic cycle shows high degrees of periodicity, which can be used to actively trigger a company's action for change, before market and environment force it to. Along an economic sinus interval of about 9 years, sub-periods are defined that trigger the re-initialization of a manufacturing system's set of FRs and thus establish the system's agility.

## 2. Agility – an Answer to Growing Environmental Complexity

The actual economic crash initiated by the subprime mortgage crisis has been leading to another global follow-up recession. Most enterprises are struggling with overcapacities caused by an abrupt decrease in market demand, and our industrial nations - traditional sources of common wealth in our "old world" - are groaning under the burden of mountains of debts. But did this crisis really come surprisingly?
The answer is no, although nobody could exactly determine its starting point in time. In fact, the economic cycle is a well-known phenomenon. Often new business opportunities created by a new technology (e.g. digital photography, GPS, smart items, photovoltaic cells, etc.) or some "hypes" such as the "dotcoms" in the late 90s may trigger an economic boom. Initially, wealth is created when growing market demand for new or "hip" products generates new jobs and promotes productivity and growth. However, quantitative economic growth is limited (Matt, 2007) and when it turns to be artificially maintained on an only speculative basis, the economic system is going to collapse.
Analyzing analogical behaviors in natural and other systems, we understand that the reason for this lays in the interaction of a system's elements in terms of causal or feedback loops (O'Connor & McDermott, 1998). System growth is driven by positive (or escalating) causal loops (Senge 1997). Even an exponential growth of a system is limited, either by the system's failure or collapse (for example, the growth of cancer cells is limited by the organism's death) or by negative feedback loops (for example, a continuous growth of an animal population is stopped by a limited availability of food, see Briggs & Peat 2006).
To maintain stability and survivability, a growing system needs to establish subsystems that are embedded in a superior structure (Vester 1999). Life on earth has not been spread all over the earth ball as a simple mash of organic cells but started to structure and differentiate, that is to grow qualitatively. A randomized cross-linking of the system components will inevitably lead to a stability loss. Thus, a system can overcome its quantitative growth limits only by qualitative growth, establishing a stabile network structure with nodes that are subject to cell division as soon as they reach a critical dimension.

### 2.1 The Mechanisms of Complexity

A system's ability to grow depends to a considerable extent on its structure and design. Its design is "good" if it is able to fulfill a set of specific requirements or expectations.
An entrepreneur or an investor for instance expects that a company makes profit and that it increases its value. The entrepreneurial risk expresses the uncertainty that these targets or expectations are fulfilled, especially over time when environmental conditions change and influence the system design. The complexity of a system is determined by the uncertainty in achieving the system's functional requirements (Suh 2005) and is caused by two factors: by a time-independent poor design that causes a system-inherent low efficiency (system design), and by a time-dependent reduction of system performance due to system deterioration or to market or technology changes (system dynamics).
To enable a sustainable and profitable system growth, its entire complexity must be reduced and then be controlled over time. To reduce a system's complexity, its subsystems should not overlap in their contribution to the overall system's functionality, they must be mutually exclusive. On the other hand, the interplay of system components must be collectively

exhaustive in order to include every issue relevant to the entire system's functionality. Finally, this procedure has to be repeated over time as changes in the system's environments might impact its original design and thus lead to a loss in efficiency and competitiveness.

The **time-independent complexity** of a system is a measure for a system's ability to satisfy a set of functional requirements without worrying about time-dependent changes that might influence the system's behavior. It consists of two components: a time-independent real **complexity** and a time-independent **imaginary complexity**. The real complexity tells if the system range is inside or partly or completely outside the system's design range. The imaginary complexity results from a lack of understanding of the system design, in other words the lack of knowledge makes the system complex. If the system is designed to always fulfill the system requirements, that is the range of the system's functional requirements (system range) is always inside the system's range of design parameters (design range), it can be defined a "good" design. This topic will be treated in more detail in a following section.

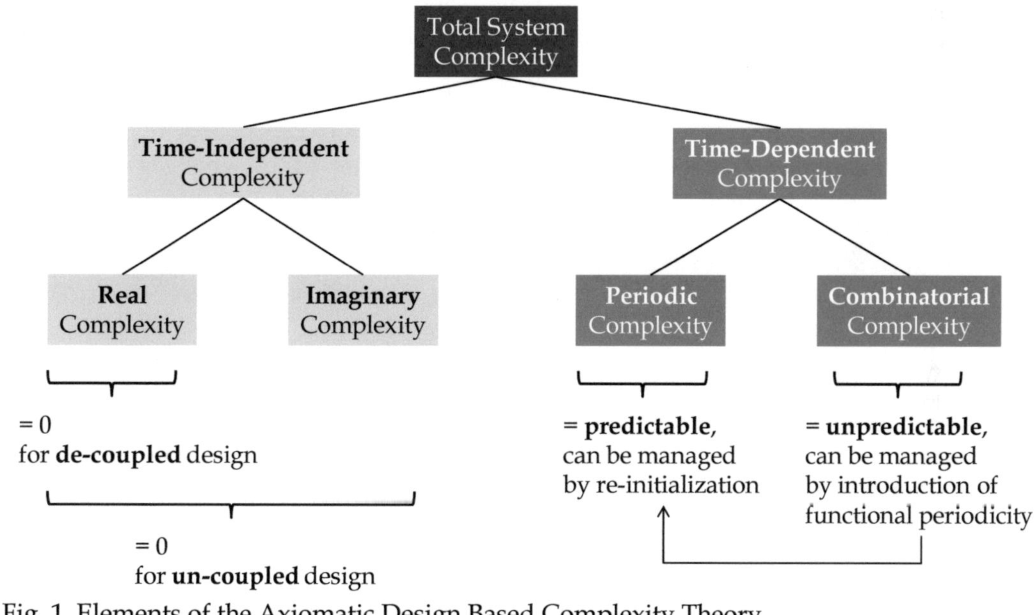


Fig. 1. Elements of the Axiomatic Design Based Complexity Theory

**Time dependent system complexity** has its origins in the unpredictability of future events that might change the current system. There are two types of time-dependent complexities (Suh 2005): The first type of time-dependent complexity is called **periodic complexity**. It only exists in a finite time period, resulting from a limited number of probable combinations. These probable combinations may be partially predicted on the basis of existing experiences with the system or with a very systematic research of possible failure sources.

The second type of time-dependent complexity is called **combinatorial complexity**. It increases as a function of time proportionally to the time-dependent increasing number of possible combinations of the system's functional requirements. This may lead to a chaotic

state or even to a system failure. The critical issue as to combinatorial complexity is that it is completely unpredictable.
According to Nam Suh, the economic cycle is a good example of time-dependent combinatorial complexity at work (Suh, 2005). To provide stabile system efficiency, the time-dependent combinatorial complexity must be changed into a time-dependent periodic complexity by introducing a functional periodicity. If the functional periodicity can be designed in at the design stage, the system will last much longer than other systems. This way the system becomes "agile".

### 2.2 Agility

In recent scientific publications, terms like flexibility (De Toni & Tonchia, 1998), reconfigurability (Koren et al., 1999), agility (Yusuf et al., 1999) and more recently changeability (Wiendahl & Heger, 2003) or mutability (Spath & Scholz, 2007) have been defined in many different contexts and often refer to the same or at least a very similar idea (Saleh et al., 2001). Nyhuis et al. (2005) even state that changeover ability, reconfigurability, flexibility, transformability, and agility are all types of changeability, enumerated in the order of increasing system level context.
Flexibility means that an operation system is variable within a specific combination of in-, out- and throughput. The term is often used in the context of flexible manufacturing systems and describes different abilities of a manufacturing system to handle changes in daily or weekly volume of the same product (volume flexibility) to manufacture a variety of products without major modification of existing facilities (product mix flexibility), to process a given set of parts on alternative machines (routing flexibility), or to interchange the ordering of operations (operation flexibility) on a given part (Suarez et al., 1991). Reconfigurability aims at the reuse of the original system's components in a new manufacturing system (Mehrabi, 2000). It is focused on technical aspects of machining and assembly and is thus limited to single manufacturing workstations or cells (Zaeh et al., 2005). Agility as the highest order of a system's changeability, in contrast, means the ability of an operation system to alter autonomously the configuration to meet new, previously unknown demands e. g. from the market as quickly as the environmental changes (Blecker & Graf, 2004).
Unlike flexible systems, agile ones are expected to be capable of actively varying their own structure. Due to the unpredictability of change, they are not limited to a pre-defined system range typical for so called flexible systems but are required to shift between different levels of systems ranges (Spath & Scholz, 2007).

### 2.3 The Principles of Axiomatic Design (AD)

The theory of Axiomatic Design was developed by Professor Nam P. Suh in the mid-1970s with the goal to develop a scientific, generalized, codified, and systematic procedure for design. Originally starting from product design, AD was extended to many different other design problems and proved to be applicable to many different kinds of systems. Manufacturing systems are collections of people, machines, equipment and procedures organized to accomplish the manufacturing operations of a company (Groover, 2001). As system theory states, every system may be defined as an assemblage of subsystems.

Accordingly, a manufacturing system can be seen as an assemblage of single manufacturing stations along the system's value stream (Matt, 2006).
The Axiomatic Design world consists of four domains (Suh, 2001): the customer domain, the functional domain, the physical domain and the process domain.
The customer domain is characterized by the customer needs or attributes (CAs) the customer is looking for in a product, process, system or other design object. In the functional domain the customer attributes are specified in terms of functional requirements (FRs) and constraints (Cs). As such, the functional requirements represent the actual objectives and goals of the design. The design parameters (DPs) express how to satisfy the functional requirements. Finally, to realize the design solution specified by the design parameters, the process variables (PVs) are stated in the process domain (Suh, 2001). For the design of manufacturing systems the physical domain is not needed (Reynal & Cochran, 1996).
Most system design tasks are very complex, which makes it necessary to decompose the problem. The development of a hierarchy will be done by zigzagging between the domains. The zigzagging takes place between two domains. After defining the FR of the top level a design concept (DP) has to be generated.
Within mapping between the domains the designer is guided by two fundamental axioms that offer a basis for evaluating and selecting designs in order to produce a robust design (Suh, 2001):

- **Axiom 1: The Independence Axiom.** Maintain the independence of the functional requirements. The Independence Axiom states that when there are two or more FRs, the design solution must be such that each one of the FRs can be satisfied without affecting the other FRs.
- **Axiom 2: The Information Axiom.** Minimize the information content I of the design. The Information Axiom is defined in terms of the probability of successfully achieving FRs or DPs. It states that the design with the least amount of information is the best to achieve the functional requirements of the design.

The FRs and DPs are described mathematically as a vector. The Design Matrix [DM] describes the relationship between FRs and DPs in a mathematical equation (Suh, 2001):

$$\{FR\} = [DM]\{DP\} \tag{1}$$

With three FRs and three DPs, the above equation may be written in terms of its elements as:

$$\begin{aligned} FR_1 &= A_{11}\, DP_1 + A_{12}\, DP_2 + A_{13}\, DP_3 \\ FR_2 &= A_{21}\, DP_1 + A_{22}\, DP_2 + A_{23}\, DP_3 \\ FR_3 &= A_{31}\, DP_1 + A_{32}\, DP_2 + A_{33}\, DP_3 \end{aligned} \tag{2}$$

The goal of a manufacturing system design decision is to make the system range inside the design range (Suh, 2006). The information content I of a system with n FRs is described by the joint probability that all n FRs are fulfilled by the respective set of DPs. The information content is measured by the ratio of the common range between the design and the system range (Suh, 2001). To satisfy the Independence Axiom, the design matrix must be either diagonal or triangular (Fig. 2). When the design matrix is diagonal, each of the FRs can be satisfied independently by means of exactly one DP. It represents the ideal case of an uncoupled system design where the design range of every single DP perfectly meets the system range of exactly one FR, irrespective of the sequence of the fulfillment of the functional requirements. This means, that the design equation can be solved without any restrictions. In this case, the above equation (2) may be written as:

$$\begin{aligned} FR_1 &= A_{11}\, DP_1 \\ FR_2 &= A_{22}\, DP_2 \\ FR_3 &= A_{33}\, DP_3 \end{aligned} \tag{2.1}$$

Both components of the time-independent complexity - the real complexity and the imaginary complexity - are zero, in other words: the total time-independent complexity of the system is zero (see also Fig. 1).

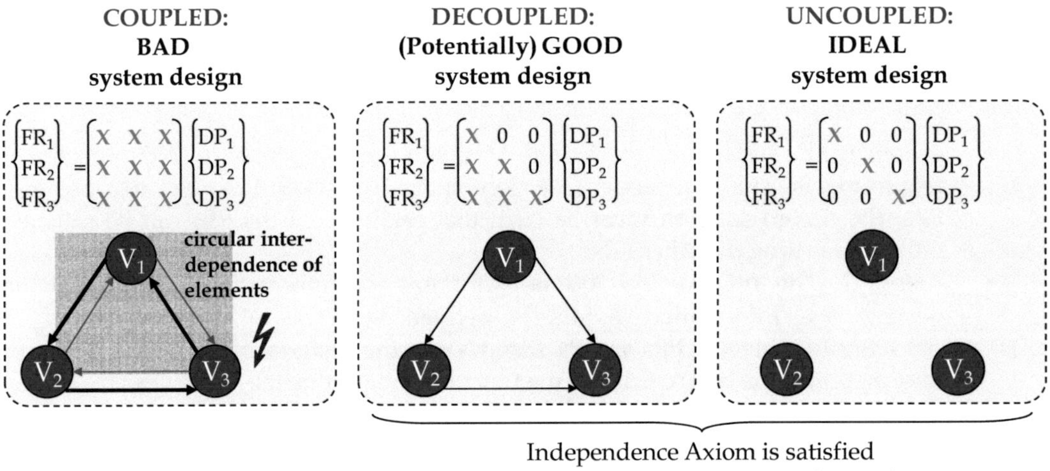


Fig. 2. Exemplary illustration of the Independence Axiom (Lee & Jeziorek, 2006)

When the matrix is triangular, the independence of FRs can be guaranteed if and only if the DPs are determined in a proper sequence. In the case of a decoupled design, which design range also fits the system range, the real complexity equals to zero, but the complexity consists in the uncertainty of fulfilling the design task due to different possible sequences. Thus, it depends on a particular sequence and represents a decoupled design creating a time-independent imaginary complexity. In terms of equation (2), this has the following consequence:

$$\begin{aligned} FR_1 &= A_{11}\, DP_1 \\ FR_2 &= A_{21}\, DP_1 + A_{22}\, DP_2 \\ FR_3 &= A_{31}\, DP_1 + A_{32}\, DP_2 + A_{33}\, DP_3 \end{aligned} \tag{2.2}$$

Any other form of the design matrix is called a full matrix and results in a coupled design.

## 3. Axiomatic Design of Agile Manufacturing Systems

A manufacturing system is a dynamic system, because it is subject to temporal variation and must be changeable on demand (Cochran et al., 2000; Matt, 2006). Market and strategy changes will influence its system range of functional requirements and therefore impact the system's design (Reynal & Cochran, 1996). Considering for example a given production program, all possible product variants that can be manufactured at a certain point in time determine the static system complexity. However, the dynamic complexity is determined by the frequency and magnitude of changes of the production program when new product variants are introduced or eliminated. When both complexities are low, then the system is simple. In the case of a high (low) structural complexity and low (high) dynamic complexity, the system is considered to be complicated (relatively complex). When both complexities are high, then the system is said to be extremely complex (Ulrich & Probst, 1995). On the basis of these definitions, every approach aiming at the reduction of a system's complexity consequently has to focus on the redesign of the system elements and their relationships.

Following the considerations made in section 2.1, two general ways to attack the problems associated with complex systems can be identified. The first is to simplify them, the second to control them. Leanness is about the former in that it advocates waste removal and simplification (Naylor et. al., 1999). It aims at the complexity reduction of a system at a certain point in time. Thus, system simplification is about eliminating or reducing the time-independent complexity of a system. Agility is the ability to transform and adapt a manufacturing system to new circumstances caused by market or environmental turbulences (Zaeh et. al., 2005). Thus, complexity control is associated with the elimination or reduction of a system's time-dependent complexity. To adopt design strategies that consider Lean and Agility principles, it is important to introduce decoupling points. A material decoupling point is the point in the value chain to which customer orders are allowed to penetrate. At this point there is buffer stock and further downstream the product is differentiated. A very helpful tool in this context is value stream mapping, a key element of the Lean toolbox, which represents a very effective method for the visualization, the analysis and the redesign of production and supply chain processes including material flow as well as information flow (Rother & Shook, 1998). The methodology provides process boxes, which describe manufacturing or assembly processes following the flow principle, with no material stoppages within their borderlines. Ideally, a continuous flow without interruptions can be realized between the various assembly modules. However, most process steps have different cycle times and thus buffers (decouplers) have to be provided at their transitions for synchronization (Suh, 2001).
To define the functional requirements of a manufacturing system and to transform them into a good system design, Axiomatic Design (AD) is proposed to be a very helpful tool (Cochran & Reynal, 1999): the authors analyse the design of four manufacturing systems designs in terms of system performance and use the methodology to design an assembly area and to improve a machining cell at two different companies. However, the lifetime of such a design varies from 3 to 18 months (Rother & Shook, 1998). During this period, the design can be supposed to behave in a nearly time-independent way. Afterwards, it is again subject to changes. Thus, to maintain the efficiency of a manufacturing system design, also the time-dependent side of complexity has to be considered.

Thus, the methodology presented in the following provides two steps based on the AD complexity theory: First, the system is designed to fulfill the time-independent requirements of **efficiency and flexibility** within a "predictable" planning horizon of 6 to 24 months (Rother & Shook, 1998; Matt, 2006). This design step uses the approach of the production module templates (Matt, 2008).
In a second step, a (time-dependent) **agility** strategy is elaborated to allow a quick shift to another (nearly) time-independent system level.

### 3.1 Efficiency and Flexibility: Reduce the Time-Independent Complexity

One of the major goals of manufacturing system design is to reduce the time-independent real complexity to zero. The real complexity is a consequence of the system range being outside of the design range. If the system design is coupled it is difficult to make the system range lie inside the design range. Therefore, the following procedure is recommended:
First, the system designer must try to achieve an uncoupled or decoupled design, i.e. a design that satisfies the Independence Axiom.
Then, every DP's design range has to be fitted and adapted into the corresponding FR's system range. This way, the system becomes robust by eliminating the real complexity. The imaginary complexity rises with the information content of the design. In an uncoupled design, the information content is zero and so an imaginary complexity does not exist. However, in the case of a decoupled design, the designer has to choose the best solution among different alternatives, which is the one with the less complex sequence.

The probably most important step in Axiomatic design is the definition of the first level of FRs. It requires a very careful analysis of the customer needs regarding the design of the manufacturing systems.
The translation of the CAs into FRs is very important and difficult at the same time, because the quality of the further design depends on the completeness and correctness of the chosen CAs. According to generally accepted notions (Womack and Jones, 2003; Bicheno, 2004) regarding a manufacturing systems objective system, the following three basic CAs can be identified:

CA1: Maximize the customer responsiveness (according to the 6 "Rs" in logistics: the right products in the right quantity and the right quality at the right time and the right place and at the right price)
CA2: Minimize the total manufacturing cost per unit
CA3: Minimize inventory and coordination related costs

Starting from these basic CAs, the following generally applicable FRs for manufacturing system design can be derived:

FR1: Produce to demand
FR2: Realize lowest possible unit cost
FR3: Realize lowest possible overhead expenses

The design parameters mapped by functional requirements are:

DP1: Only consistent increments of work demanded by customers are released
DP2: Manufacturing stations are designed for low cost production
DP3: Strategy to keep inventory and coordination related costs at the lowest level

The design matrix provides a decoupled design (triangular design matrix) as shown in the following equation:

$$\begin{Bmatrix} FR1 \\ FR2 \\ FR3 \end{Bmatrix} = \begin{bmatrix} X & 0 & X \\ 0 & X & 0 \\ 0 & 0 & X \end{bmatrix} \cdot \begin{Bmatrix} DP1 \\ DP2 \\ DP3 \end{Bmatrix} \qquad (3)$$

Since the design solution cannot be finalized or completed by the selected set of DPs at the highest level, the FRs need to be decomposed further. This decomposition is done in parallel with the zigzagging between the FRs and DPs (Suh, 2001; Cochran, et al., 2002).

| | |
|---|---|
| **FR 1** Produce to demand | **FR 11** Identify the required output rate<br>**FR 12** Create a continuous flow<br>**FR 13** Respond quickly to unplanned production problems<br>**FR 14** Minimize production disturbances by planned standstills<br>**FR 15** Achieve operational flexibility |
| **FR 2** Realize lowest possible unit costs | **FR 21** Achieve a high yield of acceptable work units<br>**FR 22** Minimize labor costs<br>**FR 23** Minimize one time expenditures |
| **FR 3** Realize lowest possible overhead expenses | **FR 31** Minimize the distance between source and process<br>**FR 32** Provide a complete order picking<br>**FR 33** Eliminate unnecessary motion and prevent defects throughout the material handling operation |

Fig. 3. Second level decomposition of the FR-tree (Matt, 2009/a)

The so developed 2nd level FR-tree is shown in Fig. 3. By doing the zigzagging between FRs and DPs, as done on the first level, the DPs for the second level corresponding to FR-2 can be identified in order to maximize independence (Matt, 2006):

DP-11 Determine and produce to takt time (for details see: Matt, 2006 and Matt, 2008)
DP-12 (a) *Single model case*: no significant variations, sufficient volumes to justify the dedication of the system to the production of just one item or a family of nearly identical items. Introduction of process-principle (multi-station system) if sequentially arranged stations can be balanced to in-line continuous flow.

(b) *Batch model case*: Different parts or products are made by the system. Batching is necessary due to long setup or changeover times
(c) *Mixed model case*: different parts or products are made by the system, but the system is able to handle these differences without the need for setup or changeover.
(c.1) Introduction of process-principle (sequential multi-station system with fixed routing) if sequentially arranged stations can be balanced to in-line continuous flow independent from product variants and their production sequence.
(c.2) Introduction of object-principle if lead times of the single process steps vary widely and cannot be balanced (single-station system, eventually parallel stations if cycle times of the single station exceed the takt time). This is usually the case with a high complexity of the production program with very different variants.

DP-13 Visual control and fast intervention strategy (Introduction of TPM – Total Productivity Maintenance)
DP-14 Reduction and workload optimized scheduling of planned standstills (TPM)
DP-15 Setup reduction (Optimization with SMED – Single Minute Exchange of Die)

The effective design parameters (DPs) for FR-21, FR-22 and FR-23 are the following (Matt, 2006):

DP-21 Production with increased probability of producing only good pieces and of detecting/managing defective parts
DP-22 Effective use of workforce
DP-23 Investment in modular system components based on a system thinking approach

For FR-31, FR-32 and FR-33, the design parameters mapped by functional requirements are (Matt, 2009/b):

DP-21: Short distances between material storage location and process
DP-22: Design equipment and methods that allow handling and transport of the complete order set
DP-23: Design equipment and methods that allow an effective and defect-free interaction between humans and material

The single level Design Matrices as well as the complete Design Matrix are decoupled. Interested readers are referred to (Matt, 2008) for more detailed information about the above described AD based template approach for manufacturing system design.

### 3.2 Agility: Control Time-Dependent Complexity

Time dependent system complexity has its origins in the unpredictability of future events that might change the current system and its respective system range. The shifting between different levels of system ranges cannot be controlled by the normal flexibility tolerances provided in a manufacturing system design. It is subject to system dynamics and thus has to be handled within the domain of time-dependent complexity. According to Suh (2005), there are two types of time-dependent complexities:
As previously outlined, the first type of time-dependent complexity is called periodic complexity. It only exists in a finite time period, resulting from a limited number of probable

combinations. These probable combinations may be partially predicted on the basis of existing experiences with the system or with a very systematic research of possible failure sources, e.g. with FMEA.

The goal of a manufacturing system design is to make the system range lie inside the design range. The information content I of a system with n FRs is described by the joint probability that all n FRs are fulfilled by the respective set of DPs. The information content is measured by the ratio of the common range between the design and the system range (Suh, 2006). However, a system might deteriorate during its service life and its design range will move outside the required system range. In this case, the system's initial state must be established by re-initialization.

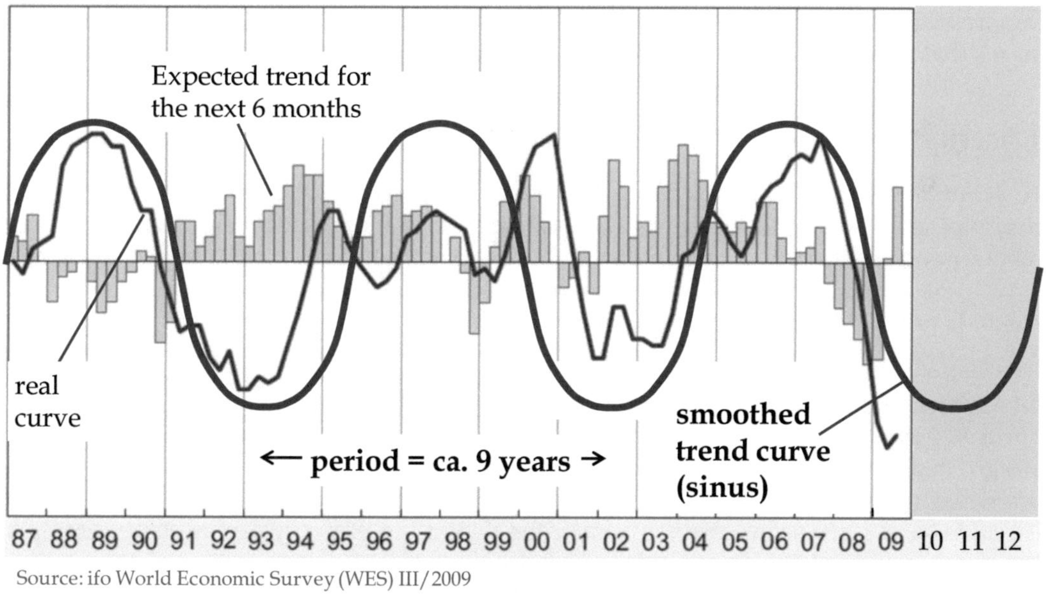


Fig. 4. The economic cycle drives an organization's functional periodicity (Matt, 2009/a)

The second type of time-dependent complexity is called combinatorial complexity. It increases as a function of time proportionally to the time-dependent increasing number of possible combinations of the system's functional requirements. It may lead to a chaotic state or even to a system failure. The critical issue with combinatorial complexity is that it is completely unpredictable. Combinatorial complexity can be reduced through re-initialization of the system by defining a functional period (Suh, 2005).

A functional period is a set of functions repeating itself on a regular time interval, like the one shown in Fig. 4 showing the periodicity of our economic system. Organizational systems - e.g. a manufacturing system - need (organizational) functional periodicity. When they do not renew themselves by resetting and reinitializing their functional requirements, they can become an entity that wastes resources (Suh, 2005).

To maximize the operational excellence of a manufacturing system in order to provide its transformability to unforeseen changes, the system must be designed to satisfy its FRs at all times. Ideally, such a system has zero total complexity, i.e. both time-independent and time-dependent complexity. Once the manufacturing system has been designed according to the

above described principles of time-independent complexity reduction, its time-dependent complexity has to be reduced in order to manage unpredictable shifts between different levels of the manufacturing system's range of functional requirements.
To design an agile manufacturing system, the time-dependent combinatorial complexity must be changed into a time-dependent periodic complexity by introducing a functional periodicity. If the functional periodicity can be designed in at the design stage, the system's changeability will be more robust than in any other system (Suh, 2005).
It is important to anticipate the economic cycle in order to maintain competitiveness (Fig. 5). However, the average period of the economic cycle (ca. 9 years) might be too long for the company specific dynamics. The current research results obtained from the observation of good industrial practice show that a possible solution might be to introduce a sinus interval compressed by a 1/n factor (stretching constant), with for example n=2 or n=3. For n=2, this means that the re-organization cycle repeats about every 4-5 years, for n=3 this is 3 years.

## 4. Illustrative Example

To illustrate the previously described approach, an industrial example of a manufacturer of electrotechnical tools and equipment is discussed. For a recently developed and presented cable scissor, an efficient and flexible assembly system has to be designed: two scissor blades have to be joined with a screw, a lining disc and a screw nut; afterwards, the assembled scissor is packaged together with some accessories.

### 4.1 Efficiency and Flexibility: Reduce the Time-Independent Complexity

The first step is the elimination or reduction of the time-independent complexity. Thus, the design must first fulfill the Independence Axiom. According to the design template presented in section 3.1, the single model case is chosen: the product has no significant variations and sufficient volumes to justify the dedication of the system to the assembly of just one item or a family of nearly identical items.
To meet the required takt time, a semi-automatic screwing device is provided as first station in a two-station assembly system. However, to create a robust system, the real complexity has to be reduced or eliminated by fitting the DPs' design range to the corresponding FRs' system range. Thus, a dynamometric screwdriver is applied which torque tolerance fits the required system range. To evade the problem of imaginary complexity, the system design has to be uncoupled. In an inline multi-station assembly system, this requirement can be achieved by introducing de-couplers (buffers) between the single stations.
However, buffers have the negative effect to create an increase of handling and therefore a loss in the system's efficiency. A possible solution to decouple an assembly system and at the same time maintain a low level of non value adding activities is the so called "moving fixture" for workpieces (Lotter et al., 1998).
It consists of a base plate with holding fixtures to clamp the single workpieces and is manually or automatically moved on a belt conveyor from one to the next station. To decouple the line, several of these moving fixtures form a storage buffer between the single assembly stations.

## 4.2 Agility: Control Time-Dependent Complexity

The next step is to reduce and control the system's time-dependent complexity. The new designed system might deteriorate during its service life and its design range will move outside the required system range. In this case, the system's initial state must be established by re-initialization. This can be done by defining fixed maintenance intervals or by regular or continuous tool monitoring, where the status of the screwing unit is determined and the decision is taken whether to continue production, to maintain or even substitute the tool. In the specific case of the electrotechnical device manufacturer, the design range of the dynamometric screwdriver moves out of the scissors' system range and thus creates quality problems. To reduce or even eliminate this periodically appearing complexity (**periodic complexity**), regular checks of the screwing device are introduced.

However, the most critical aspect in system design is the **combinatorial complexity**. Being completely unpredictable, this type of complexity can be just controlled by transforming it into a periodic complexity. Combinatorial complexity mostly results from market or environmental turbulences that create extra organizational efforts.

As a socioeconomic system, a company is embedded in general economic cycles of upturn and downturn phases (Fig. 4).

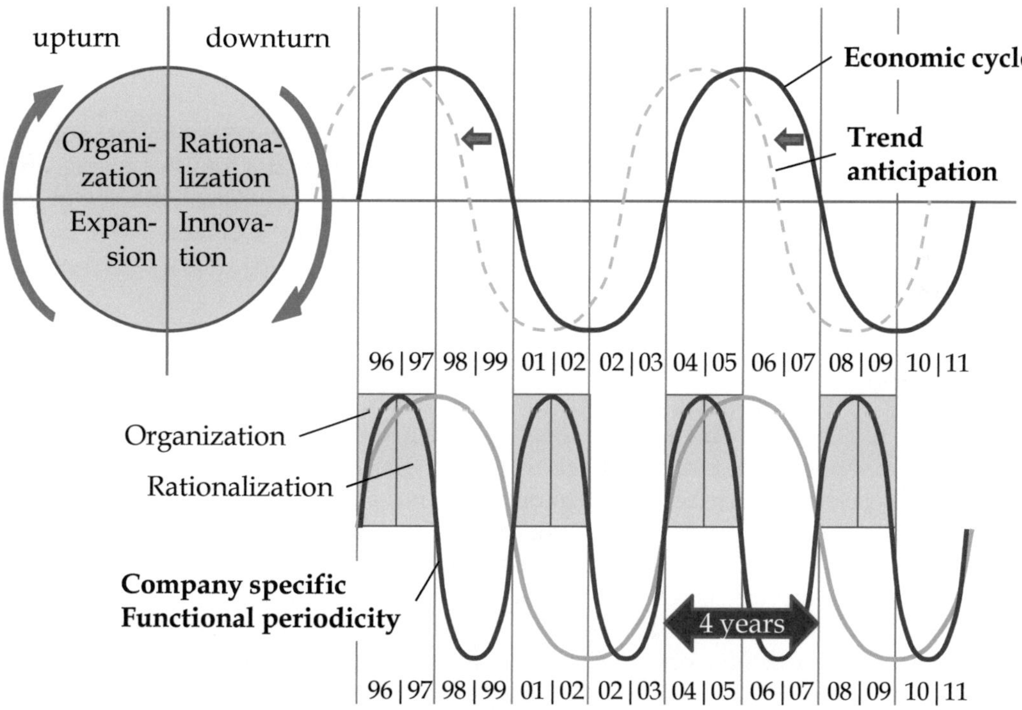


Fig. 5. Company specific functional periodicity of the manufacturing system

Obviously, every economic sector or even every single company has a different cyclic behavior regarding the timeline (Fig. 5). It passes always the following four stages: rationalization, innovation, expansion and organization. The company individual adaptation is given by the mapping of this generally applicable cycle along the timeline as a sinus curve (Matt, 2009/a). The company individual interval can be determined heuristically, i.e. based on data and experiences from past. In our example, the company specific ideal sinus-interval of the manufacturing system's functional periodicity is 4 years (n=2). As far as research showed, it is determined very much by the average product life cycle and the related company specific innovation cycles.

Fig. 6. One-set flow with moving fixtures plates in different flow-variants

Knowing the rhythm of change within a specific industry, suitable strategies for fast volume and variant adaptation can be developed, transforming combinatorial into the manageable periodic complexity. Fig. 6 shows for the present example the re-initialization strategy for the current process-oriented manufacturing system design (Spath & Scholz, 2007): as the number of variants shows a significant increase, a switch of DP-12 towards a mixed-model case c.1 or c.2 is possible.

## 5. Conclusion

In this chapter, a concept for the integrated design of efficient, flexible and changeable manufacturing systems was discussed. Starting from the AD based complexity theory, a procedure was presented that helps system designers not only to design assembly systems with low or zero time-independent complexity (focus: flexibility and efficiency), but also to prevent the unpredictable influences of the time-dependent combinatorial complexity by transforming it into a periodic review and adaptation of the system's volume and variant capabilities (focus: agility). Future research will concentrate on a more sophisticated determination of the stretching constant in the company individual sinus-curve-model.

## 6. References

Blecker, T.; Graf, G. (2004) Changeability in Operations: A Critical Strategic Resource for European Manufacturing? *Proceedings of the 2nd International Conference "An Enterprise Odyssee: Building Competitive Advantage"*, Galetic L (Ed.), Zagreb/Croatia: pp. 904-917.

Briggs, J., Peat, F. D. (2006) *Die Entdeckung des Chaos - Eine Reise durch die Chaos-Theorie.* Ungekürzte Ausgabe März 1993, 9. Aufl., dtv, ISBN-13: 978-3-423-33047-3, München.

Cochran, D. S.; Arinez, J. F.; Duda, J. W.; Linck, J. (2002) A decomposition approach for manufacturing system design. *Journal of Manufacturing Systems*, Vol. 20, No. 6, pp. 371-389

Cochran, D. S.; Eversheim, W.; Kubin, G.; Sesterhenn, M. L. (2000) The Application of Axiomatic Design and Lean Management Principles in the Scope of Production System Segmentation. *The International Journal of Production Research*, Vol. 38, No. 6, pp. 1377-1396.

Cochran, D. S.; Reynal, V. A. (1996) Axiomatic Design of Manufacturing Systems. *The Lean Aircraft Initiative*, Report Series, #RP96-05-14.

De Toni, A.; Tonchia, S. (1998) Manufacturing flexibility: a literature review. *Journal of Production Research*, 36, pp. 1587-1617.

Groover, M. P. (2001) *Automation, Production Systems and Computer Integrated Manufacturing*, Prentice Hall, ISBN 0-13-089546-6, New Jersey.

Koren, Y.; Heisel, U.; Jovane, F.; Moriwaki, T.; Pritschow, G.; Ulsoy, G.; Van Brussel, H.; (1999) Reconfigurable Manufacturing Systems. *Annals of the CIRP*, Vol. 48, No. 2, pp. 527-540.

Lee, T.; Jeziorek, P. N. (2006) Understanding the value of eliminating an off-diagonal term in a design matrix. *Proceedings of 4th International Conference on Axiomatic Design, ICAD06*, Florence/Italy, June 2006.

Lotter, B.; Hartel, M.; Menges, R. (1998) *Manuelle Montage wirtschaftlich gestalten*, Expert Verlag, Renningen-Malmsheim/Germany.

Matt, D. T. (2005). Design of self-contained, adaptable factory modules. *Proceedings of CARV 05 International Conference on Changeable, Agile, Reconfigurable and Virtual Production.* Garching/Germany, September 2005.

Matt, D. T. (2006). Fast Configuration of Lean and Customer Driven Production Systems with Axiomatic Designed Production Module Templates. *Proceedings of CIRP-ICME06 – International Conference on Intelligent Computation in Manufacturing Engineering*, pp. 373-378, Ischia/Italy, July 2006.

Matt, D. T. (2007). Reducing the Structural Complexity of Growing Organizational Systems by Means of Axiomatic Designed Networks of Core Competence Cells. *Journal of Manufacturing Systems*, 26, pp. 178-187, doi:10.1016/j.jmsy.2008.02.001

Matt, D. T. (2008). Template based Design of Lean Production Systems. *Journal of Manufacturing Technology Management*, Vol. 19, No. 7, pp. 783-797, doi:10.1108/17410380810898741

Matt, D. T. (2009/a). (Re-)Design to Agility using the Concept of Organisational Periodicity. *Proceedings of CARV 2009 – 3rd International Conference on Changeable, Agile, Reconfigurable and Virtual Production*, pp. 683-692, ISBN 978-3-8316-0933-8, Munich/Germany, October 2009.

Matt, D. T. (2009/b). Design of lean manufacturing support systems in make-to-order production. *Key Engineering Materials*, Vol. 410-411, pp. 151-158, doi:10.4028/www.scientific.net/KEM.410-411.151

Mehrabi, M. G.; Ulsoy, A. G.; Koren, Y. (2000) Reconfigurable manufacturing systems: Key to future manufacturing. *Journal of Intelligent Manufacturing*. Kluwer Academic Publishers, 11, pp. 403-419.

Naylor, J. B.; Naim M. M.; Berry, D. (1999) Leagility: integrating the lean and agile manufacturing paradigms in the total supply chain. *International Journal of Production Economics*, 62, pp. 107-118.

Nyhuis, P.; Kolakowski, M.; Heger, C. L. (2005) Evaluation of Factory Transformability. *Proceedings of CIRP 3rd International Conference on Reconfigurable Manufacturing*, Ann Arbor/MI/USA.

O´Connor, J.; McDermott, I. (1998) *Die Lösung lauert überall - Systemisches Denken verstehen und nutzen*. VAK Verlags GmbH, ISBN 3-932098-29-3, Kirchzarten bei Freiburg.

Reynal V. A., Cochran D. S. (1996) Understanding Lean Manufacturing According to Axiomatic Design Principles. *The Lean Aircraft Initiative*, Report Series, #RP96-07-28.

Rother, M.; Shook, J. (1998) *Learning to See. Value Stream Mapping to Create Value and Eliminate Muda*, The Lean Enterprise Institute, Brookline, MA.

Senge, P. M. (1997) *Die Fünfte Disziplin. Kunst und Praxis der lernenden Organisation*. 4. Aufl., Klett-Cotta, ISBN: 3-608-91379-3, Stuttgart.

Spath, D.; Scholz, O. (2007) Mutability for a Profitable Assembly in Germany (in German). *Industrie Management*, Vol. 23, No. 2, pp. 61-64.

Suarez, F.; Cusumano, M.; Fine, C. (1991) Flexibility and Performance: A Literature Critique and Strategic Framework. *Sloan School WP 3298-91-BPS*, MIT/Cambridge/MA

Suh, N. P. (2001) *Axiomatic Design – Advances and Applications*. Oxford University Press, ISBN 978-0-19-513466-7, New York

Suh, N. P. (2005) *Complexity – Theory and Applications*. Oxford University Press, ISBN 0-19-517876-9, New York.

Suh, N. P. (2006) Application of Axiomatic Design to Engineering Collaboration and Negotiation. *Proceedings of 4th International Conference on Axiomatic Design, ICAD06*, Florence/Italy, June 2006.

Ulrich, H.; Probst, G. (1995) *Anleitung zum ganzheitlichen Denken und Handeln – Ein Brevier für Führungskräfte*, 4. Aufl., Paul Haupt Verlag, ISBN : 978-3-258-05182-6, Bern.

Vester, F. (1999) *Die Kunst, vernetzt zu denken. Ideen und Werkzeuge für einen neuen Umgang mit Komplexität*. Deutsche Verlags-Anstalt (DVA), ISBN-10 3421053081, Stuttgart.

Wiendahl, H.-P., Heger, C. L. (2003) Justifying Changeability - A Methodical Approach to Achieving Cost Effectiveness. *Proceedings of CIRP 2nd International Conference on Reconfigurable Manufacturing*, Ann Arbor/MI/USA.

Zaeh, M. F.; Moeller, N.; Vogl, W. (2005) Symbiosis of Changeable and Virtual Production – The Emperor's New Clothes or Key Factor for Future Success? *Proceedings of CARV 05 International Conference on Changeable, Agile, Reconfigurable and Virtual Production*. Garching/Germany, September 2005.

# 10

# A blended process model for Agile software development with lean concept

Indika Perera
*University of Moratuwa*
*Sri Lanka*

## 1. Introduction

This research addresses a known set of issues with Agile Software Development through a unique form of solution. In fact, the research approach can be considered as one of the first ever research to propose a hybrid process paradigm to overcome Agile process issues with the assistance of Lean manufacturing principles. Research results show a significant improvement for the normal Agile practices, which indeed a unique and worthy finding for Agile practitioners. After years of being practiced in the industry the Agile software development process possesses standard characteristics of a process paradigm (Perera, 2009). However, due to the inherited higher degree of flexibility and the exceptional abstract nature of the process principles, Agile process heavily depends upon the project and people norms once it is implemented. Having more flexibility is a better attribute for a process, if it is used by competent experts who can take productive decisions at right moments. However, depending too much on expert knowledge to process and product adjustments is a questionable concern to a growing project with rapid changes to its development and releases.

Software applications are complex and intangible products, which are difficult to manage. Hence Software Lifecycle management becomes one of the key research areas in software engineering. Due to the nature of the software, software researchers and practitioners are focused on improving the software processes which are used to develop software. The underline assumption is that there is a direct correlation between the quality of the process and the quality of the developed software (Fuggetta, 2000). A software process can be considered as a set of tools, methods and practices, we use to produce a software product (Humphrey, 2006). These are the prime parameters, also known as Key Process Areas (KPAs), that differentiate the process based software development from ad-hoc programming. Identifying KPAs is one of the main considerations when a certain process model to be improved (Fatina, 2005). In this research, the KPAs of Agile practice were studied and reviewed for required improvements, considering criticisms on those. Specially, the driving KPAs of Agile practice such as, the non-standardized process flow, reliance on key people, and immense flexibility were the considerations for this study. Then the Lean principle for possible key practices to incorporate with classical Agile practice was examined.

The remainder of this chapter is arranged into 8 sections as follows: section 2 provides some background literature on the major areas of interest with respect to this research; section 3 describes the research problem this research worked on as an extension to the literature review. The section 4 presents the proposed blended process model as a solution to the research problem being considered. Section 5 and section 6 elaborate the detail on the experiment conducted to evaluate the proposed process model and the analysis of the results obtained, respectively. The Section 6 describes possible policy implications and future work before concluding. Finally, the section 8 with references completes the chapter.

## 2. Background

This section includes a comprehensive synopsis of the literature referred for the study. In fact, the main emphasis was given on the topics; the Agile software development, the Lean principle, and the Lean software development. Therefore, this section is divided into three main areas of literature, representing the focus of the study. There is a plethora of case studies and application stories on Agile software practice and Lean principle in an isolated manner. As the paper explains in the problem section, most of those cases do not emphasize the possible improvements to the two practices to overcome their weaknesses. Further, there are concerns of using these two practices in certain applications and specific cases, considering their weaknesses. For this study, it was considered that the proposed blended process model should be derived upon the fundamental parameters of the two practices, for the simplicity and to obtain a generic process model as the outcome. Hence, the literature focus was decided to be on fundamental concepts of the selected two practices than their applications or customized models.

### 2.1 Agile Software development

Agile software process was introduced and defined by a group of experts in a collective nature, to overcome issues with the traditional software processes. Agile Manifesto was the proper introduction of the Agile methods to the software industry. According to the Agile Manifesto the following four norms are the basics of the Agile methods.

- Individuals and interactions over processes and tools
- Working software over comprehensive documentation
- Customer collaboration over contract negotiation
- Responding to change over following a plan (Agile Manifesto, 2001).

Most of the traditional software processes suffer from having heaps of documents once the project finishes. Despite from those most obvious differences between plan-driven life-cycle models and Agile development is that Agile models are fewer documents oriented and place more emphasis on code development (Perera & Fernando, 2007). By the nature of this paradigm, it also provides some other benefits like, flexible project management, cost effective adaptability (Perera & Fernando, 2009), increase communication and ultimately increased customer satisfaction (Perera & Fernando, 2007). Agile Methods are a reaction to traditional ways of developing software and acknowledge the need for an alternative to documentation driven, heavyweight software development processes (Cohen, *et al.*, 2003). Augustine (2005) has defined Agile Software Development as the work of energizing, empowering, and enabling project teams to rapidly and reliably deliver business value by

engaging customers and continuously learning and adapting to their changing needs and environments.
Agile software development which emphasizes sense-and-respond, self-organization, cross-functional teams, and continuous adaptation, has been adopted by an increasing number of organizations to improve their software development (Lee and Xia, 2010). However, it was observed that when applied to large scale industrial projects, Agile practices fail to keep their stability and performance measures within the expected norms (Perera & Fernando, 2007). This also confirms the fact of the unbalanced number of many successful Agile projects teams with few developers, ideally less than 10 persons. Agile techniques have demonstrated immense potential for developing more effective, higher-quality software. However, scaling these techniques to the enterprise presents many challenges (Shalloway, *at el.*, 2009). One of the main objectives of this study to raise the stability of Agile process with Lean principle, which may help to sustain with large development teams, even though the experiment environment of this research does not incorporate such teams. Moreover, in the Agile world, requirements change rapidly developers expect this and are not fazed by the possibility of having to discard their work and start over (Black, *at el.*, 2009). However, the software process and productivity standards and norms believe that such level of work discard and alterations are essentially impact to the end productivity; more or less it will be compensated either by compromising the customer expectations or more frequently, at the expense of the developer time. This factor was considered as a prime motive when the experiments were designed to assess the productivity of the proposed process model. More detailed analysis on Agile process issues is included in the section 3.

## 2.2 Lean Principle

The Lean principle has a long history of application in Japanese automobile industry, especially within the Toyota manufacturing process. Taiichi Ohno has done a pioneering work to introduce the Toyota Production System with based on Lean concept. Taiichi Ohno and Shigeo Shingo introduced their new concept to the Toyota Production System in result in a significant productivity boost in early 1950 (Ohno, 1988). After few decades of the Lean concept introduction in Japan, many researchers around the world began to investigate possible applications of this concept as a generic production model. The early articles were named as Toyota Production System instead of the name Lean concept/principle, and the first English article was published by Sugimori, *et al.* (1977) on the principles of the Toyota Production System. However, with the book on 'Lean Thinking' by Womack and Jones in 1996 has triggered the momentum on Lean applications to various industries and relevant researches (Womack and Jones, 2003). More interestingly, software development (Poppendieck, 2007) and pharmaceutical (Petrillo, 2007) industries were the early adapters of this new manufacturing concept, spanning across two segments of the commercial arena; the service sector and manufacturing sector, respectively. More discussion on the Lean Software Development is included in the next section.
The Lean principle is based on two practices: the elimination of the waste (Muda) from the production process, and the continuous quality inspection, known as Jidoka, within the production process (Danovaro, *at. el*, 2008). In fact, Jidoka is an operational process which ensures the waste is within the expected amounts or at zero levels. Therefore, essentially, the elimination of waste is the fundamental objective of the Lean principle, although, there are derived and extended applications can be seen, to date.

There are five basic principles of Lean manufacturing: as Specify value, Identify all the steps in the value stream, Flow smoothly, Pull value, and Pursue perfection are the five principles in Lean practice (Womack & Jones, 2003).

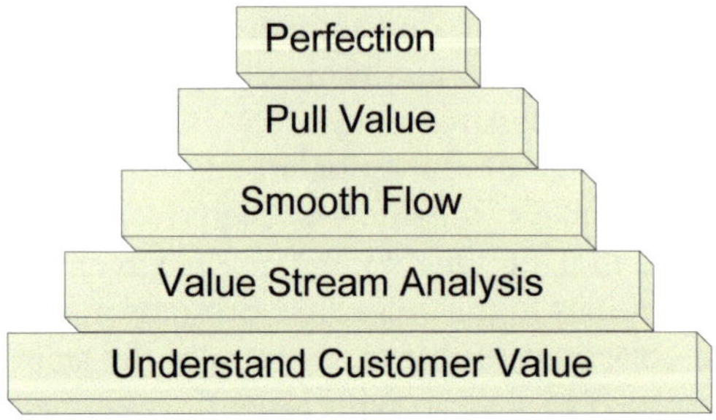


Fig. 1. Lean Principle Model - Five Basic Principles and Their Relationship

**Step 1 - Understand Customer Value**—Value must be externally focused. Only what customers perceive as value is important for the development.
**Step 2 - Value Stream Analysis**— Once the value that is required to deliver to the customers has been identified, you need to analyze all the steps in your business processes to determine which ones actually add value. If an action does not add value, you should consider changing it or removing it from the process.
**Step 3 - Smooth Flow**—Instead of moving the product from one work centre to the next in large batches, production should flow continuously from raw materials to finished goods in dedicated production cells.
**Step 4 - Pull Value** —Rather than building goods to stock, customer demand pulls finished goods through the system. Work is not performed unless the part is required downstream.
**Step 5 - Perfection**—As you eliminate waste from your processes and flow product continuously according to the demands of your customers, you will realize that there is no end to reducing time, cost, space, mistakes, and effort (Womack and Jones, 2003).
Lean principle is composite with unique methodologies to perform the operational activities. Kanban (Pull) production system is one important method (Gross, 2003). In that approach, throughout the production lines one can schedule the value flow process efficiently, and activities flow using signalling to each other with respect to the workflow. In 1953 Toyota applied this logic in their main plant machine shop (Ohno, 1988). Just In Time (JIT) is the basic process Toyota used, and the Kanban is an improved process of the JIT (Kupanhy, 1995). For this research Kanban was identified as a key element of the proposed blended process model to facilitate value flow without an overhead burden to the Agile software practitioner.

### 2.3 Lean Software development

Applying Lean principles to software development projects has been advocated for over ten years, and it will be shown that the extensive Lean literature is a valuable source of ideas for software development (Middleton, *at el.*, 2007). One of the domains affected by the Lean Thinking was the Software Development, which generated the term Lean Software Development (Udo, *at el.*, 2008). The first glimpse of Lean Software Development was appeared with the research work done by Middleton (2001), on two industry case studies of software engineering with Lean implementation. However, Mary Poppendiek and Tom Poppendiek (2003) were the pioneers of introducing a more enhanced software development practice based on Lean principle, which was branded as Lean Software Development.

Lean Software Development mainly focuses on defect minimization within the software development activities. Effectively, the Lean Software Development model has been able to map seven wastes in production systems to software domains; a brief overview of this mapping is shown in the following table 1. It is adapted from the work of Poppendiek & Poppendiek (2003), and Poppendiek (2007).

| Wastes in Production Domain | Corresponding wastes in Software Domain |
|---|---|
| Overproduction | Extra Features |
| Inventory | Requirements |
| Extra processing steps | Extra steps |
| Motion | Finding Information |
| Defects | Defects not caught by tests |
| Waiting | Waiting, including customers |
| Transportation | Handoffs |

Table 1. Corresponding seven types of waste in software development

There are some criticisms on this way of thinking with the software development activities. Specially, even though these seven waste types and the Lean Software Development practices are successful enough to minimize defects of the software development, this abstract way of process mapping does not provide a sufficient level of information for a comprehensive software process practice. In deed that is the key issue with using Lean Software Development practice in large scale industry projects. In fact, Lean Software Development does not incorporate the key software lifecycle activities, such as Requirement Engineering, Software Design, Software Testing and Software Deployment, but the Software Development. Without any of these key steps it is rather inappropriate to consider Lean Software Development as a software process model for generic use.

## 3. The Research Problem

The main purpose of this research was to formulate a new software process paradigm model and evaluate its success. Having said so, let's consider the research problem that this research tries to address. In fact, this research mainly considers Agile practice and Lean concept, as the research's basis. Agile development has significantly impacted industrial software development practices; though it's wide popularity, there's an increasing perplexity on software architecture's role and Agile approaches (Abrahamsson, *at el.*, 2010). The team lead engineers and the software development managers may be unsure how to adopt Agile methods incrementally, which situational practices should perform, and how to engender enthusiasm in team members (Syed-Abdullah, *et al.*, 2007). Chow and Cao (2008) have done a survey study to identify critical success factors in Agile software projects which they have categorized into four major aspects; Quality, Scope, Time, and Cost. This also indicates that precise settings for quality, scope, time and cost will result in a successful Agile software project. The Agile development literature is largely anecdotal and prescriptive, lacking empirical evidence and theoretical foundation to support the principles and practices of Agile development (Lee and Xia, 2010); this also indicates that there is a concern on standard practice of Agile principles and norms across the industry. Mainly, due to the high flexibility and lack of awareness, Agile practitioners interpret their own forms of Agility, where in most of the cases deviate heavily from the optimum Agile best practices. In terms of scalability Agile practices are usually applied to projects with smaller teams of ten or fewer people (Deek, *at el.*, 2005). This limitation is also considered due to the uncertain and highly expert based interpretations and implementations of the basic Agile principles.

There have been many efforts to improve Agile practices but to date some of the Agile process based software projects suffer due to the weaknesses inclusive to the process practices and individual forms of Agile applications. However, there is no successful method to address the behavioural issues with Agile practices and standardizing it for a uniform practice independent from expert judgment, unfortunately. Process improvement offers a sustainable method of making project success probability a significantly higher value irrespective of individual dependencies (Jacobs, 2006). Salo and Abrahamsson (2005) have done an empirical study on Agile software development integration with software process improvement, where they state continuous improvement of Agile software development processes is important in enhancing the capabilities of the project members and the organization as a whole. Miller and Sy (2009) have indicated an extensive summary of factors that affects Agile software processes, where the major concerns are concentrated on aspects such as poor communication, lack of expertise for autonomous development, weak value flow, dependency issues, etc. Essentially, this provides an excellent arena to perform Lean practices along with Agile process as a remedy; The Lean principle more or less address most of these issues in a more flexible manner where agility could not be successful. Lean manufacturing has a proven set of records for flexible and productive manufacturing in many industries. However, Leanness alone may not be appropriate for software development, instead of agility, as the basic Lean model focuses on defect minimization as the prime objective. Narasimhan and others (2006) have indicated that Agile practice could presume leanness but leanness might not presume Agile nature. This is an interesting claim. However, there are no significant further studies done afterwards.

Therefore, as the basic problem domain of this research, the above mentioned Agile process weaknesses have been considered. The research has successfully tried to formulate a blended process model with combining appropriate Lean principles with the Agile software process to improve the Agile process stability, certainty, and productivity without compromising its advantages.

## 4. The Blended Process: Lean-Agile hybrid model

Yusuf and Adeleye (2002) have compared the effectiveness of Lean and Agile manufacturing in UK. Even though, the conclusions were derived that Agile manufacturing slightly outperform Lean manufacturing, there is no comprehensive study have been done on software development context. Further, the research focus on agility and Lean practices in software development have been more or less equally supportive observations for both Agile and Lean software development approaches. Therefore, this research was driven with the prime motivation of combining the best aspects of the both processes to form a blended process model.

Santana (*at el.,* 2009) have indicated that the focus of Agile project learning should be on improving the performance of the ongoing project. This continuous learning with an ongoing project will effectively increase the quality and productivity of the produce being developed. It is important to have a parallel vigilance over the Agile activities, as they are more or less implemented according to the individual project and people norms. Prince and Kay (2003) combined Agile manufacturing with Lean concept to achieve better production flow. Integrating Lean and Agile characteristics becomes an important study on how these philosophies can assist business to prosper (Naylor, *at el.,* 1999), although, software process development researches have not been guided to a reasonable extent so far, unfortunately. The solution for the Agile process poor scalability is to integrate the principles and practices of Lean with Agile ideology and methods (Shalloway, *at el.,* 2009). Moreover, to combine Agile process with Lean practices, it is required to ensure that there will not be redundant process steps in the outcome due to the higher degree of similarity between the two processes being considered. Though Agile and Lean practices are appeared to be similar, there are basic differences between the two in the context they are applied; the difference is in the underlying perspective and philosophy and the mindset (Hibbs, *at el.,* 2009).

As briefly explain in the section 3 above, with this blended process model, the identified key Agile weaknesses were addressed through Lean practices. In that sense, the prime objective of this proposed model was to increase the process stability, developer autonomy, and higher degree of productivity over the classical Agile practices. It was hypothesised that incorporating Lean streamlined routine based activities within Agile development phase would help to achieve these objectives. Specially, the main hypothesis was that through Lean incorporation, developers get more time to focus on their development work than worrying about the process management activities.

### 4.1 Lean-Agile Hybrid Model

As the first step towards the model development, a simple value stream map was developed respective to the Agile process based on the basic classical Agile principles and standard practices. According to Oppenheim (2004), the current-state of the value stream map enables the identification of wastes and possible improvements; hence, using this value stream, possible weak spots were identified comparing the Lean value stream against the Agile. Even though this was a trivial activity, it was one of the crucial steps for the success of the model. One of the major drawbacks with classical Agile value stream map was the higher degree of developer effort to streamline the development process. Literary, this is an overhead task for a developer. Unfortunately, the role of the project manager is not significant at the grass root level of development in an Agile team; hence development team members should perform relevant scheduling and workflow management, individually. This can also affect to their decision making on the technical designs as well. In the proposed model, these possible overhead points were incorporated with relevant Lean steps at the micro level. The underline hypothesis was to inject routine practises of Lean steps into flexible yet weak Agile process points where the blended process will have more certain and stable process points over classical Agile, respectively.

With this understanding of the Agile process weak points, the proposed blended process model was developed with a 4 step Lean model; in fact, one step of the altered Lean model is a combination of the basic steps 'Smooth Flow' and 'Pull Value'. It was identified that these two core Lean principles can be combined and practiced along with the Agile Development phase. Literally, the Development phase of the Agile practice overwhelms significantly the other phases; it further justifies the decision taken to merge these two Lean steps. The proposed model and the classical Agile model are shown in the following figure 2.

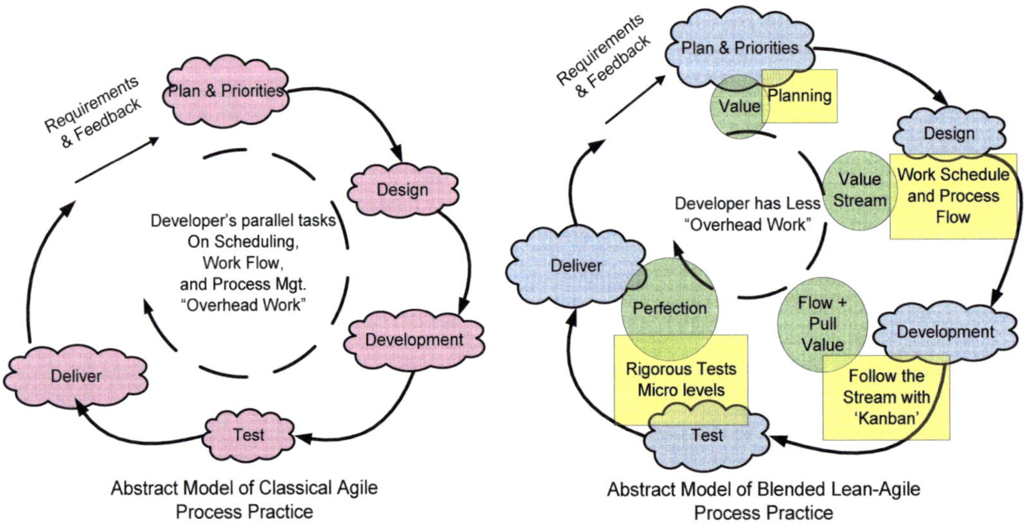


Fig. 2. The Classical Agile Process Model (Left) and the Proposed Lean-Agile Blended Process Model (Right)

The proposed model mainly tries to address the issue of overhead work on an Agile developer, despite the expected Development work. Specially, this is a significant issue which is the fundamental cause for many Agile weakness described above. With the value stream map, it was identified that due to the nature of Process Micro Management by the individual developer this overhead work can be significant, and affects the developer productivity in a substantial manner. With the incorporation of Lean steps the process expects to routines most of the trivial work parallel with the development work, without much effort from the developer end. However, the mere incorporation of the altered Lean steps with relevant Agile activities would not give a practicable model with realistic steps. Therefore, a further step towards process implementation was incorporated in a more tangible manner. The rectangular boxes in the figure represent these steps which ware the linkages with abstract Lean principles and Agile phases as appropriately. The first step - Planning represents the initial action towards the respective iteration of the Agile process, where it covers the value understanding and development priorities for the iteration. The second step - the work schedule and process flow is the effective starting point of the proposed model. There the developers decide how their development work (Value Stream) should be, and they decide the process flow. The most significant improvement can be seen with third step – following stream with 'Kanban', where each developer (or pair) should follow the decided the value stream based work through 'Kanban' cards. A Kanban card is a simple form of signalling mechanism between the workers of a Lean practice. Anybody can define their own Kanban cards according to their requirements. Though it is a simple technique, it has a proven record of success. Specially, a person who follows the process with Kanban does not have to think about the process at all, but the work assigned with that Kanban card. A simple, yet sufficient Kanban card was designed for the experiments. One of the used Kanban cards is shown in the following figure 3.

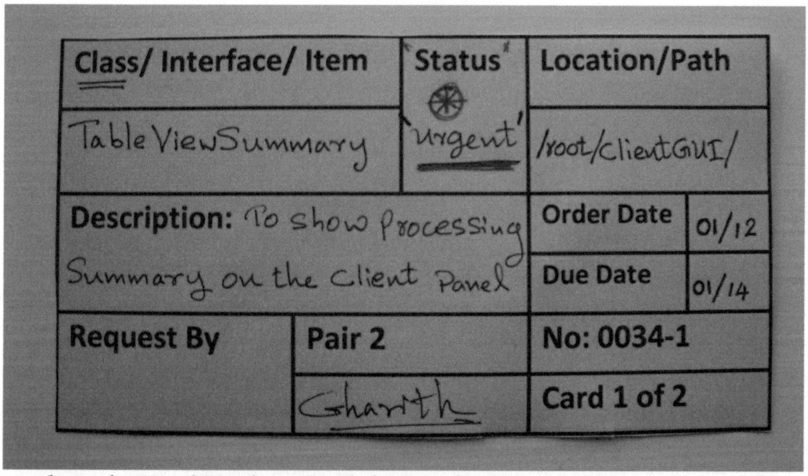


Fig. 3. A Snapshot of a Used Kanban Card during the Experiment

As the final step of the proposed model, a rigorous testing at the micro level was introduced as a perfection norm. This testing effectively higher granular than unit testing, making lesser load on unit testing and sub system testing activities, while giving more opportunities to find errors in the code, specially before they are being camouflaged. Beyond this step, the blended process reiterates with the next cycle of the development similar to the classical Agile process.

## 5. The Experiment Methodology

Conducting an experiment to evaluate a software process is not an easy task. There are various study types that can be performed. In many cases, the type of study will depend on the circumstances. Much of what we do in the software engineering domain is opportunistic, and we are often limited by the situation available (Basili, 2007). However, "The approaches vary in cost, level of confidence in the results, insights gained, and the balance between quantitative and qualitative research methods" (Basili, 1993). First, the experiment methodology should comply with the objectives of the underlined study. Further, the experimental paradigms require an experimental design, observation, data collection and validation on the process or product being studied (Basili, 1993). With these objectives in mind, following experiment environment was designed to evaluate the proposed process model's success over classical Agile process.

### 5.1 The Experiment Environment

This experiment was designed to evaluate the introduced new hybrid software process paradigm's appropriateness in the practical environments. The experiment environment and the performance measures were carefully selected in accordance with the prime objective of the experiment and to suit with the research hypothesis.

The experiment environment was selected in a way to practice the two software paradigms, collect their respective measures, and analyze the collected results. Therefore, it was decided to conduct the experiment with a controlled sample.

The final year student projects of the Department of Computer Science and Engineering (CSE), University of Moratuwa, were the best possible test samples available for this study. Those final year projects created a homogeneous experiment platform among each project. For this study, 10 final year project groups were selected to participate in this experiment in voluntarily basis. Each project group was consisted of 4 final year CSE students.

Before the experiment 4 mini workshops were conducted for the entire sample (10 groups) on the Agile practice of software development. This was to ensure to diminish the knowledge gap on the Agile process between the groups as well as between the group members within a group. After that 5 groups were separated from the rest to follow the new Lean-Agile blended process model, which is considered as the experiment sample. These 5 groups were selected entirely upon the voluntarily basis. For this experiment sample, 3 additional mini workshops were conducted to familiarize the Lean principle and the new process model. However, extra measures were taken on planning and delivering of these 3 mini workshops to ensure no additional Agile process skills were developed on the experiment sample students over the controlled sample. The other five groups were asked to follow the classical Agile practice, which was considered as the controlled experiment sample. While leaving both samples a fortnight to practice their process methods, the experiment phase started. Data gathering was done thereafter for 10 weeks, on per group basis, within the Software Development phase of their projects. A typical working instance model of a group which practiced the proposed blended process is shown in the figure 4 below.

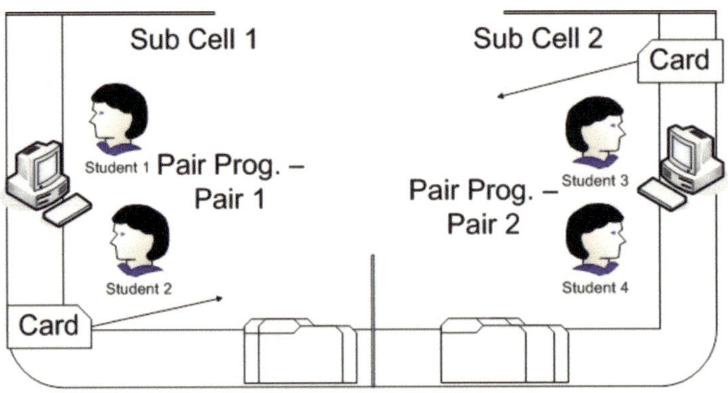


Fig. 4. The Lean-Agile Value Cell - An Instance of a Student Group Development Work

### 5.2 Performance measures

Since the experiment was focused on comparing the two samples to find out which one is better with respect to the software development, a set of essential measurement parameters were identified. One important fact to mention here is that the division of the Lines Of Codes (LOC) parameter to three parts as New LOC, Changed LOC, and Removed LOC. If only total LOC is measured, changes to where the LOC comes from may go undetected (Rozum, 1991). These changes may reflect significant differences in the effort and schedule required to complete the project. Performance measures are shown in the following table 2.

| Measurement | Amount | Work level (Human hours) |
|---|---|---|
| New LOC | N_LOC | H_NLOC |
| Changed LOC | C_LOC | H_CLOC |
| Removed LOC | R_LOC | H_RLOC |
| Defects fixed | D | H_D |
| Expected work | -- | H_EW |
| Actual work | -- | H_AW |

Table 2. Measurement Parameters for the Experiment

The number of defects fixed is important to understand the difference between the two practices in the context of quality enhancements to the developed software. Expected work amount and the actual work amount values were used to compare the effective work completion rate between two paradigms.

All these performance attributes were measured as quantitative values during the examined time period along with their respective work amount in human hours. This human hour measure is very important to understand the difference of work loads in the classical Agile and the improved Agile process model. In addition to that, the human hour values were used to identify the weighted factors for the statistical analysis on the three types of LOC; New, Changed and Removed, measured during the experiment. Furthermore, in some cases, human hour values were used significantly to verify the respective LOC values for their accuracy. One could question the appropriateness of the selected experiment environment and the measurement parameters for this study. However, due to the following reasons, the experiment methodology can be justified steadily.

### 5.3 Experiment Rationalization

Software process related experiments are heavily suffered by the people factor. Process can provide a useful framework for groups of individuals to work together, but process per se cannot overcome a lack of competency (Cockburn, 2001). Individual competency discrepancies on software development can cause varying results in the experiments. But, the selected participants of this experiment have the least skill differences when compared with the other possible participant samples. The same year (final year) students, who have followed a more or less similar set of courses and projects, can be considered as equally skilled developers, compared to generic sample of developers. Industrial software firms always have different competent software developers for their projects, and organization to organization, people competencies differ significantly. It affects to the homogeneity of the sample participants to a significant extent.

Another aspect is the scope of the selected projects. All these projects are worth of 10 GPA credits for the students' graduation. Therefore, initial guidance on project scope was given to the students. In addition to that at the beginning of the project work, a set of experts analyzed those project proposals and ensured to keep the projects within the expected scope for a final year project. If the industrial projects were taken into the experiment, this type of similar scoped projects may not be available. Therefore, it is reasonable to assume all projects have a similar work level required for their completion.

Furthermore, the final year students have equal time and resource constraints while practicing their project with their other academic activities. Specially, this was a key success factor to constrain student development work to have a uniform experiment environment.

All these create the best experiment platform one could ever find to this type of an experiment. If the samples were taken from the industry, this kind of uniformity would not be possible, since different organizations have different resource levels and different schedules for the completion of their projects. As the experiment is sensitive to relative measures, that kind of project environment can create too much deviated results from the tolerant levels.

## 6. Results and Analysis

### 6.1 Analysis – Hypothesis Testing 1

Though the main objective for this research was defined at the beginning, for this analysis a derived hypothesis from the main objective was used. In fact, what has been evaluated in this analysis was the main objective of the study, but using a slightly different hypothesis, merely because to align the analysis with the data and the objectives of the study.

In this analysis, the software development productivity rate was considered as a measure of the effective Line of Codes (LOC) being produced. With that perspective, following hypotheses were defined for the analysis.

**Null hypothesis ($H_0$)** - Agile software development productivity cannot be improved by combining Lean principles

**Alternative hypothesis ($H_1$)** - Agile software development productivity can be improved by combining Lean principles

Since the analysis is based on LOC and the collected data samples have three different LOC values, i.e. New LOC, Changed LOC, and Deleted LOC, weighted summations of those three were derived on per group, per week basis. Considering the human work hours spent on each category and the usefulness to the final code, following three weights were identified. $W_N = 1$ for New LOC, $W_C = -0.5$ for changed LOC, and $W_R = -1$ for removed LOC. Using these weights, 50 data values were derived for a sample and the values are shown in the following tables for the two samples. For a given Week (W*i*) and for a given Group (G*i*), the LOC value was obtained as the equation (1).

$$LOC = W_N*N\_LOC + W_C*C\_LOC + W_R*R\_LOC \quad (1)$$

| | W1 | W2 | W3 | W4 | W5 | W6 | W7 | W8 | W9 | W10 |
|---|---|---|---|---|---|---|---|---|---|---|
| **G1** | 55.5 | 80.5 | 191.5 | 320 | 365 | 409.5 | 373 | 557 | 462 | 668 |
| **G3** | 67 | 107.5 | 145 | 275 | 375 | 377.5 | 420 | 585 | 744.5 | 575 |
| **G4** | 66 | 112 | 168 | 301 | 345 | 397 | 336.5 | 192 | 580.5 | 442.5 |
| **G9** | 52.5 | 187 | 200.5 | 475.5 | 676.5 | 581.5 | 444.5 | 567 | 745 | 944 |
| **G10** | 48 | 203.5 | 364.5 | 481 | 551.5 | 659 | 672.5 | 751.5 | 386 | 908.5 |

Table 3. Sample A (Classical Agile Process) weighted sums of LOC

| | W1 | W2 | W3 | W4 | W5 | W6 | W7 | W8 | W9 | W10 |
|---|---|---|---|---|---|---|---|---|---|---|
| **G2** | 102.5 | 204 | 240 | 391.5 | 486 | 542 | 533 | 533.5 | 680.5 | 540.5 |
| **G5** | 107.5 | 170 | 375 | 494 | 574.5 | 470.5 | 220 | 610 | 501 | 755 |
| **G6** | 97 | 221 | 388.5 | 751 | 692 | 812 | 594.5 | 939 | 688.5 | 779 |
| **G7** | 130.5 | 380.5 | 570.5 | 695.5 | 844.5 | 919 | 879.5 | 1081 | 1088 | 699.5 |
| **G8** | 49 | 342 | 435 | 86 | 920.5 | 1097 | 1084 | 968 | 943 | 981.5 |

Table 4. Sample B (Lean-Agile Process) weighted sums of LOC

To compare the samples with above data and test the hypothesis, Analysis of Variance (ANOVA) statistical method was selected. Instead of manual calculation, the Minitab© statistical application was used. Minitab release 13.20 was the application version which used for this analysis. As the name implies, ANOVA is based on variance analysis between the samples, and it is a widely used statistic model for comparing two or more samples for their means. Actually, the Analysis of Variance (or ***F*-test**), as with Student's ***t*-test**, is fairly robust with respect to violations of the assumptions of normality and homogeneity of variance, so the primary claim is that of equality of means; the alternative hypothesis, then, is that at least one of the population means is different from the others. The ***p*-values** derived from its use are not strongly affected by such violations, as long as the violations are not too extreme (Vokey and Allen, 2002). ANOVA uses the following equation (2).

$$SST = SSW + SSB \tag{2}$$

This is the fundamental equation of ANOVA; the unique partitioning of the total sum of squares (*SST*) into two components: the sum of squares within groups (*SSW*) plus the sum of squares between groups (*SSB*). This is a very abstract model, though the computations of those values are somewhat complex, however, further detail of ANOVA is beyond the scope of this research. The Minitab output for the ANOVA on this hypothesis test is shown in the figure 5, below.

```
Welcome to Minitab, press F1 for help.
Retrieving project from file: I:\PROGRESS2ND\MINITAB.MPJ

One-way ANOVA: C1, C2

Analysis of Variance
Source     DF        SS         MS        F        P
Factor      1    755856     755856    10.23    0.002
Error      98   7238383      73861
Total      99   7994239
                                     Individual 95% CIs For Mean
                                     Based on Pooled StDev
Level       N      Mean      StDev   --------+---------+---------+-------
C1         50     399.9      233.0   (-------*-------)
C2         50     573.8      305.6                     (------*-------)
                                     --------+---------+---------+-------
Pooled StDev =    271.8                   400       500       600
```

Fig. 5. Minitab output - ANOVA output for hypothesis test on effective LOC

Sample A: The mean value $\mu_A$ = 399.9 LOC/week and the standard deviation $\sigma_A$ = 233.0.
Sample B: The mean value $\mu_B$ = 573.8 LOC/week and the standard deviation $\sigma_B$ = 305.6.

From the average developed LOC values for a week, clearly, the blended Agile practice is capable of producing more effective LOC than the classical Agile practice; hence, a higher productivity. However, just by considering the means of the samples, hypothesis tests cannot be done, statistically. With the ANOVA test, the p-value or the significance level was 0.002 for the groups A and B. In ANOVA, to reject the $H_0$ the p-value should be less than 0.05 and if not the $H_0$ will be accepted. In this case, the p-value is 0.002 i.e. $p < 0.05$; therefore, reject the Null hypothesis ($H_0$) with 95% confidence level. This implies that the Agile process development productivity can be improved by applying Lean practices.

### 6.2 Analysis – Hypothesis Testing 1

This analysis is similar to the previous one where in this case, the same derived hypothesis was used. The only difference in this analysis is the data samples were derived using the collected two parameters of expected work and the actual work. In this analysis, the successful achievement levels of the scheduled workloads were used to evaluate the two methods. Once again, the same hypothesis with the LOC analysis was used as follows.

**Null hypothesis ($H_0$)** – Agile process developer productivity cannot be improved through applying Lean principle

**Alternative hypothesis ($H_1$)** – Agile process developer productivity can be improved by using Lean practice techniques

For a given week (Wi) and a given group (Gi), the actual work level was calculated considering the work hour measures. For this analysis, the work done on defect fixing time was not considered, since that time was not scheduled explicitly, in the expected work norms. However, with the results it was obvious that there had been a direct impact from the defect fixing work to the actual work level, making it further deviate from the expected work norm.

Since the analysis was based on achievement level of the scheduled work for a given week, the following model (3) was used to derive the required sample information for the hypothesis testing.

$$\frac{(\text{Actual work level} - \text{Expected work level})}{\text{Expected work level}} * 100\% \qquad (3)$$

Having a positive value as the output from this equation means, in that particular week, the actual work done has exceeded the scheduled or expected work amount. A negative value indicates that the actual work done is less than that of the expected. The values were considered as percentages for the comparison and analysis convenience.

| | W1 | W2 | W3 | W4 | W5 | W6 | W7 | W8 | W9 | W10 |
|---|---|---|---|---|---|---|---|---|---|---|
| **G1** | -39% | -27.5% | -17% | -28% | -9.6% | -7% | -34% | -18.7% | -24.7% | -22.3% |
| **G3** | -30% | -23.3% | -31.4% | -27.7% | -16% | -16% | -38% | -28% | -16% | -13% |
| **G4** | -29% | -25.8% | -34% | -32% | -23.1% | -17.3% | -16.2% | -12% | -15% | -17.6% |
| **G9** | -17.3% | -18% | -35.6% | -22.7% | -18% | -29.3% | -16% | -4% | -15.1% | -27.2% |
| **G10** | -14.7% | -16% | -33.3% | -9.3% | -21% | -29.2% | -12% | -1.2% | -23.2% | -17.7% |

Table 5. Sample A – deviation percentage from expected work amount

| | W1 | W2 | W3 | W4 | W5 | W6 | W7 | W8 | W9 | W10 |
|---|---|---|---|---|---|---|---|---|---|---|
| **G2** | -42% | -34% | 8% | -24% | -21.6% | -8% | -15.2% | -19% | -2.7% | -40% |
| **G5** | -20% | -20% | -2.7% | 1% | -18% | -1.5% | -14% | -24% | -16 | -13.8% |
| **G6** | -14.4% | -2% | -12% | -9% | -3.8% | -5.7% | -26% | -58.7% | -5.3% | -11.7% |
| **G7** | -10.6% | -28% | -9.3% | -18% | -5% | -13.3% | -17.3% | -5.3% | -30% | -22.2% |
| **G8** | -16% | -16% | -7% | -12% | -15.3% | -20% | -26.7% | -17.3% | -32% | -13.3% |

Table 6. Sample B – deviation percentage from expected work amount

The above two tables (Table 5 and 6) show the deviation percentages from the expected work amount for the two samples. These data samples used to test the hypothesis using ANOVA method as done in the previous analysis.

```
Welcome to Minitab, press F1 for help.
Retrieving project from file: C:\DOCUME~1\A\DESKTOP\MINITABACT.MPJ

One-way ANOVA: C1, C2

Analysis of Variance
Source     DF        SS        MS        F        P
Factor      1    0.0672    0.0672     6.07    0.016
Error      98    1.0864    0.0111
Total      99    1.1536
                                Individual 95% CIs For Mean
                                Based on Pooled StDev
Level       N      Mean     StDev  --+---------+---------+---------+----
C1         50   -0.2140    0.0882   (---------*---------)
C2         50   -0.1621    0.1199             (---------*---------)
                                   --+---------+---------+---------+----
Pooled StDev =   0.1053          -0.240    -0.210    -0.180    -0.150
```

Fig. 6. MINITAB output - ANOVA output for hypothesis test on work levels

According to the obtained results from the ANOVA test, following information was obtained for the analysis.

For the sample A: The mean value $\mu_A$ = -21.4% and the standard deviation $\sigma_A$ = 8.82.
For the sample B: The mean value $\mu_B$ = -16.21% and the standard deviation $\sigma_B$ = 11.99.

According to the used equation model, having a higher (towards to the positive values) mean value is better since the deviation from the expected work level is lesser. Furthermore, statistically the ANOVA test resulted in the p-value as 0.016; i.e. $p< 0.05$. This means that the Null hypothesis (**$H_0$**) can be rejected with 95% confidence level, based on achieving the expected effective work level for a given time period. For the work hour measures, since the additional tasks were not incorporated, clearly, the blended process practice allows higher productivity with a higher effective work level. This implies that the Agile process developer productivity can be improved by incorporating Lean practices along with developers' usual Agile process tasks.

### 6.3 Analysis – Defect Rate Behaviour

Apart from the hypothesis testing, the defect fixing rate was also analyzed for the two samples through the examined time period. A defect was classified as an unexpected or erroneous behaviour of a selected piece of module or component, which has been already compiled successfully and committed to the project. With that respect, compilation errors of the code were not considered as defects. The main intention for this analysis was to examine the supportability level towards the work perfection of the two paradigms.

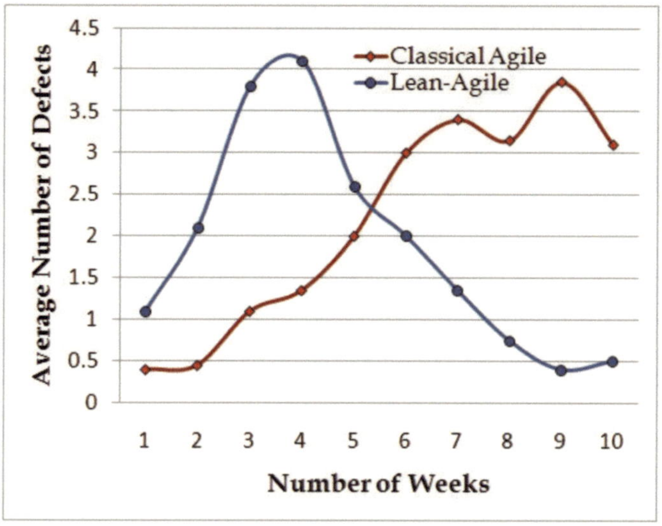


Fig. 7. Average defects rates in the examined time period for the two samples

A significant pattern difference between the defect rates of the two samples was observed during the experiment period, as shown in Figure 7. A higher rate for Lean-Agile sample at the early stages of development was due to that their autonomous and value perfection norms with the development. On the other hand, the classical Agile groups did not find many defects during the early weeks, since they did not pay much attention to the perfection of what they were developing. At the later stages, this situation swapped between the two samples and the Lean-Agile practice seems to have a stable minimal defect rate, where as the classical Agile practice has experienced a high and varying defect rate. A possible explanation to this behaviour is that the unfixed hidden defects in components from early developments would cause emerging defects once they integrated each other. Importantly, having lesser defects in the later stages is very essential for the stability of the project and to be aligned with the project schedule. Furthermore, defects emerge in later stages are relatively expensive to fix than that of the early stages. However, the average defect rates per week for the two samples were close to each other as (Sample A) $\mu_A$ = 2.2, and (Sample B) $\mu_B$ = 1.92. If the study was extended further, this closeness would have changed since the group A is getting further defects with its trend. However, based on the available information it can be concluded that applying Lean principles stabilizes the Agile development phase with respect to quality and perfection, especially in later stages of the development phase.

### 6.4 Experiment Limitations

There have been few experimental limitations were identified with this research, which are mentioned below. However, their impact to the result was not significant enough to create externalities among the data samples; hence to the outcome. One of the key limitations observed with the experiment was that the assumed identical skill level among the students in the sample. It is a known fact that no two humans can have identical skill levels. However, this fact is a generic limitation to all the experiments, which involve human skill based activities. The best possible scenario one could achieve is to have nearly similar skilled people within the sample, i.e. minimize the skill differences as much as possible. In that regard, the selected experiment population is one of the best cases one could find for such an experiment. The reasons for such a strong statement were discussed under the experiment rationalization section. Therefore, the impact of this limitation was minimal to the study.

Another limitation with the study was the truncation errors of the collected data. Literally, what have happened to be the developers were confident on expressing their values with integer figures of hours without the decimal or fractional values. For an example, they might have said their actual work amount as 23 hours, but the precise value may be 23.2 hours or 22.7 hours, etc. This was with the LOC measures as well. If there were extreme cases, which questioned the accuracy of the data additional parameters such as compile time and codebase log files, were used to cross validate the claimed figures, as a sanity check. However, since this is common to both samples of the experiment this was nullified at the end. Furthermore, this type of truncation errors have the normal distribution behaviour where the standard error mean is 0; i.e. the impact at the population level is insignificant.

Another limitation was the domain differences between the projects. Sometimes, domain specific knowledge can be a significant factor for project success. Some of the projects were in different domains, which introduced some impact to the experiments. However, since students have already followed their literature survey and background studies, at the time they engage with software development, every group had a sufficient level of competence on their respective domains, resulted in lesser impact to the experiment outcome.

## 7. Conclusion

This research has introduced significant policy implications to Agile practitioners. First of all, software development activities which follow Agile process, can be considerably benefited through using the proposed process model. In fact, the proposed process model successfully, creates more value oriented, certain, value streamed, and productive software development environment over the classical Agile approach. The research results also reveal a more defect free development activity, essentially in the crucial stages of the development. Importantly, the proposed blended process shows more stability over frequent requirement changes, which is inevitable within an Agile process based software development. The used Lean principles have acted as stabilizing agents within certain Agile practices.

Another possible implication derives from this study is that, like the proposed process practice improves the development works within the software development phase, there is a significant potential to improve the other software lifecycle phases, such as, Requirement Engineering, Design, Testing, and Deployment, even though they are less visible within the Agile practices. In fact, more dominancy on development phase alone, has made the Agile practices more vulnerable to process instability, frequent changes and overhead development works. With the Lean practices, Agile process can have short yet steady Requirement Engineering, Design and Testing phases without affecting to the main development works.

Moreover, the recent hype on Agile manufacturing can also be benefited from the amalgamation of suitable Lean concepts as required. This means, though this study was mainly focused on software industry, it is possible to extend the proposed process model as required for other industries of interest. Specially, the industries of promising future with

Agile manufacturing, could be enhanced the process potentials resulting in fruitful returns. Moreover, the flexibility given in the proposed process model allows practitioners to customize their practices as per the industry norms without reducing the benefits.
It is required a further examine on this proposed process model in a broad spectrum of industrial environments and formulate a standardized process practice for the proposed model. It is crucial to substantially practice the model in a wider range of projects in diversified environments to fine tune the proposed practices. Therefore, it is expected, thus encourage industrial practitioners to use this model widely while interested researchers to research further to improve, standardize and make popular for the benefit of Agile practitioners.

## 8. References


Abrahamsson, P., Babar, M. A., Kruchten, P., (2010), Agility and Architecture: Can They Coexist?, *IEEE Software*, Vol. 27, No. 2, March/April 2010, pp. 16-22, IEEE Press

Agile Manifesto, (2001), Manifesto for Agile software development, [available at] http://Agilemanifesto.org/, [accessed on 19th December 2009]

Augustine, S., (2005), *Managing Agile Projects*, Robert C. Martin series, Prentice Hall Publishers

Basili, V., (1993) , The Experimental Paradigm in Software Engineering," *in LNCS 706, Experimental Software Engineering Issues: Critical Assessment and Future Directives*, H.D. Rombach, V. Basili, and R. Selby, eds., *Proceedings of Dagstuhl-Workshop, September 1992*, Springer-Verlag,.

Basili, V., (2007), The Role of Controlled Experiments in Software Engineering Research, *in Empirical Software Engineering Issues, LNCS* 4336, V. Basili et al., (Eds.), Springer-Verlag, pp. 33-37

Black, S., Boca, P.P., Bowen, J.P., Gorman, J., Hinchey, M., (2009), Formal Versus Agile: Survival of the Fittest?, *Computer, IEEE Press*, Vol. 42, pp. 37-45

Cockburn, A., Highsmith, J., (2001), Agile software development: the people factor, *IEEE Computer*, pp 131-133.

Chow, T., Cao, D., (2008), A survey study of critical success factors in Agile software projects, *Journal of Systems and Software*, Vol. 81, Issue 6, pp. 961-971

Cohen, D., Lindvall, M., Costa, P. (2003), *A State of the Art Report: Agile Software Development*, Data and Analysis Center for Software 775 Daedalian Dr. Rome, New York 13441-4909, p. 01

Danovaro, E., Janes, A., Succi, G. (2008), Jidoka in software development, *In Companion To the 23rd ACM SIGPLAN Conference on Object-Oriented Programming Systems Languages and Applications, OOPSLA Companion '08. ACM*, pp. 827-830.

Deek, F. P., McHugh J. A. M., O. M. Eljabiri, (2005), *Strategic Software Engineering an Interdisciplinary Approach*, Auerbach Publications, FL, USA, p. 94

Fatina, R., (2005), *Practical Software Process Improvement*, Artech House, Boston, p. 06

Fuggetta, A., (2000), Software Process: A Roadmap, *in Proc. of the Conference on the Future of Software Engineering, ICSE*, Limerick, pp. 25-34

Gross, J. M., McInnis, K. R., *Kanban Made Simple: Demystifying and Applying Toyota's Legendary Manufacturing Process*, AMACOM, 2003

Hibbs, C., Jewett, S., Sullivan, M., (2009), *The Art of Lean Software Development: A Practical and Incremental Approach*, O'reilly Media, CA, USA

Humphrey, W. S., (2006), *Managing the Software Process*, SEI, Pearson Education, India, p. 03

Jacobs D., (2006), *Accelerating Process Improvement Using Agile Techniques*, Auerbach Publications, FL, USA

Kupanhy, L., (1995), Classification of JIT techniques and their implications, *Industrial Engineering*, Vol. 27, No.2

Lee, G., Xia, W., (2010), Toward Agile: An Integrated Analysis of Quantitative and Qualitative Field Data, *MIS Quarterly*, Vol.34, No.1, pp.87-114

Middleton, P., (2001), Lean Software Development: Two Case Studies. *Software Quality Journal*, Vol.9, No.4, pp. 241-252

Middleton, P., Taylor, P. S., Flaxel, A., Cookson, A., (2007), Lean principles and techniques for improving the quality and productivity of software development projects: a case study, *International Journal of Productivity and Quality Management*, Vol. 2, No. 4, Inderscience publishers, pp. 387-403

Miller, L. Sy, D. 2009. Agile user experience SIG, *In Proc. of the 27th International Conference Extended Abstracts on Human Factors in Computing Systems, CHI '09*. ACM, New York, NY, pp. 2751-2754

Narasimhan, R., Swink, M., Kim, S.W., (2006), Disentangling leanness and agility: An empirical investigation, *Journal of Operations Management*, Vol. 24, No.5, pp. 440–457

Naylor, J.B., Naim, M.M., Berry, D., (1999), Leagility: Integrating the Lean and Agile manufacturing paradigms in the total supply chain, *International Journal of Production Economics*, Vol. 62, No. (1/2), pp. 107–118.

Ohno, T. (1988), *Toyota Production System: Beyond Large-Scale Production*, Productivity Press, Cambridge, MA, USA

Oppenheim, B. W., (2004), Lean product development flow, *Systems Engineering*, Vol.7, No. 4, pp. 352-376

Perera, G.I.U.S., (2009), Impact of using Agile practice for student software projects in computer science education, *International Journal of Education and Development using Information and Communication Technology (IJEDICT)*, Vol. 5, Issue 3, pp.83-98

Perera, G.I.U.S. and Fernando, M.S.D. (2007), Bridging the gap - Business and information systems: A Roadmap, *In Proc. of 4th ICBM conference*, pp. 334-343.

Perera, G.I.U.S. and Fernando, M.S.D. (2007), Enhanced Agile Software Development — Hybrid Paradigm with LEAN Practice, *In Proc. of 2nd International Conference on Industrial and Information Systems, ICIIS 2007*, IEEE, pp. 239 – 244.

Perera, G.I.U.S. & Fernando, M.S.D., (2009) Rapid Decision Making For Post Architectural Changes In Agile Development – A Guide To Reduce Uncertainty, *International Journal of Information Technology and Knowledge Management*, Vol. 2, No. 2, pp. 249-256

Petrillo, E. W., (2007), Lean thinking for drug discovery - better productivity for pharma. *DDW Drug Discovery World*, Vol. 8, No.2, pp. 9–16

Poppendieck, M., (2007), Lean Software Development, *29th International Conference on Software Engineering (ICSE'07)*, IEEE Press

Poppendieck, M., Poppendieck, T., (2003), *Lean Software Development: An Agile Toolkit* (The Agile Software Development Series), Addison-Wesley Professional

Prince, J., Kay J.M., (2003), Combining Lean and Agile characteristics: Creation of virtual groups by enhanced production flow analysis, *International Journal of Production Economics*, Vol. 85, No. 3, pp. 305–318

Rozum, J. A., (1991), Defining and understanding software measurement data, *Software Engineering Institute,*

Salo, O., Abrahamsson, P., (2005), Integrating Agile Software Development and Software Process Improvement: a Longitudinal Case Study, *2005 International Symposium on Empirical Software Engineering, IEEE press,* pp. 193-202

Santana, C., Gusmão, C., Soares, L., Pinheiro, C., Maciel, T., Vasconcelos, A., and A. Rouiller, (2009), Agile Software Development and CMMI: What We Do Not Know about Dancing with Elephants, P. Abrahamsson, M. Marchesi, and F. Maurer (Eds.): *XP 2009, LNBIP 31*, Springer-Verlag, Berlin Heidelberg, pp. 124 - 129

Shalloway, A., Beaver, G., Trott, J. R., (2009), *Lean-Agile Software Development: Achieving Enterprise Agility*. 1st. Addison-Wesley Professional

Sugimori, Y., Kusunoki, K., Cho, F., Uchikawa, S., (1977), Toyota production system and Kanban system: materialisation of just-in-time and respect-for-human system, *International Journal of Production Research,* Vol. 15, No.6, pp.553–564.

Syed-Abdullah, S., Holcombe, M., Gheorge, M., (2007), The impact of an Agile methodology on the well being of development teams, *Empirical Software Engineering*, 11, pp. 145–169

Udo, M., Vaquero, T. S., Silva, J. R., and Tonidandel, F., (2008) Lean software development domain, *In Proc. of ICAPS 2008 Scheduling and Planning Application workshop,* Sydney, Australia

Vokey, J. R., Allen S. W., (2002), *Thinking with Data*, 3rd Ed., PsyPro, Alberta

Womack J. P., Jones, D.T., (2003), *Lean Thinking: Banish Waste and Create Wealth in Your Corporation*, New Ed., Free Press, UK

Yusuf, Y.Y., Adeleye, E.O., (2002), A comparative study of Lean and Agile manufacturing with a related survey of current practices in the UK, *International Journal of Production Research*, Vol. 40, No.17, pp. 4545–4562.

# 11

# Process Monitoring Systems for Machining Using Audible Sound Energy Sensors

Eva M. Rubio and Roberto Teti
*National Distance University of Spain (UNED)*
*Spain*
*University of Naples Federico II*
*Italy*

## 1. Introduction

In the last fifty years, many manufacturers have chosen the implementation of Flexible Manufacturing Systems (FMS) or Computer Integrated Manufacturing (CIM) in their shop floor or, at least, the automation of some of the operations carried out therein with the intention of increasing their productivity and becoming more competitive (Shawaky, 1998; Sokolowski, 2001; Cho, 1999; Govekar, 2000; Brophy, 2002).

With reference to machining operations, the implementation of these systems requires the supervision of different aspects related to the machine (diagnostic and performance monitoring), the tool or tooling (state of wear, lubrication, alignment), the workpiece (geometry and dimensions, surface features and roughness, tolerances, metallurgical damage), the cutting parameters (cutting speed, feed rate, depth of cut), or the process itself (chip formation, temperature, energy consumption) (Byrne, 1995; D'Errico, 1997; Tönshoff, 1988; Grabec, 1998; Inasaki, 1998; Kopac, 2001; Fu, 1996; Masory, 1991; Huang, 1998; Teti, 1995; Teti, 1999).

For the monitoring and control of the above mentioned aspects, it has been necessary to make notable efforts in the development of appropriate process monitoring systems (Burke & Rangwala, 1991; Chen *et al.*, 1994; Chen *et al.*, 1999; Chen, 2000). Such systems are typically based on different types of sensors such as cutting force and torque, motor current and effective power, vibrations, acoustic emission or audible sound (Desforges, 2004; Peng, 2004; Lin, 2002; Sokolowski, 2001; Ouafi *et al.*, 2000; Karlsson *et al.*, 2000; Chen & Chen, 1999; Jemielniak *et al.*, 1998; Byrne, 1995; Dornfeld, 1992; Masory, 1991). However, despite all the efforts, standard solutions for their industrial application have not been found yet. The large number and high complexity of the phenomena that take place during machining processes and the possibility to choose among numerous alternatives in each implementation step of the process monitoring system (e.g. cutting test definition, type and location of sensors, monitoring test definition, signal processing method or process modeler selection) are the main responsible for the existence of more than one solution.

The review and analysis of the relevant literature on this topic revealed that it is necessary to develop and implement an experimental system allowing for the systematical characterizarion of the different parameters that influence the process before realizing a process monitoring system applicable to industry (Hou, 2003; Jin & Shi, 2001; Hong, 1993; Malakooti *et al.*, 1995; Venkastesh *et al.*, 1997; Xiaoli *et al.*, 1997; Xu & Ge, 2004). This will allow to establish an adequate knowledge and control of the critical factors involved in the process monitoring system by means of single factor variations. Moreover, it will be also possible to identify the variations produced by potential spurious sources when the process monitoring system is applied to real situations in the shop floor.

This work reports on the approach for the development of a machining process monitoring system based on audible sound sensors. Audible sound energy appears as one of the most practical techniques since it can serve to replace the traditional ability of the operator, based on his experience and senses (mainly vision and hearing), to determine the process state and react adequately to any machine performance decay (Lu, 2000). This technique has been attempted for decision making on machining process conditions but it has not been extensively studied yet for applications in industrial process monitoring (Teti, 2004; Teti & Baciu, 2004). The main critical issues related to the employment of this technology in industry are the need to protect the sensor from the hazardous machining environment (cutting fluids and metal chips) and the environment noise (from adjacent machines, motors, conveyors or other processes) that may contaminate the relevant signals during machining (Lu, 2000; Teti & Baciu, 2004; Teti *et al.*, 2004; Wilcos, 1997; Clark, 2003).

The principal benefits of audible sound sensors for machining process monitoring are associated with the nature of the sensors employed in the acquisition of the signals. These are, in general, easy to mount on the machine tool, in particular near the machining point, with little or no interference with the machine, the tool, the workpiece or the chip formation. Besides, these sensors, basically microphones, are easy to use in combination with standard phonometers or spectrum analysers. These characteristics of audible sound sensors make the realization of the monitoring procedure quite straightforward. In addition, their maintenance is simple since they only require a careful handling to avoid being hit or damaged. Accordingly, they usually provide for a favourable cost/benefit ratio.

The key novelties of the approach proposed in this work are, on the one hand, the application of a systematic methodology to set up the cutting trials allowing for a better comparison with other similar experimental works and, as a result, the advance in the standardization for the development of such systems. On the other hand, the independent signal analysis of the noise generated by the machine used for the cutting trials and by the working environment allows to filter this noise out of the signals obtained during the actual material processing. Lastly, the possibility has been verified to apply the results of this approach for the development of process monitoring procedures based on sensors of a different type, in particular acoustic emission sensors, where the stress waves produced within the work material do not travel through air but only in the work material itself. The combined application of audible sound energy sensors and acoustic emission sensors could allow for the acquisition of more exhaustive information from both low frequency (audible sound) and high frequency (acoustic emission) acoustic signal analysis. This would decidedly contribute to the realization of the concept of sensor fusion technology for process monitoring (Emel, 1991; Niu *et al.*, 1998).

The described methodology was applied to characterize the audible sound signals emitted by different cutting conditions during milling processes. The classification of audible sound signal features for process monitoring in milling was carried out by graphical analysis and parallel distributed data processing based on artificial neural networks. In the following sections, the methodology, the experimentation, the sensor signal detection and analysis methods, and the obtained results are reported and critically assessed.

## 2. Methodology

The methodology proposed for the design and implementation of a process monitoring system based on audible sound energy sensors includes the steps described below.

*Cutting tests definition.* All the elements involved in the cutting tests, along with their basic characteristics and properties, should be defined in this step, as reported in the systematic methodology proposed in (Rubio & Teti, 2005) for the establishment of tool condition monitoring systems. In particular, the cutting operation, the machine tool, the workpiece (material and size), the tools (type, material, coating, dimensions and fresh/worn state), the cutting parameters (cutting speed, feed rate, depth of cut) and the possible use of cutting fluid, should be defined. Although this seems obvious and there are in the literature works that report thorough descriptions of the cutting tests (Teti & Buonadonna, 1999), most of the authors do not provide, or not with the desired detail, all the necessary information to allow for a correct analysis of the results and an adequate comparison with the results obtained by other authors.

*Process monitoring tests definition.* The monitoring tests dealt with in this work are based on the use of audible sound energy sensors. The broadband sound pressure level of the audible signals is detected by means of sensing devices dedicated to the measure and display this type of signals. All detected audible sound signals are transferred on PC and off-line analysed. In order to verify the repeatability of the monitoring tests, the audible sound signal specimens should be recorded several times ($\geq$ 3) for each cutting condition. The noise of the machine tool running unloaded should be recorded as well in order to be able, later, to characterise the audible sound signals from the cutting process deprived of the disturbing noise generated by both machine and working environment.

*Selection of signal processing and decision making methods.* To select the most adequate signal processing and decision making methods, a review of the main advanced signal processing (Rubio *et al.*, 2006a) and decision making procedures (Rubio *et al.*, 2006b) used in machining process monitoring based on acoustic sensors was carried out. As a result, the Fast Fourier Transform (FFT) was selected for signal processing and feature extraction whereas supervised Neural Network (NN) paradigms were adopted for signal feature pattern recognition and process conditions decision making.

*Experimental layout.* The most essential aspects of the experimental layout concern the audible sound sensor location and protection: firstly, the selection of the distance between sensor and cutting point in order to detect the signals correctly, and, secondly, the way to protect the sensor from the chips, the cutting fluid and other pollutants during machining. Besides these actions, particular attention must be paid to isolate the experiments from environmental noise that could seriously contaminate the signal detection.

*Performance of the cutting and process monitoring tests.* Once all the previous steps have been completed, the machining tests with process monitoring must be carried out. As stated earlier, the tests should be rehearsed several times in order to verify their repeatability. Furthermore, the noise of the machine tool running unloaded should be recorded for its later subtraction from audible sound signals detected during the material removal process.

*Signal processing and decision making.* After the sensor monitoring tests, the processing and analysis of the recorded signals by means of the methods selected earlier must be carried out together with the decision making procedure applied to significant signal features: in this work, the FFT for signal processing and supervised NN paradigms for decision making.
*Design and implementation of the process monitoring system.* On the basis of the issues of the previous steps, the implementation procedure for an on-line machining process monitoring system based on audible sound energy sensors can be proposed.

## 3. Application

According to the methodology described in the previous section, experimental applications were carried out as outlined below.
*Cutting tests definition.* Following the methodology for the definition of the cutting tests (Rubio & Teti, 2005), the machining operation was defined as a milling process carried out on a conventional DORMAC FU-100 milling machine. The workpiece was a plate of size of 100 x 200 x 40 mm made of T4-6056 Al alloy. The tool was a fresh 5-teeth milling cutter of 12.16 x 8.18 x 5.16 mm, made of WC-Co inserts coated with TiN. The cutting conditions were: spindle speed, $S$ = 800 and 1000 rpm; feed rate, $f$ = 40, 80 and 160 mm/min and depth of cut, $d$ = 0.5 and 1 mm. The tests were conducted under dry cutting conditions. Table 1 summarizes the cutting test description.

| Element | Type/ Characteristics/Properties |
|---|---|
| Cutting operation | Milling |
| Machine Tool | Conventional: DORMAC FU-100 milling machine |
| Workpiece | Material: 6056 aluminium alloy with T4 thermal treatment<br>Dimensions: 100 x 200 x 40 mm |
| Tool | Type: 5-teeth milling cutter<br>Material: tungsten particles and cobalt matrix carbide (WC-Co)<br>Coat material: titanium nitride (TiN)<br>Dimensions: 12,16 x 8,18 x 5,16 mm<br>State: Fresh |
| Cutting conditions | Cutting speed, $S$ = 800 - 1000 rpm<br>Feed rate, $f$ = 40 – 80 - 160 mm/min<br>Depth of cut, $d$ = 0.5 - 1 mm |
| Coolant | No |

Table 1. Summary of the cutting test description.

*Process monitoring tests definition.* The audible sound energy monitoring system was composed of a Larson Davis 2800 Spectrum Analyser, a standard Larson Davis preamplifier model PRM 900B, a ½" free field high sensitivity sensor and a ½" pre-polarized microphone (Fu, 1996). All audible sound signals detected by the Larson Davis 2800 Spectrum Analyser were transferred on PC for off-line analysis.

*Selection of signal processing and decision making methods.* The selected signal processing and feature extraction method was the FFT and the signal features pattern recognition for decision making was based on supervised NN data processing since this approach had been used in previous works with satisfactory results (Teti, 2004; Teti & Baciu, 2004).
*Experimental layout.* Figure 1 shows the experimental layout. The distance between the microphone and the cutting point was set in such a way that, during each machining operation, was approximately equal to 85 mm. Particular attention was paid to protect the microphone from the chips by means of a plastic mesh and to isolate the experimental area from environment noise that could contaminate the detected signals.

Fig. 1. Experimental layout.

*Performance of the cutting and process monitoring tests.* The experimental tests carried out with the different cutting conditions are reported in Table 2. Each test was rehearsed 3 times in order to check for repeatability. Simultaneously, the sensor monitoring procedure was applied during each test.
*Signal processing and decision making.* The spectrum analyser was set to 800 lines acquisition mode and a FFT zoom was set equal to 2. In this way, as the capture interval was from 0 to 10000 Hz, by dividing this frequency interval into 800 lines, a step of 12.5 Hz was achieved. Besides the audible sound signal detected in sound Level Meter mode, a series of signal parameters (SUM (LIN) SUM (A), SLOW, SLOW MIN, SLOW MAX, FAST, FAST MIN, FAST MAX, IMPULSE, LEQ, SEL, PEAK, Tmax3 and Tmax5) were obtained and recorded as well. The option "by time" allowed to save the measurements automatically, with end time equal to 10 seconds and step equal to 1 second. The transfer velocity was set at 9600 Baud, which was the same as the velocity imposed to the PC for file transfer. For graphical data processing and display, Spectrum Pressure Level-Noise (Spectrum Pressure Lave, 1998) and Vibrations Works (OS Windows) (Noise and Vibrations Works, 1998) and CA Cricket Graph III (OS Mac) (CA-Cricket Graph III,1992) software packages were used. For NN data processing, the Neural Network Explorer software package was used (Masters, 1993).

| Test Id. | *S (rpm)* | *f (mm/min)* | *d (mm)* |
|---|---|---|---|
| 1 | 800 | --- | --- |
| 2 | 800 | --- | --- |
| 3 | 800 | --- | --- |
| 4 | 1000 | --- | --- |
| 5 | 1000 | --- | --- |
| 6 | 100 | --- | --- |
| 7 | 800 | 40 | 0.5 |
| 8 | 800 | 40 | 0.5 |
| 9 | 800 | 40 | 0.5 |
| 10 | 800 | 80 | 0.5 |
| 11 | 800 | 80 | 0.5 |
| 12 | 800 | 80 | 0.5 |
| 13 | 800 | 160 | 0.5 |
| 14 | 800 | 160 | 0.5 |
| 15 | 800 | 160 | 0.5 |
| 16 | 800 | 40 | 1 |
| 17 | 800 | 40 | 1 |
| 18 | 800 | 40 | 1 |
| 19 | 800 | 80 | 1 |
| 20 | 800 | 80 | 1 |
| 21 | 800 | 80 | 1 |
| 22 | 800 | 160 | 1 |
| 23 | 800 | 160 | 1 |
| 24 | 800 | 160 | 1 |
| 25 | 1000 | 40 | 0.5 |
| 26 | 1000 | 40 | 0.5 |
| 27 | 1000 | 40 | 0.5 |
| 28 | 1000 | 80 | 0.5 |
| 29 | 1000 | 80 | 0.5 |
| 30 | 1000 | 80 | 0.5 |
| 31 | 1000 | 160 | 0.5 |
| 32 | 1000 | 160 | 0.5 |
| 33 | 1000 | 160 | 0.5 |
| 34 | 1000 | 40 | 1 |
| 35 | 1000 | 40 | 1 |
| 36 | 1000 | 40 | 1 |
| 37 | 1000 | 80 | 1 |
| 38 | 1000 | 80 | 1 |
| 39 | 1000 | 80 | 1 |
| 40 | 1000 | 160 | 1 |
| 41 | 1000 | 160 | 1 |
| 42 | 1000 | 160 | 1 |

Table 2. Cutting test parameters.

*Design and establishment of the process monitoring system.* Once the audible sound signals have been fully characterized for each of the diverse cutting conditions, it becomes possible to compare these reference signals with the new ones detected during the normal process operation in such a way that the differences between reference signals and current signals allow for the reliable sensor monitoring and control of the machining process. The target is to achieve an on-line monitoring system using as reference the signals conditioned through machine tool and working environment noise filtering and suppression.

## 4. Results

After audible sound signals detection, the repeatability of the tests was verified by calculating the differences between recorded signals and dividing the result by 800 (number of acquisition lines of the spectrum analyser). All the computed values were less than 5%. Then, a reference signal for the machine and environment noise was established as the average of the 3 signals obtained from each of the unloaded machine tool running tests. Figure 2 shows the reference signal in terms of amplitude, *Sa* (dB), versus frequency, *f* (Hz), for the 5th second of the cutting test with *S* = 800 rpm and *f* = 80 mm/min. Along with the reference signal for the machine and environment noise, the average signals for *d* = 0.5 mm and *d* = 1 mm under the same *S* and *f* conditions were plotted as well.

The reference signal was subtracted from the audible sound signals detected during the actual machining tests to obtain a "difference signal" for classification analysis. All further analyses were carried out using these difference signals (Figure 3).

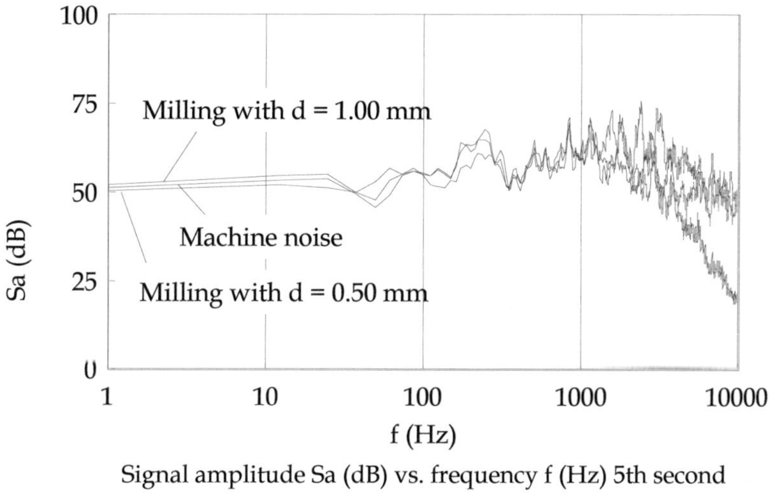


Fig. 2. Signal amplitude *Sa* (dB) vs. frequency *f* (Hz) of the audible sound signals for the 5th second of each test. Namely, milling with *S* = 800 rpm, *f* = 80 mm/min, *d* = 0.5 mm; milling with *S* = 800 rpm, *f* = 80 mm/min, *d* = 1 mm, and machine tool running unloaded at *S* = 800 rpm.

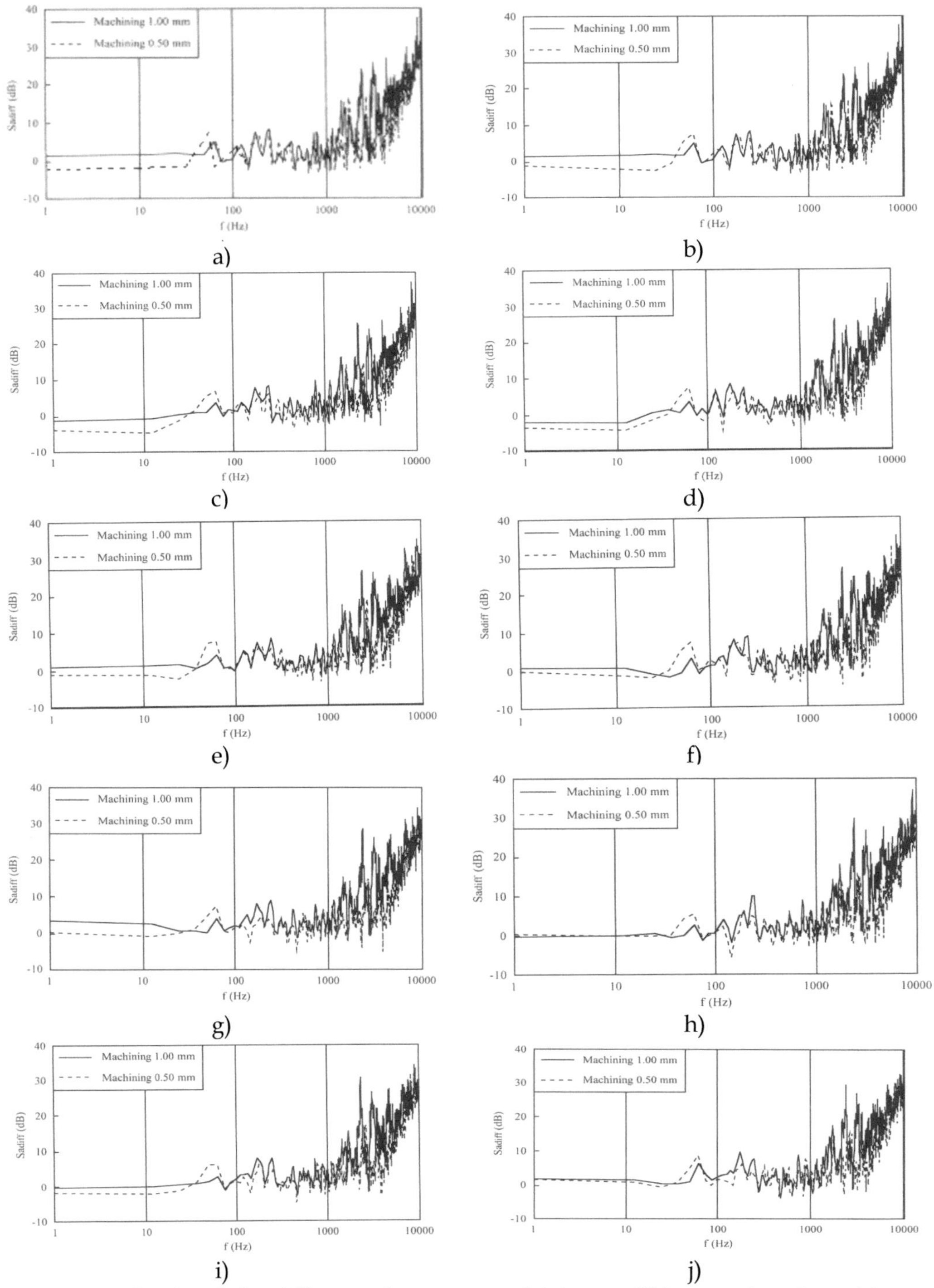


Fig. 3. Amplitude of the difference between machining audible sound and machine tool noise ("difference signal") for each of the ten seconds of cutting test: a) first; b) second; c) thrird; d) fourth; e) fifth; f) sixth; g) seventh; h) eighth; i) ninth; j) tenth second.

The maximum amplitude of the "difference signal" was evaluated for each frequency interval and for each second of cutting test. The six frequency intervals selected for audible sound signal processing were: 0-0.25, 0.25-0.5, 0.5-1, 1-2.5, 2.5-5, 5-10 kHz. Figure 4 reports examples of the "difference signal" maximum amplitude $Sa\ diff_{MAX}$ (dB) versus frequency intervals $\Delta f$ (Hz) for cutting tests with $S$ = 800 rpm, $f$ = 80 mm/min and $d$ = 0.5 mm or 1 mm cases, for each of the ten seconds of each cutting test. The figure shows that for frequency values higher than 1 kHz it is possible to discriminate audible sound signals obtained from machining with different depth of cut values.

Graphical representation of data in high dimensions (> 3) feature spaces is not feasible. Thus, the results are presented in a 2 dimensions feature space by pair-wise plotting of frequency intervals maximum signal amplitude as shown in Figure 5 for two low frequency intervals, in Figure 6 for two medium frequency intervals, and in Figure 7 for two high frequency intervals. The figures show that for the two high frequency intervals the separation between cluster points characteristic of the two depth of cut values is very good. The same can be seen if the "difference signal" maximum amplitude is plotted versus depth of cut as shown in Figure 8 for low, medium and high frequency intervals.

At low frequencies (0-0.25 kHz; 0.25-0.5 kHz), the $Sa\ diff_{MAX}$ value is around 10 dB for both depth of cut values (0.5 and 1 mm). In this case, depth of cut discrimination is unfeasible. However, at high frequencies (1-2.5 kHz; 2.5-5 kHz) the $Sa\ diff_{MAX}$ value is around 10 dB for depth of cut 0.5 mm and around 30 dB for a depth of cut 1 mm and recognition becomes feasible.

A supervised NN data processing was utilized for pattern recognition using the 6-component feature vectors made of the "difference signal" maximum amplitudes for the 6 frequency intervals. A three-layers feed-forward back-propagation NN was built with the following configuration: input layer with 6 nodes; hidden layer with 3 nodes determined by the cascade learning procedure (Teti & Buonadonna, 1999); output layer with 1 node.

The 6-3-1 NN was trained and tested according to the leave-k-out procedure with k = 2 (Teti & Buonadonna, 1999), using a number of learning steps comprised between 1000 and 14000. In Figure 9, the NN output is reported versus the number of input patterns for 12000 and 14000 learning steps. From this figure, it can be seen that the NN Success Rate (SR) in the identification of depth of cut becomes 100% after 14000 learning steps.

Figure 10 reports the NN SR versus learning steps for different treshold values. From the figure, it can be noted that the NN SR is 85% as early as 2000 learning steps.

Figure 11 reports the NN SR versus threshold value for variable numbers of learning steps. From the figure, it can be observed that the NN SR starts decreasing gradually only for threshold values < 0.3, except in the case of the lowest number of learning steps (i.e. 1000) for which a rapid SR reduction is expectedly verified.

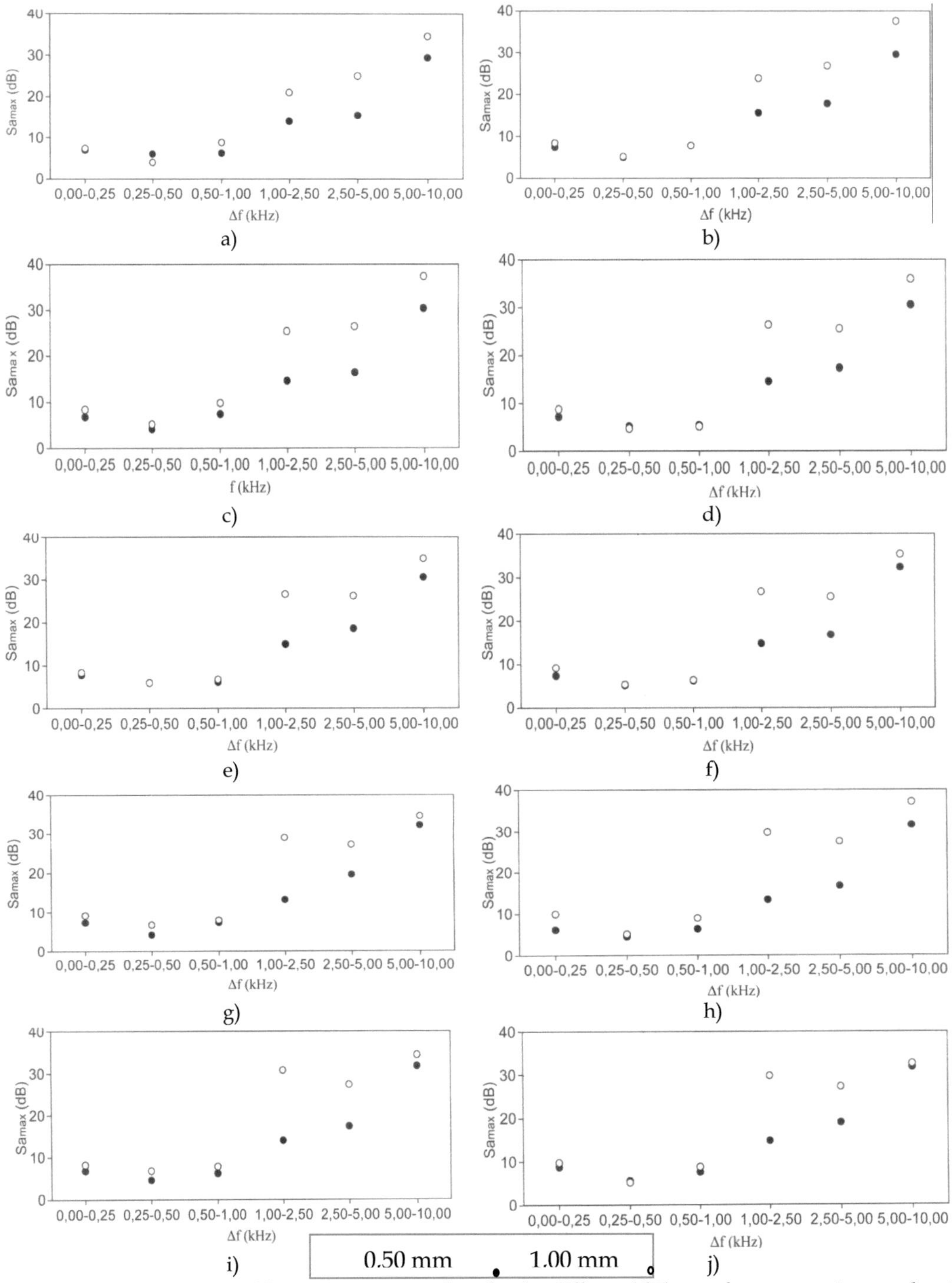


Fig. 4. "Difference signal" maximum amplitude *Sa* $diff_{MAX}$ (dB) vs. frequency intervals $\Delta f$ (Hz) for the $S$ = 800 rpm, $f$ = 80 mm/min, and $d$ = 0.5 or 1 mm cases, for each of the ten seconds of cutting test: a) first; b) second; c) third; d) fourth; e) fifth; f) sixth; g) seventh; h) eighth; i) ninth; j) tenth second.

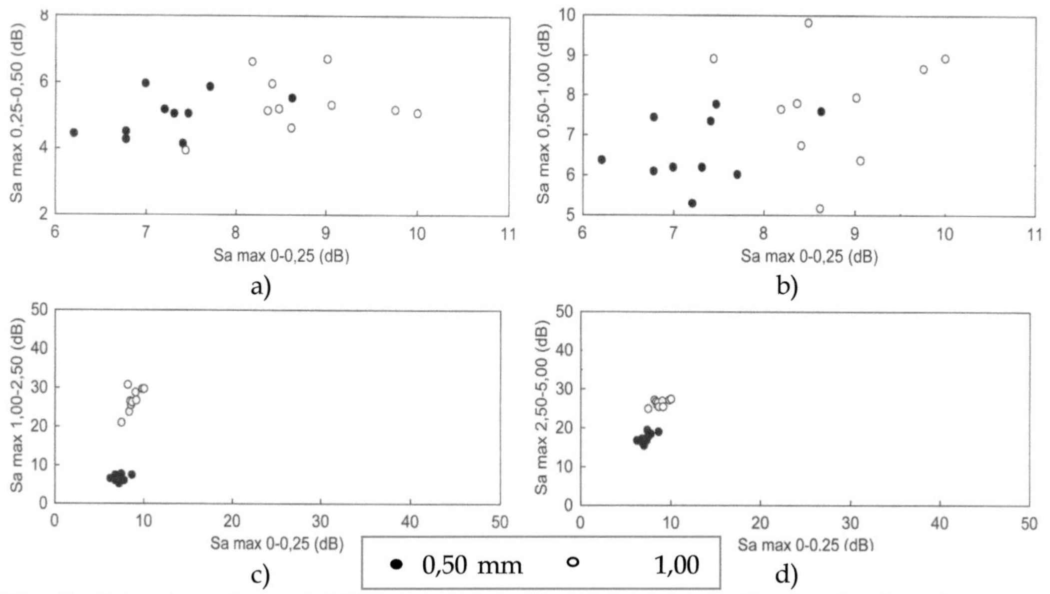

Fig. 5. Pair-wise plots of "difference signal" maximum amplitudes for low frequency intervals.

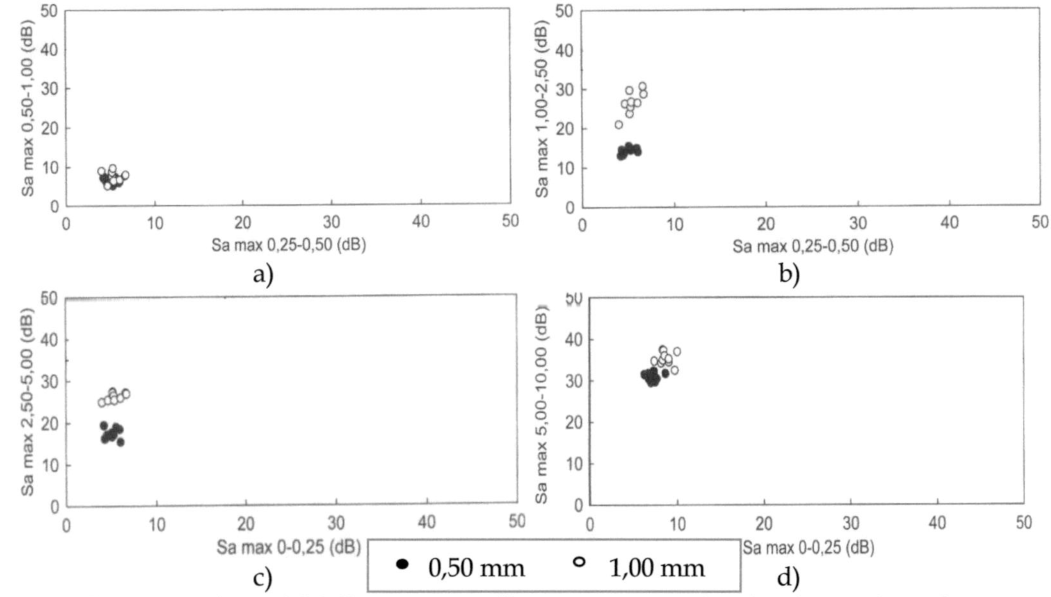

Fig. 6. Pair-wise plots of "difference signal" maximum amplitudes for medium frequency intervals.

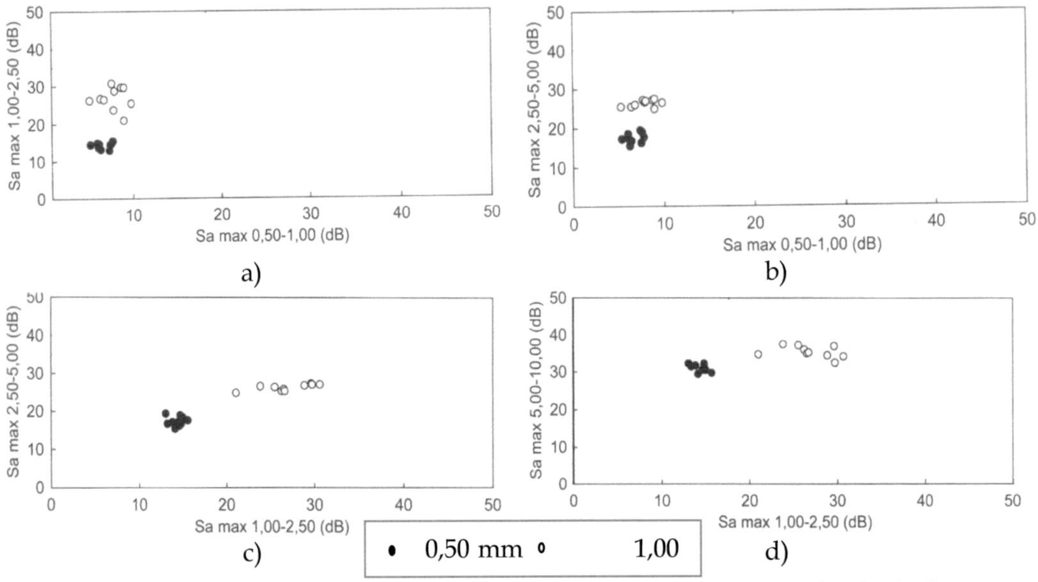


Fig. 7. Pair-wise plots of "difference signal" maximum amplitudes for high frequency intervals.

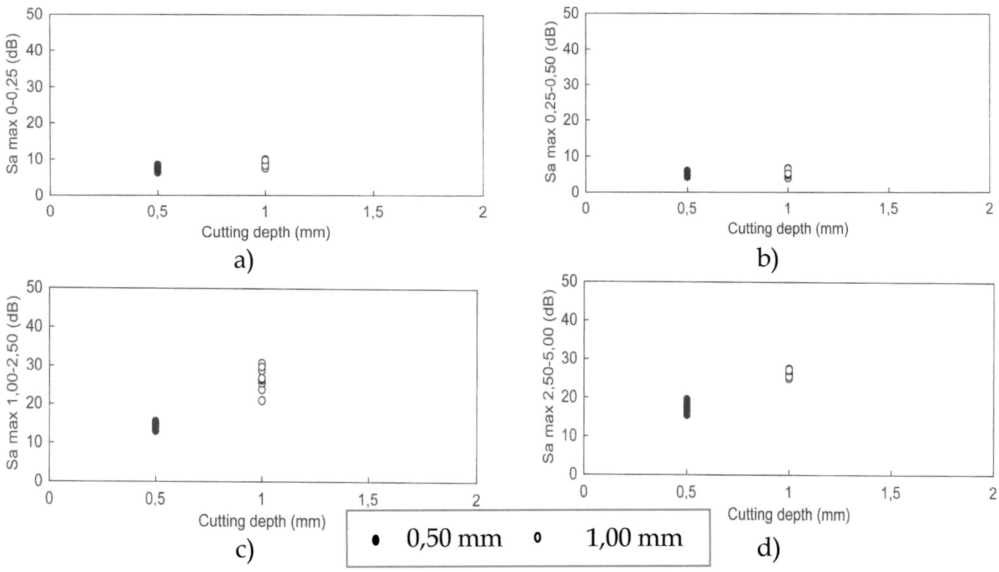


Fig. 8. "Difference signal" maximum amplitudes vs. depth of cut.

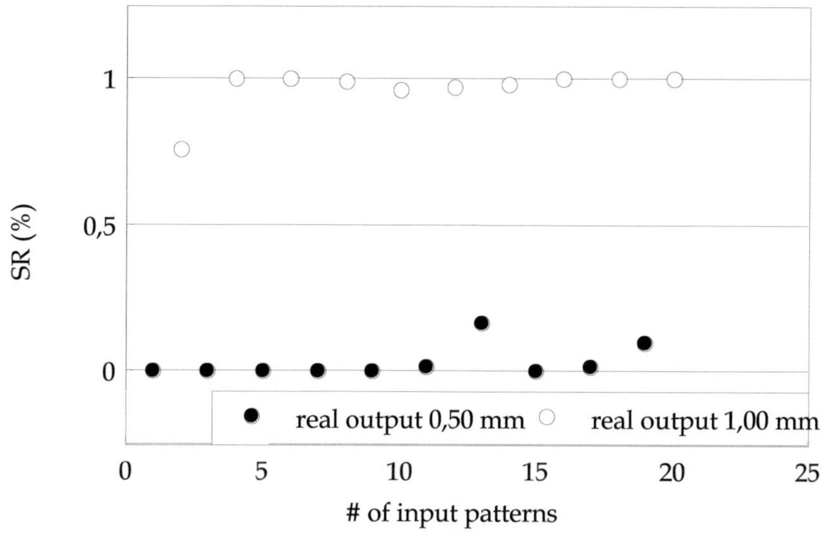


a)

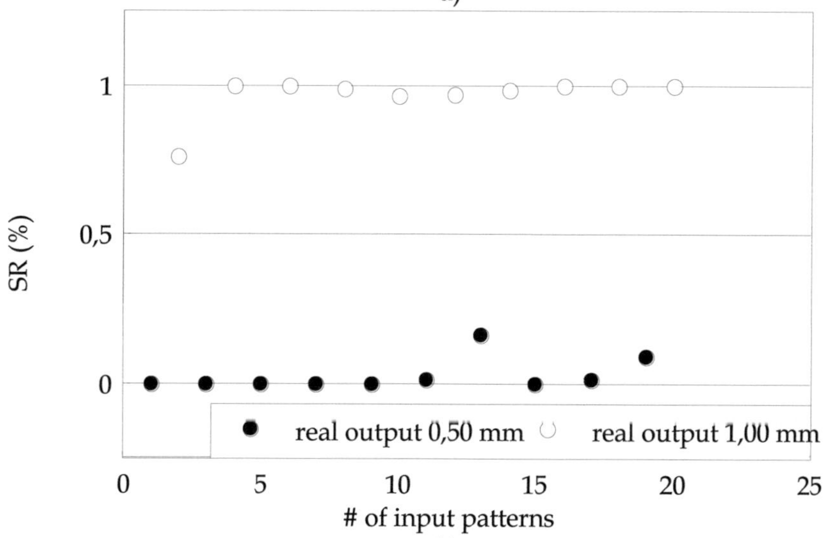


b)

Fig. 9. Neural Network output vs. number of input patterns for: a) 12000 and b) 14000 learning steps.

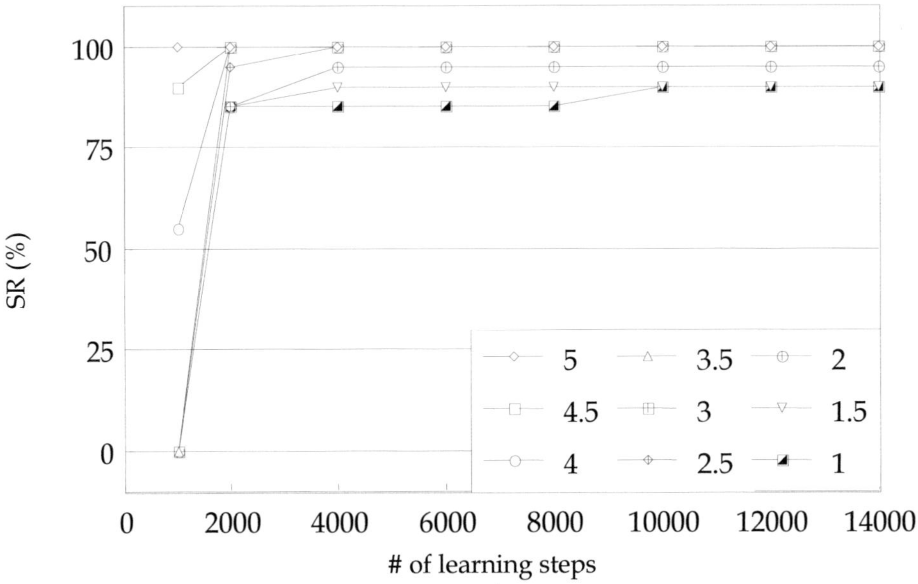


a)

Fig. 10. Neural Network Success Rate vs. number of learning steps for different threshold values.

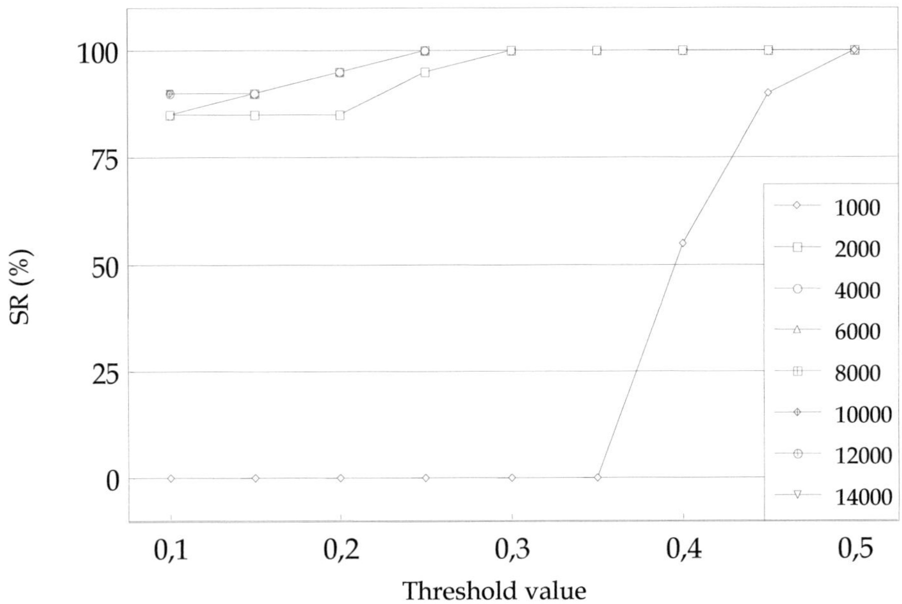


Fig. 11. Neural Networks Success Rate vs. threshold value for different numbers of learning steps.

## 5. Conclusion

During the last years, notable efforts have been made to develop reliable and industrially applicable machining monitoring systems based on different types of sensors, especially in production environments that require fully unmanned operation such as Flexible Manufacturing Systems (FMS) or Computer Integrated Manufacturing (CIM).

The main focus of this work is the establishment of a methodology to implement a process monitoring system based on audible sound energy sensors for application to milling operations.

In order to characterise the audible sound energy signals emitted by different cutting conditions during milling of T4-6056 Al alloy plates, machining parameters were varied and the corresponding acoustic signals were detected and processed in the frequency domain by a real-time spectrum analyser.

The classification of audible sound signal features was performed in two-dimensional space by graphical analysis and in multi-dimensional spaces by parallel distributed data processing using a supervised Neural Network paradigm.

The experimental results showed that the identification of depth of cut variation can realised only with reference to high frequency ranges. Besides, the supervised Neural Network data processing proved that the recognition of depth of cut value can be reliably achieved independently of the frequency range.

The proposed approach allows to state that: (1) the application of a systematic methodology to set up the cutting tests permits a more thorough comparison with other similar experimental works; (2) sensor signal analysis independent of the noise generated by the machine tool and the working environment is obtainable by subtracting the noise characteristic signal from the signals detected during the cutting tests; (3) the results obtained in this approach can be utilized for the development of process monitoring procedures based on sensors of different types, such as acoustic emission sensors where the high frequency (> 20 kHz) stress waves produced within the work material do not travel through air but only in the material itself. The combined application of audible sound energy sensors and acoustic emission sensors could make available more comprehensive information on process conditions through both low frequency (audible sound) and high frequency (acoustic emission) signal analysis, realizing the concept of sensor fusion technology.

## 6. Acknowledgements

Funding for this work was partly provided by the Spanish Ministry of Education and Science (Directorate General of Research), Project DPI2008-06771-C04-02.

The activity for the preparation of this work has received funding support from the European Community's Seventh Framework Programme FP7/2007-2011 under grant agreement no. 213855.

## 7. References

Brophy, C., Kelly, K., Byrne, G. (2002) AE-based condition monitoring of the drilling process. *Journal of Materials Processing Technology*, 124, 3, 305-310, ISSN: 0924-0136.

Burke, L.I., Rangwala, S. (1991) Tool condition monitoring in metal cutting. A neural network approach. *Journal of Intelligent Manufacturing*, 2, 5, 269-280, ISSN: 0956-5515.

Byrne, G., Dornfeld, D., Inasaki, I., Ketteler, G., König, W., Teti, R. (1995) Tool Condition Monitoring (TCM) - The Status of Research and Industrial Application. *Annals of the CIRP*, 44, 2, 541-567, ISSN: 0007-8506.

CA-Cricket Graph III For Macintosh: Version 1.0, *User Guide*, 1992.

Chen, C., Lee, S., Santamarina, G. (1994) An object-oriented manufacturing control system. *Journal of Intelligent Manufacturing*, 5, 5, 315-321, ISSN: 0956-5515.

Chen, F.F., Huang, J., Centeno, M.A. (1999) Intelligent scheduling and control of rail-guided vehicles and load/unload operations in a flexible manufacturing system. *Journal of Intelligent Manufacturing*, 10,5, 405-421, ISSN: 0956-5515.

Chen, J.C. (2000) An effective fuzzy-nets training scheme for monitoring tool breakage. *Journal of Intelligent Manufacturing*, 11,1, 85-101, ISSN: 0956-5515.

Chen, J.C., Chen, W.L. (1999) A tool breakage detection system using an accelerometer sensor. *Journal of Intelligent Manufacturing*, 10, 2, 187-197, ISSN: 0956-5515.

Cho, D.W., Lee, S.J., Chu, C.N. (1999) The state of machining process monitoring research in Korea. *International Journal of Machine Tools and Manufacturing*, 39, 11, 1697-1715, ISSN: 0890-6955.

Clark, W.I., Shih, A.J., Hardin, C.W., Lemaster, R.L., McSpadden, S.B. (2003) Fixed abrasive diamond wire machining—part I: process monitoring and wire tension force. *International Journal of Machine Tools Manufacturing*, 43, 5, 523-532, ISSN: 0890-6955.

D'Errico, G.E. (1997) Adaptive systems for machining process monitoring and control. *Journal of Materials Processing Technology*, 64, 1-3, 75-84, ISSN: 0924-0136.

Desforges, X., Habbadi, A., Geneste, L., Soler, F. (2004) Distributed machining control and monitoring using smart sensors/actuators. *Journal of Intelligent Manufacturing*, 15, 1, 39-53, ISSN: 0956-5515.

Dornfeld, D.A. (1992) Monitoring of machining process - Literature Review. *Annals of the CIRP*, 41, 1, 93-96, ISSN: 0007-8506.

Emel, E. (1991) Tool wear detection by neural network based acoustic emission sensing. *ASME, Dynamic Systems and Control Division Publication*, 28, 79–85, ISSN: 0022-0434.

Fu, J.C., Troy, C.A., Mori, K. (1996) Chatter classification by entropy functions and morphological processing in cylindrical traverse grinding. *Precision Engineering*,18, 2-3, 110-117, ISSN: 0141-6359.

Govekar, E., Gradišek, J., Grabec, I. (2000) Analysis of acoustic emission signals and monitoring of machining processes. *Ultrasonics*, 38, 1-8, 598-603, ISSN: 0041-624X.

Grabec, I., Govekar, E., Susic, E., Antolovic, B. (1998) Monitoring manufacturing processes by utilizing empirical modelling. *Ultrasonics*, 36, 1-5, 263-271, ISSN: 0041-624X.

Hong, S.Y. (1993) Knowledge-based diagnosis of drill conditions. *Journal of Intelligent Manufacturing*, 4, 3, 233-241, ISSN: 0956-5515.

Hou, T.H., Liu, W.L., Lin, L. (2003) Intelligent remote monitoring and diagnosis of manufacturing processes using an integrated approach of neural networks and rough sets. *Journal of Intelligent Manufacturing*, 14, 2, 239-253, ISSN: 0956-5515.

Huang, P.T., Chen, J.C. (1998) Fuzzy logic-base tool breakage detecting system in end milling operations. *Computers and Industrial Engineering*, 35, 1-2, 37-40, ISSN: 0360-8352.

Inasaki, I., (1998) Application of acoustic emission sensor for monitoring machining processes. *Ultrasonics*, 36, 1-5, 273-281, ISSN: 0041-624X.

Jemielniak, K., Kwiatkowski, L., Wrzosek, P. (1998) Diagnosis of tool wear based on cutting forces and acoustic emission measures as inputs to a neural network. *Journal of Intelligent Manufacturing*, 9, 5, 447-455, ISSN: 0956-5515.

Jin, J., Shi, J. (2001) Automatic feature extraction of waveform signals for in-process diagnostic performance improvement. *Journal of Intelligent Manufacturing*, 12, 3, 257-268, ISSN: 0956-5515.

Karlsson, B., Karlsson, N., Wide, P. (2000) A dynamic safety system based on sensor fusion. *Journal of Intelligent Manufacturing*, 11, 5, 475-483, ISSN: 0956-5515.

Kopac, J., Sali, S. (2001) Tool wear monitoring during the turning process. *Journal of Materials Processing Technology*, 113, 312-316, ISSN: 0924-0136.

Larson Davis Laboratory, 2800 Manual, Preliminary Documentation 1/27/93.

Lin, B., Zhu, M.Z., Yu, S.Y., Zhu, H.T., Lin, M.X. (2002) Study of synthesis identification in the cutting process with a fuzzy neural network. *Journal of Materials Processing Technology*, 129, 1-3, 131-134, ISSN: 0924-0136.

Lu, M.C., Kannatey-Asibu, E. Jr. (2000) Analysis of sound signal generation due to flank wear in turning. *International ME2000 Congress & Exposition*, Orlando, FL.

Malakooti, B.B., Zhou, Y.Q., Tandler, E.C. (1995) In-process regressions and adaptive multicriteria neural networks for monitoring and supervising machining operations. *Journal of Intelligent Manufacturing*, 6, 1, 53-66, ISSN: 0956-5515.

Masory, O. (1991) Monitoring machining processes using multi-sensor readings fused by artificial neural network. *Journal of Materials Processing Technology*, 28, 1-2, 231-240, ISSN: 0924-0136.

Masters, T. (1993) *Practical Neural Networks Recipies in C++*, Academic Press, San Diego, CA.

Niu, Y., Wong, Y., Hong, G. (1998) Intelligent sensor system approach for reliable tool flank wear recognition. *International Journal of Advanced Manufacturing Technology*, 14, 2, 77-84, ISSN: 0268-3768.

Noise and Vibrations Works, References Manual version 1.22.

Okafor, C., Adetona, O. (1995) Predicting quality characteristics of end-milled parts based on multi-sensor integration using neural networks, individual effects of learning parameters and rules. *Journal of Intelligent Manufacturing*, 6, 6, 389-400, ISSN: 0956-5515.

Ouafi, A.E., Guillot, M., Bedrouni, A. (2000) Accuracy enhancement of multi-axis CNC machines through on-line neurocompensation. *Journal of Intelligent Manufacturing*, 11, 6, 535-545, ISSN: 0956-5515.

Peng, Y. (2004) Intelligent condition monitoring using fuzzy inductive learning. *Journal of Intelligent Manufacturing*, 15, 3, 373-380, ISSN: 0956-5515.

Rubio, E.M., Teti, R. (2004) Cutting tests definition for effective tool condition monitoring. *IFAC-MIM '04, Int. Conf. on Manufacturing, Modelling, Management and Control*, Athens, 21-22 Oct.

Rubio, E.M., Teti, R., Baciu, I.L. (2006) Advanced signal processing in acoustic emission monitoring systems for machining technology. *2nd Int. Virtual Conf. on Intelligent Production Machines and Systems - IPROMS 2006*, 3-14 July: 189-192.

Rubio, E.M., Teti, R., Baciu, I.L. (2006) Main decision making procedures used in the monitoring systems of machining processes based on acoustic emission sensors. *5th CIRP Int. Sem. on Intelligent Computation in Manufacturing Engineering CIRP ICME '06*, Ischia, 25-28 July: 189-192.

Shawaky, A., Rosenberger, T., Elbestawi, M. (1998) In process monitoring and control of thickness error in machining hollows shafts. *Mechatronics*, 8, 301-322.

Sokolowski, A., Kosmol, J. (2001) Selected examples of cutting process monitoring and diagnostics. *Journal of Materials Processing Technology*, 113, 1-3, 322-330, ISSN: 0924-0136.

Spectrum Pressure Lavel (Spl 3100), Manual Version 0.98, Program Version 1.10.

Teti, R., Buonadonna, P. (1999) Round Robin on Acoustic Emission Monitoring of Machining. *Annals of the CIRP*, 48, 3, 47-69, ISSN: 0007-8506.

Teti, R., (1995) A Review of Tool Condition Monitoring a Literature. *Annals of the CIRP*, 44, 2, 659-666, ISSN: 0007-8506.

Teti, R., Baciu, I.L. (2004) Neural network processing of audible sound signal parameters for sensor monitoring of tool conditions. *4th CIRP Int. Sem. on Intelligent Computation in Manufacturing Engineering – CIRP ICME '04*, Sorrento, 30 June - 2 July: 385-390.

Teti, R., Baciu, I.L., Rubio, E.M. (2004) Neural network classification of audible sound signals for process monitoring during machining. *Annals of DAAAM* for 2004 & Proc. 15th Int. DAAAM Symp. on Intelligent Manufacturing Systems: Globalisation-Technology-Man-Nature, Ed. B. Katalinic, DAAAM International, ISSN 1726-9679, ISBN 3-901509-42-9: 459-460.

Tönshoff, H.K., Wulsferg, J.P., Kals, H.J., Köning, W., Van Luttervelt, C.A. (1988) Developments and Trends in Monitoring and Control of Machining Processes. *Annals of the CIRP*, 37, 2, 611-622, ISSN: 0007-8506.

Venkatesh, K., Zhou, M., Caudill, R.J. (1997) Design of artificial neural networks for tool wear monitoring. *Journal of Intelligent Manufacturing*, 8, 3, 215-226, ISSN: 0956-5515.

Wilcos, S.J., Reuben, R.L., Souquet, P. (1997) The use of cutting force and acoustic emission signals for the monitoring the tool insert geometry during rough face milling. *International Journal of Machine Tools Manufacturing*, 32, 4, 481-494, ISSN: 0890-6955.

Xiaoli, L., Yingxue, Y., Zhejun, Y. (1997) On-line tool condition monitoring system with wavelet fuzzy neural network. *Journal of Intelligent Manufacturing*, 8, 4, 271-276, ISSN: 0956-5515.

Xu, Y., Ge, M. (2004) Hidden Markov model-based process monitoring system. *Journal of Intelligent Manufacturing*, 15, 3, 337-350, ISSN: 0956-5515.

# 12

# Hybrid particle swarm algorithm for job shop scheduling problems

Xiaoyu Song
*School of Information and Control Enineering, Shenyang Jianzhu University*
*China*

## 1. Introduction

Particle swarm optimization (PSO) algorithm is a kind of random optimization algorithm based on swarm intelligence. Swarm intelligence of PSO is produced by cooperation and competition between particles, which is used for guiding optimization search. PSO has been studied widely in many applications due to its good global searching ability. Currently PSO has been widely used in function optimization, neural network training, pattern classification, system control and other applications. The research on PSO in recent years indicates that PSO has fast convergence speed and good quality in solutions and fine robustness on optimization in multidimensional space functions or in dynamic objectives, which is suitable for project applications. In this chapter, we firstly introduce searching mechanism and algorithm processes of PSO. Then, some important problems are solved when PSO is used for job shop scheduling problems (JSSP), such as hybrid algorithms between particle swarm and other algorithms (HPSO), its deadlock issues, and the proof of PSO and HPSO convergence. This chapter can provide guides effectively for readers who apply particle swarm optimization algorithm.

## 2. Particle Swarm Optimization Algorithm for JSSP

Particle swarm optimization (PSO) is an evolutionary computation technique developed by Kennedy and Eberhart in 1995. The particle swarm concept was motivated by the simulation of social behaviors. PSO algorithm constitutes the simple conduct rules for each particle, remembers the best position of the particles, and shares the information between particles. That is, PSO algorithm achieves the optimization through cooperation and competition between the individuals of population. Comparing with other evolutionary algorithms, PSO algorithm retains the global search strategy based on population, and belongs to the simple model of movement and velocity. PSO algorithm can dynamically adjust the current search with unique memory. Considering the currency and validity of the algorithm, PSO algorithm has been studied in many applications.
Job shop scheduling problem (JSSP) is the simplification model of an actual problem, and among the most typical and hardest combinatorial optimization problems, which is a NP

complete problem. JSSP is often used to test the performance of the intelligent algorithms, which has important research and actual engineering meanings.

### 2.1 Introduction of PSO

PSO algorithm simulates the prey behavior of a bird flock. We can imagine a scene, a group of birds are random searching the food, and there is only a piece of food in this region. All the birds don't know the place of food, but they know distance from the current location to the place of food. What is the optimal strategy of searching the food? The most simple and effective strategy is to search the areas where are close to the birds.

PSO algorithm is motivated from the model, and is used to solve optimization problems. Each optimization is considered as a bird in the search space called a particle. Each particle has a fitness value that is decided by an optimization function, has a velocity to determine its flight direction and distance. PSO algorithm constructs an initial particle swarm (random solutions), then find the optimal solution through iterations. In each iteration, particles update their velocities and positions by tracking the two extreme values. An optimal solution is the individual extremum *pBest* that is found by the particle itself, and another optimal solution is the global individual extremum *gBest* that is found by the current population.

In traditional PSO algorithm, the particle swarm searches results in space of m*n dimension, each particle position means a result of the problem. The particle continuously adjusts itself position $X$ to search new results. Let $P_{id}$ denote the optimal result that the particle obtains. Let $P_{gd}$ denote the optimal position that the particle swarm passed, the best total result in the search domain. Let $V$ denote the speed of the particle.

$$V_{id}(t+1)= \omega \times V_{id}(t) + \eta_1 \times rand() \times (P_{id} - X_{id}(t)) + \eta_2 \times rand() \times (P_{gd} - X_{id}(t)) \quad (1)$$

Let $V_{id}(t)$ denote the speed of $d$ dimension of particle $i$ in generation $t$, $\omega$ denote inertia weight, and '-' denote distance. Let $\eta_1$ and $\eta_2$ denote parameter, which can adjust $P_{id}$ and $P_{gd}$ respectively. *rand()* is the random number generation function. Therefore, we can get the next particle position.

$$X_{id}(t+1) = X_{id}(t) + V_{id}(t+1) \quad (2)$$

Considering the formula (1) and (2), we can find that the moving direction of particle is decided by three parts. That is, the initial speed $X_{id}(t)$ of the particle, and the optimum distance $P_{id} - X_{id}(t)$ that the particle passed, and the optimum distance $P_{gd} - X_{id}(t)$ that the particle swarm passed. The relative importance of three parts is decided by weighting coefficient $\omega$, $\eta_1$, $\eta_2$。

The traditional PSO algorithm is described as follows.

STEP 1: Construct an initial particle swarm, that is randomly set the initial position $X$ and the initial velocity $V$ of each particle;

STEP 2: Calculate fitness value of each particle;

STEP 3: Compare each particle fitness value and its best position fitness value $P_{id}$, if better, update $P_{id}$;

STEP 4: Compare each particle best position $P_{id}$ and the best position of particle swarm $P_{gd}$, if better, update $P_{gd}$;
STEP 5: adjust the velocity and position according the formula (1) and (2);
STEP 6: If termination conditions are satisfied (good enough position or the maximum number of iterations), then end; otherwise, go to 2.
PSO algorithm is a kind of evolutionary algorithm, which has several typical characteristics. First, the individual of population has been randomly initialized a random solution in the initialization process. Secondly, the better solutions of a new generation are obtained by searching the solution space. At last, a new generation of population is produced on the basis of the previous generation.

### 2.2 Convergence of PSO

The convergence of intelligence optimization algorithm is an important problem for the application of intelligent optimization algorithms. It is necessary that we discuss the convergence of PSO algorithm before solving a practical problem.

#### 2.2.1 Convergence of Traditional PSO

It is a difficult problem to prove the convergence for an intelligent optimization algorithm. Two assumptions H1 and H2 proposed by Solis and Wet were introduced, which were used to prove the global convergence of the pure optimization algorithm with probability *1*. General requirements of stochastic optimization algorithm convergence are described as follows.
An optimization problem $\langle A, f \rangle$ and stochastic optimization algorithm $D$ are given. $x_k$ is the results of the *k-th* iterations, and results of the next iteration is $x_{k+1}$ ($x_{k+1} = D(x_k, \zeta)$), where $\zeta$ is the solution that has been searched by algorithm $D$.
**Condition H1:** $f(D(x, \zeta)) \le f(x)$, if $\zeta \in \mathbf{A}$, set $f(D(x_k, \zeta)) \le f(\zeta)$, where $A$ is the subset of the $R^n$, and $A$ denotes the constraint space of the problem.
Conditions H1 random algorithm can guarantee the correctness; their objective is to ensure optimization of the solution to the fitness value of f (x) non-incremental.
A global convergence of the algorithm, which means sequence $\{f(x_k)\}_{k=0}^{\infty}$ can reach infimum $inf(f(x) : x \in \mathbf{A})$ in the feasible solution $A$. Because it is possible that the feasible solution $A$ of optimization problem exist discontinuity spaces or isolated spots, infimum and other fitness value is incontinuous. Considering this potential problem, search infimum is defined in Lebesgue measure space as shown in formula 3, where $v[X]$ denotes Lebesgue measure in set $X$.

$$\Psi = inf(t : v[x \in A \mid f(x) < t] > 0) \tag{3}$$

Formula (3) implies that non-empty set of the search space is existent, where the fitness of its members infinitely are close to $\Psi$. The definition of $v[X]$and $A$ guarantee that nonempty point does not exist in set $A$. So the algorithm can reach or be close to the infimum without searching all points of set $A$.
Therefore, the optimal region can be defined as the following formula, where $\varepsilon > 0$, $M \to \infty$.

$$R_{\varepsilon,M} = \begin{cases} \{x \in A \mid f(x) < \Psi + \varepsilon\}, & \text{finite}\Psi \\ \{x \in A \mid f(x) < -\text{M}\}, & \Psi = -\infty \end{cases}$$

If the optimization algorithm find a point among $R_{\tau,\ M}$, the point is an acceptable global optimal or near to the global optimum.

**Condition H2:** $\prod_{k=0}^{\infty}(1 - v_k[B]) = 0$, $\forall B \subseteq A$, s.t.$v[B] > 0$

Where $v_k[B]$ is probability measure in set $B$, and $B$ is the *kth* iteration set of algorithm $D$. Algorithm $D$ satisfies condition H2. It means, it is impossible that algorithm $D$ searches the points among set B, and let $v[B] > 0$. Because $R_{\tau,\ M} \subseteq$ A, it is possible that the global optimum can be found.

**Theorem 1 (Global Convergence):** Supposing that $f$ is measurable and feasible solution space $A$ is measurable subset of $R^n$, algorithm $D$ satisfies condition H1 and H2. And algorithm $D$ generates series $\{x_k\}_{k=0}^{\infty}$. Then

$$\lim_{k \leftarrow +\infty} P[x_k \in R_{\varepsilon,M}] = 1$$

$P[\, x_k \in R_{\tau,\ M}]$ is probability measure in $R_{\tau,\ M}$, and $R_{\tau,\ M}$ is the *kth* iteration set of algorithm $D$. Considering theorem 1, the global convergence of stochastic algorithms must satisfy the conditions H1 and H2. Because each iteration of PSO algorithm has kept the best position, conditions H1 must be satisfied. However, utilizing Markov chain theory and mathematical theory of real variable, Dr. Van den Bergh has proved that PSO algorithm does not satisfy conditions H2.

### 2.2.2 Convergence of Improved PSO

Because the traditional PSO algorithm does not guarantee global convergence, the position and velocity update equations are improved for solving JSSP. Considering the formula (1) and (2), although $v_k$ and $x_k$ is multidimensional variable, each dimension is independent. Therefore the convergence analysis can be simplified to the one-dimensional. In order to expand the solution space of PSO algorithm, we adopt the velocity update equation and position update equation of particle $i$ as follows:

$$v_i(t+1) = \alpha(P_i - x_i(t)) + \beta(P_g - x_i(t)) \quad (4)$$

$$x_i(t+1) = x_i(t) + v_i(t+1) \quad (5)$$

In formula (4) and (5), $\alpha$, $\beta(\alpha, \beta \in [0,\ 1])$ are random numbers. The part $\alpha(P_i - x_i(t))$ represents that the best private distance experience of the particle $i$ in group t is inhabited by probability $\alpha$; And the part $\beta(P_g - x_i(t))$ represents that the best group distance experience of all the particles in group $t$ is inhabited by probability $\beta$.

It can be obtained from formula (4) and (5) that the bigger the value of $\alpha$ is the larger impact of $P_i$ is, and the greater possibility of particle's moving to the local optimum will become;

Similarly, the bigger the value of $\beta$ is the larger impact of $P_g$ is and the greater possibility of particle's moving to the global optimum will become.
The particle $i$ will stop moving when $x_i(t) = P_i = P_g$. Namely $x_i(t+1) = x_i(t)$. In order to expand the solution space of PSO algorithm, we save $P_g$ as historical global best position and regenerate position $x_i(t+1)$ of particle $i$ randomly in the solution space, which will make the particle $i$ continue to search.
Through the operation, equation (5) can be deformed as follows:

$$x_i(t+1) = (1-c)\,x_i(t) + c_1 p_i + c_2 p_g \tag{6}$$

When $p_i$, $p_g$ fixed, equation (6) is a simple linear difference equation, when $x_i(0) = x_{i\,0}$, its solution is:

$$\begin{aligned} x_i(t) &= \mathrm{k} + (x_{i\,0} - k)(1-c)\,t \\ k &= \frac{c_1 p_i + c_2 p_g}{c}, \quad c = c_1 + c_2 \end{aligned} \tag{7}$$

Considering the formula (7), the formula (6) has convergence if $|1 - c| < 1$. That is, if $t \to \infty$, then $x_i(t) \to \frac{c_1 p_i + c_2 p_g}{c}$. If $|1 - c| < 1$, then $0 < c_1 + c_2 < 2$. That is, if $0 < c_1 + c_2 < 2$, the evolution equation of improved PSO algorithm is asymptotic convergence. The convergence region shown in Fig. 1.

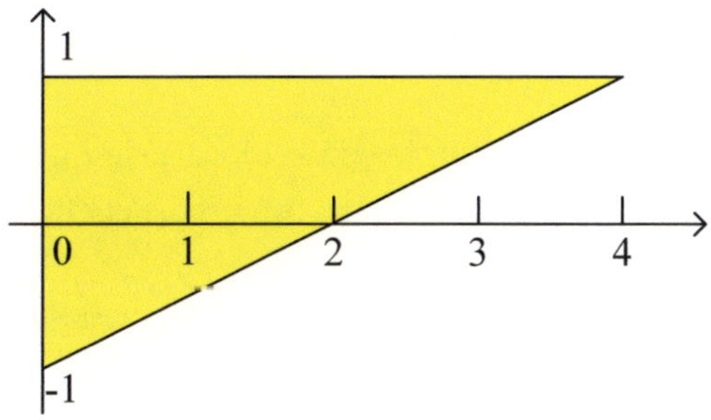


Fig. 1. Convergence region

**Theorem 2:** if $|1 - c| < 1$, $\lim_{t\to\infty} x_i(t) = p_g$.

See the formula (7), if $|1 - c| < 1$, then

$$\begin{aligned} \lim_{t\to\infty} x_i(t) = k = \frac{c_1 p_i + c_2 p_g}{c} \\ x_i(t+1) = x_i(t) - (c_1 + c_2)\,x_i(t) + c_1 p_i + c_2 p_g \end{aligned} \tag{8}$$

if $t \to \infty$,

$$\lim_{t\to\infty} x_i(t+1) = \lim_{t\to\infty} x_i(t) \tag{9}$$

Then

$$-(c_1+c_2)\lim_{t\to+\infty} x_i(t) + c_1p_i + c_2p_g = 0 \tag{10}$$

Considering theorem 1, if a random optimization algorithm satisfies condition H1 and H2, we can guarantee that the algorithm can converge to global optimal solution with probability *1*. We will discuss the problem whether the improved PSO algorithm is able to satisfy condition H1 and H2.

In the improved PSO algorithm, the solution sequence is$\{p_{g,t}\}$, where $t$ denotes evolutional generation, and $p_{g,t}$ denotes the best position of particle swarm in generation $t$. The function $D$ is redefined by the improved PSO algorithm.

$$D(p_{g,t} x_i(t)) = \begin{cases} p_{g,t}, & f(p_{g,t}) \le f(x_i(t)) \\ x_i(t), & f(p_{g,t}) > f(x_i(t)) \end{cases} \tag{11}$$

It is easy to prove that the condition H1 is satisfied.

In order to satisfy the conditions H2, the sample space of particle swarm $A$ must contain $A$. Namely,

$$A \subseteq \bigcup_{i=1}^{a} M_{i,t}, \tag{12}$$

Where $M_{i,t}$ is the support set of the sample space of particle $i$ in generation $t$. Considering particle $j$, let $M_{i,t} = A$ when $x_j(t) = p_i = p_g$. Let the other particle $i$:

$$M_{i,t} = x_i(t-1) + \varphi_1(p_i - x_i(t-1)) + \varphi_2(p_g - x_i(t-1)) \tag{13}$$

Because of $0 \le \varphi_1 \le c_1$ and $0 \le \varphi_2 \le c_2$, $M_{i,t}$ is a super rectangle with vertices, where $\varphi_1 = \varphi_2 = 0$, $\varphi_1 = c_1$, and $\varphi_2 = c_2$. Let $v[M_{i,t} \cap A] < v(A)$, when $max(c_1 \mid p_i - x_i(t-1) \mid, c_2 \mid p_g - x_i(t-1) \mid) < 0.5 \times \mathrm{diam}(A)$, where $\mathrm{diam}(A)$ denotes the length of $A$. Considering condition H2, the length of $M_{i,t}$ is near to *0* when $t \to \infty$. Therefore, the measure $v[M_{i,t}]$ of each $M_{i,t}$ is decreasing with the growth of generation $t$. And the measure $v[\bigcup_{i\neq j} M_{i,t}]$ of union $\bigcup_{i\neq j} M_{i,t}$ is decreasing. Therefore there is $N$, let $v[\bigcup_{i\neq j} M_{i,t} \cap A] < v[A]$ when $t > N$.

*Because of* $M_{i,\ t} \subseteq A$, Let $\bigcup_{i=1}^{A} M_{i,t} = A$. In order to satisfy the conditions H2,we define *A* is the Borel subset($S = M_{i,\ t}$), then $v[\ S\ ] > 0$ and $v_t[\ S\ ] = \sum_{i=0}^{a} v_{i,t}[S] = 1$. Considering theorem 1, the improved algorithm can converge to a global optimal solution with probability *1*.
There is almost identical convergence between the improved PSO algorithm and the traditional PSO algorithm. That is, the parameter $x_i(t)$ can converge to the best location within the finite range. The traditional PSO algorithm does not guarantee global convergence, but the improved PSO algorithm can converge to a global optimal solution with probability *1* when generation *t* is near to ∞.

### 2.3 Convergence of PSO

Job shop scheduling problems (JSSP) is an important part of production scheduling of an enterprise, which is one kind of the most typical and hardest combinatorial optimization problems, an NP complete problem. The main task in scheduling, in terms of production targets and constraints, is to determine the precise process route, time, machine and operation for every process object. JSSP is often used to test the performance of the intelligent algorithms, which is significant for research and actual engineering.
Particle swarm optimization (PSO) algorithm is a kind of random optimization algorithm based on continuous optimization problems. PSO algorithm is less studied to solve JSSP. The PSO algorithm design of solving JSSP is difficult, and the efficient PSO algorithm design of solving JSSP is more difficult.
Leticia etc. construct the single machine scheduling algorithm based on random coding of JSSP, and the algorithm is a kind of retardation minimum time algorithm. The algorithm utilizes the dynamic mutation operators to ensure the diversity of particle populations. The algorithm has been tested respectively with 40 jobs and 50 jobs, and the algorithm achieves good results. Lina etc. construct PSO algorithm based on operation code to solve JSSP. They apply the crossover and mutation operation of GA in place of the update operations of velocity and position of PSO algorithm.
In the hybrid particle swarm optimization, Jerald.J etc. apply GA, SA and PSO algorithm to solve scheduling problems of flexible manufacturing systems. The hybrid algorithm optimizes machine idle time and reduces the cost of production tardiness. Liu etc. combine PSO algorithm and VNS. The hybrid algorithm minimizes the makespan of the flexible JSSP. Xia etc. design the hybrid PSO algorithm based on SA local search algorithm. The hybrid algorithm can solve multi-objective flexible JSSP. In order to minimize the makespan, Sha etc. construct the hybrid algorithm based on Hash table to solve JSSP. In the hybrid algorithm, Giffler-Thompson (G&T) algorithm is adopted to construct the feasible solution from the particle location of Hash table, and SWAP operation updates the particle velocity. The hybrid algorithm combines with TS algorithm based on block structure.

#### 2.3.1 JSSP Description

Each instance of the problem $J/\ /C_{max}$ is defined by a set of jobs, a set of machines and a set of operations. Each job consists of a sequence of operations, each of which has to be performed on a given machine for a given time. A schedule is an allocation of the operations to time intervals on the machines. The problem is to find the schedule that minimizes the

makespan subject to the following constraints: (i) the precedence of operations given by each job are to be respected, (ii) each machine can perform at most one operation at a time and (iii) the operations cannot be interrupted.
Let:

- $J = \{1, \dots, n\}$ denote the set of jobs;
- $M = \{1, \dots, m\}$ denote the set of machines;
- $V = \{0, 1, \dots, \tilde{n}+1\}$ denote the set of operations, where 0 and $\tilde{n}+1$ represent the dummy start and finish operations, respectively;
- A be the set of pair of operations constrained by the precedence relations, as in (i);
- $V_k$ be the set of operations to be performed by the machine k;
- $E_k \subset V_k \times V_k$ be the set of pair of operations to be performed on the machine k and which therefore have to be sequenced, as specified in (ii);
- $p_v$ and $t_v$ denote the (fixed) processing time and the (variable) start time of the operation v, respectively. The processing time of the 0 and $\tilde{n}+1$ operations is equal to zero, i.e., $p_0 = p_{\tilde{n}+1} = 0$.

Given the above assumptions, the problem can be stated as searching minimize $t_{\tilde{n}+1}$ subject to

$$\begin{array}{ll} t_j - t_i \geq p_i, & (i, j) \in A, \\ t_j - t_i \geq p_i \vee t_i - t_j \geq p_j, & (i, j) \in E_k, k \in M, \\ t_i \geq 0, & i \in V. \end{array} \tag{14}$$

The first set of constraints represents the precedence relations among the operations of the same job, whereas the second set of constraints describes the sequencing of the operations on the same machine. These constraints impose that either $t_j - t_i \geq p_i$ or $t_i - t_j \geq p_j$. Any feasible solution of the problem (1) is called a schedule.
In this framework, it is useful to represent the job shop scheduling problem in terms of a disjunctive graph $G := (V, A, E)$, where $V$ is the set of nodes, $A$ the set of ordinary arcs (conjunctive) and $E$ the set of disjunctive arcs. The nodes of $G$ correspond to operations, the directed arcs correspond to precedence relations, and the disjunctive arcs correspond to operations to be performed on the same machine. More precisely, $E = \bigcup_{k=1}^{m} E_k$, where $E_k$ is the subset of disjunctive arcs is related to a machine $k$; each disjunctive arc of $E$ can be considered as a pair of opposite directed arcs. The length of an arc $(i,j) \in A$ is $p_i$, the length of an disjunctive arc $(i,j) \in E$ is either $p_i$ or $p_j$ depending on its orientation. The selection of a processing order on each machine involves the orientation of the disjunctive arcs, in order to produce an acyclic directed graph. A schedule on a disjunctive graph $G$ consists in finding a set of orientations that minimizes the length of the longest path (*critical path*) in the resulting acyclic directed graph.
According to the Adams et al. method, the graph $G$ can be decomposed into one direct subgraph $D = (V, A)$, by deleting disjunctive arcs, and in $m$ cliques $G_k = (V_k, E_k)$, obtained from $G$ by deleting both the conjunctive arcs and the dummy nodes 0 and $\tilde{n}+1$. A selection $S_k$ in $E_k$ contains exactly one arc between each pair of opposite arcs in $E_k$. A selection is acyclic since it does not contain any directed cycle. Moreover, sequencing the operations on

the machine $k$ is equivalent to choosing an acyclic selection in $E_k$ . A complete selection $S$ is the union of selections $S_k$ , one for each $E_k$ , $k \in M$. $S$ generates the directed graph $D_S =(V, A \cup S)$;$S$ is acyclic if the associated directed graph $D_S$ is acyclic. An acyclic complete selection $S$ infers a schedule, i.e., a feasible solution of Problem.

In order to solve the job shop scheduling problem the best acyclic complete selection $S^*$ that minimizes the value of the length of the longest critical path in the direct graph $D_{S^*}=(V,A \cup S^*)$must be determined.

The neighbourhood of the current solution can be formed by the solutions generated by inverting the direction of a disjunctive arc in the critical path of $D_S$ . To this end, as stated by other authors, it is useful to decompose the critical path into a sequence of $r$ blocks $(B_1,B_2, \ldots ,B_r)$. Each block contains the operations processed on the same machine; for each pair of consecutive blocks $B_j,B_{j+1}$ with $1 \le j \le r$ the last operation of $B_j$ and the first of $B_{j+1}$ belong to the same job but are performed on different machines.

### 2.3.2 PSO for Solving JSSP

As for applying PSO to the job shop scheduling problem, the problem can be described as that $n$ jobs are processed by $m$ machines. A certain list such as $S_m = (O_i)$, $i=1, \ldots, n$, demonstrates the list of jobs processed on a machine, then the amount of possible lists is $n!$, list set $S = \{ S_k \mid k = 1, 2, \ldots, m\}$ is used to express the process that $n$ jobs done by $m$ machines, the whole possibility of solutions is $(n!)^m$. As job shop scheduling problem, when all the operations in the solution is configured, the best processed list that satisfies the efficiency index can be seeked. Therefore, for solving job shop scheduling problem by PSO, we only need to change $m$ encoding of each particle to seek optimal solution. According to the above, definition of operating list in job shop problem is given here.

**Definition 1**: exchanging operation. In the operation list, operation $O_i$ on position $i$ and operation $O_j$ on position $j$ change their positions each other. This behavior is called exchanging operation, the operator is denoted as $\otimes$. For the list S, the exchange of $O_i$ and $O_j$ is expressed as $S\otimes(O_i,O_j)$, where, $(O_i,O_j)$ denotes the operation exchange, which can be simply expressed as $O_{i\otimes j}$. Then $S'=S\otimes(O_i,O_j)=S\otimes O_{i\otimes j}$, $S'$ denotes the list which has been disposed.

Example 1: with regard to the job shop problem in which 6 jobs are processed on $m$ machines, the list done on machine $m$ is $S_m=(2\ 4\ 6\ 1\ 3\ 5\ )$, for the list $S_m$, if operation 2 and operation 6 exchanges, their position are respectively 1 and 3, the exchange process can be described as following formula.

Here, $S'_m = S_m\otimes (O_1,O_3) =(2\ 4\ 6\ 1\ 3\ 5) \otimes (2,6) =(\ 6\ 4\ 2\ 1\ 3\ 5)$.

**Definition 2**: exchange list. The operation list composed of no less than one exchanges among operations is named as exchange list, which is denoted as CS, and $CS = (O_{1i\otimes 1j}, O_{2k\otimes 2l}, \ldots, O_{np\otimes nq}\ )$. When the list only have one time exchange operate, $CS = (O_{1i\otimes 1j},)$, where the sequence $O_{1i\otimes 1j}$, $O_{2k\otimes 2l}, \ldots, O_{np\otimes nq}$, denotes the sequence of exchanging operations in the list $S$.

Exchange list acts on certain fraction of $S_m$, and it means that all the exchange operation in the list acts on $S_m$ one by one, namely $S'_m = S_m\otimes CS = S_m \otimes (O_{1i\otimes 1j}, O_{2k\otimes 2l}, \ldots, O_{np\otimes nq}) = [[S_m \otimes (O_{1i},O_{1j})\ ] \otimes (O_{2k},O_{2l})\ ]\ldots\otimes (O_{np},O_{nq})$.

**Definition 3**: Equal set of exchange list. Different exchange list acts on the same solution, maybe the same solution is obtained. All the exchange lists which products the same solution is called the equal set.

**Definition 4**: united operation of exchanged list. When more than two exchanging lists such as $CS_1$, $CS_2$, ..., $CS_n$, which act on one list according to the sequence is named as united operate, moreover, the operator is denoted as $\oplus$, the unit of exchange list is expressed as HS, HS= $CS_1 \oplus CS_2 \ldots \oplus CS_n$. Through the principle stated above, it can be described as $S' = S \otimes HS = S \otimes (CS_1 \oplus CS_2 \ldots \oplus CS_n)$ , where $S'$ denotes the new operation list that $S$ had been exchanged according to the exchange list.

Through definition 3 and definition 4, a new solution $S'$ can be obtained after acting on solution $S$ with $CS_1$ and $CS_2$. Supposed that there is another exchange list that acts on the solution $S$, if a same solution $S'$ can be obtained, then the unite of $CS_1$ and $CS_2$ is equal to $CS$, which can be expressed as $CS = CS_1 \oplus CS_2$, $CS$ and $CS_1 \oplus CS_2$ belong to the same equal set, generally speaking, $CS$ is not sole.

**Definition 5:** Basic exchange list. In the equal set $\{CS_i\}$ of exchanging list, exchange list $BS$ with least exchange operators is called basic exchange list of this equal set. Supposed $X$ and $Y$ are operation list on machine $m$, constructing an exchange list $BS$ which satisfies $X = Y \otimes BS$, if $BS$ is of least exchange operation, then $BS$ is a basic exchange list, which is denoted as $BS = Y \rightarrow X$.

According to the following method, a basic exchange list can be constructed, supposed that two solutions of problem FT06 are given, the lists on machine $m$ are respectively $X$ and $Y$.

Eg: X= ( 1 2 3 4 5 6 ), Y= ( 2 6 3 1 5 4).

It can be seen that, in the operation $X$, $O_1 = 1$. and in operation $Y$, $O_4 = 1$, let Y's first operate exchanging $O_{1i\otimes 1}$ be $Y \otimes (O_1, O_4)$, then $Y1 = Y \otimes (O_1, O_4)$, there exists $Y1 = (1\ 6\ 3\ 2\ 5\ 4)$; in X, $O_2 = 2$, and in $Y1$, $O_4 = 2$, let the second exchange operate $O_{2k\otimes 2l}$ be $Y1 \otimes (O_2, O_4)$, then $Y2 = Y1 \otimes (O_2, O_4)$, there exists $Y2 = (1\ 2\ 3\ 6\ 5\ 4)$. Similarly, the third exchange operate $O_{3p\otimes 3q}$ is $Y2 \otimes (O_4, O_6)$, there exists $Y3 = Y2 \otimes (O_4, O_6) = X$. Spontaneously, $BS = (O_{1i\otimes 1j}, O_{2k\otimes 2l}, O_{3p\otimes 3q})$ is of the minimal exchange operations, which is named as a basic exchange list, namely, $BS = Y \rightarrow X$. Here, $BS = Y \rightarrow X = (O_{1i\otimes 1j}, O_{2k\otimes 2l}, O_{3p\otimes 3q}) = ((O_1, O_4) \oplus (O_2, O_4) \oplus (O_4, O_6))$ .

Aiming at PSO used to solve job shop problem, formula of basic PSO is not fit for this new type algorithm, so the formulas are recreated as follows:

$$v_{id} = \alpha( X_{id} \rightarrow P_{id}) \oplus \beta( X_{id} \rightarrow P_{gd}) \tag{15}$$

$$X'_{id} = X_{id} \otimes V_{id} \tag{16}$$

Where $\alpha$, $\beta$ are random number and ($\alpha$, $\beta \in [0, 1]$). $\alpha( X_{id} \rightarrow P_{id})$ expresses that all the exchange operations of basic exchange list $(X_{id} \rightarrow P_{id})$ are withheld by the probability $\alpha$. Similarly, $\beta( X_{id} \rightarrow P_{gd})$ expresses that all the exchange operations of basic exchange list $( X_{id} \rightarrow P_{gd})$ are withheld by the probability $\beta$.

According to the formula (15) and (16), it can be seen that, the greater $\alpha$ is, the stronger $P_{id}$ affects, the probability of moving towards to the local optimization is magnified. In the same way, the greater $\beta$ is, the stronger $P_{gd}$ effect, the probability of moving towards to the global optimization is magnified.

Due to the regularity of object functions, the optimal solution must be in the active scheduling set, so PSO uses the solution produced with G&T as the initial solution. For the random and widespread searching ability, the exchanging list based PSO is used to search globally. In the process of running PSO algorithm, if any infeasible solution appears, it must be adjusted. When there exists $P_i(t) = P_{id} = P_{gd}$ for the particle $P_i(t)$ of generation $t$, then recreate this particle, so that PSO algorithm for job shop problem is constructed.
The steps of solving JSSP by PSO are described as following:
Step1: Use G&T algorithm to produce an initial solution, initialize $P_{id}$ with the initial solution, initialize $P_{gd}$ with the best $P_{id}$;
Step2: If the end condition is satisfied, go to Step6;
Step3: According to the position of $X_{id}$, calculate $X_{id}$ 's next position $X'_{id}$, namely new solution;

a) $A = X_{id} \rightarrow P_{id}$ denotes that $A$ acts on $X_{id}$ to get $P_{id}$ , where, $A$ is a basic exchange list,;
b) $B= X_{id} \rightarrow P_{gd}$, where $B$ is also a basic exchange list;
c) Calculate validity $V_{id}$ of particle according to formula (15);
d) Calculate new position $X'_{id}$ (solution) according to formula (16);

Step4: Adjust infeasible solution;
Step5: Calculate fitness:

a) If a better solution is got, then update $P_{id}$;
b) If a better solution of the whole swarm is searched out, then update $P_{gd}$, simultaneously adopt G&T to recreate a new particle instead. Go Step2;

Step6: Show the optimal solution obtained by this algorithm ($P_{gd}$).
Adjustment of infeasible solution is described in hybrid PSO algorithm.

### 2.4 Summary

Particle swarm optimization (PSO) is an evolutionary computation technique developed by Kennedy and Eberhart in 1995. The particle swarm concept was motivated by the simulation of social behaviors. The original intent was to graphically simulate the graceful but unpredictable choreography of bird flock. In the section, we introduce search mechanisms and processes of PSO, and analyze the convergence of PSO theoretically. A new PSO algorithm is proposed based on exchanged factors and exchanged lists, which is put the PSO idea into the discrete field of JSSP.

## 3. Hybrid Particle Swarm optimization Algorithm for JSSP

Recently, the theorem of No Free lunch (NFL) is proposed for evaluating optimization algorithms by professor Wolpert and Macready of Stanford University. It is shown that there isn't a single solution that adapts to all problems effectively. Radcliffe and Surry have the same conclusion.
For example, if GA algorithm is better than SA algorithm when solving the problem set $A$, then SA algorithm must be better than GA algorithm when solving the problem set $B$. Considering all the circumstances, two algorithms have the same performance. Therefore, there is no kind of intelligent optimization algorithm better than the other intelligent optimization algorithms. That is, every method has its corresponding application circumstances.

In theory and practice, adopting a single intelligent algorithm is not enough for solving JSSP. The hybrid algorithm is an effective method, which enlarges the application domain and improves their performance. A hybrid algorithm combines effectively some features of several algorithms, such as optimization mechanism, process, search behavior, operation, and so on. The hybrid algorithm will have better optimization efficiency.

### 3.1 HSPSO

If adopting a single algorithm to solve job shop problems, it is hard to improve the local optimization after some running time of the algorithm, it is necessary to find out a method to escape from this local optimization. Therefore, a hybrid PSO algorithm based on exchanging list is proposed.

The design ideas of hybrid optimization algorithm HPSO are as follows: (1) Due to the regularity of object function, the optimal solution must be in the active scheduling set, so HPSO uses the solution produced with G&T as the initial solution. (2) For the randomly and widespread searching ability, the exchanging list based PSO is used to search globally. (3) In the process of running PSO algorithm, if an infeasible solution appears, it must be adjusted. (4) In order to avoid algorithm falling in a local optimization too early, TS exploiting strategy embedded critical operations based on exchanging neighbors is adopted to realize local parallel search, simultaneously improve the local search ability.

When there exists $P_i(t) = P_{id} = P_{gd}$ for the particle $P_i(t)$ of generation $t$, then adopt G&T algorithm to regenerate the particle, so that hybrid PSO algorithm for solving JSSP is constructed. The arithmetic frame is shown as Fig. 2.

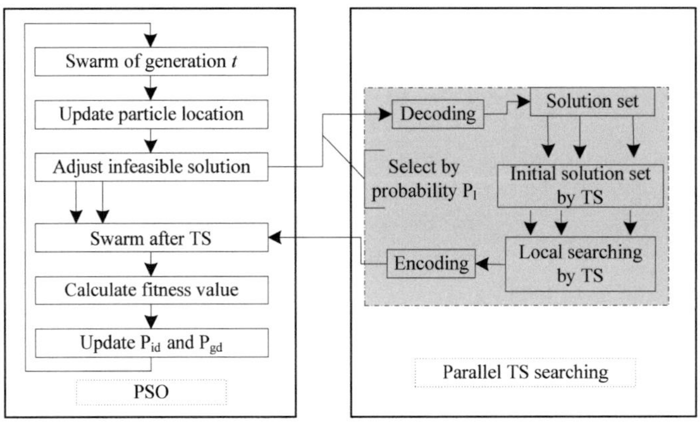


Fig. 2. Frame of the hybrid PSO algorithm

The steps of solving job shop problem by HPSO are described as following:

Step1: Use G&T algorithm to produce initial solution, initialize $P_{id}$ with an initial solution, initialize $P_{gd}$ with the best $P_{id}$;

Step2: If the end condition is satisfied, go to Step7;

Step3: According to the position of $X_{id}$, calculate $X_{id}$ 's next position $X'_{id}$, namely a new solution;

a) $A = X_{id} \rightarrow P_{id}$ denotes that $A$ acts on $X_{id}$ to get $P_{id}$, where, $A$ is a basic exchange list,;

b) $B = X_{id} \rightarrow P_{gd}$, where $B$ is also a basic exchange list;

c) Calculate validity $V_{id}$ of particle according to formula (8);
d) Calculate new position $X'_{id}$ (solution) according to formula (9);

Step4: Adjust infeasible solutions;
Step5: Select some solutions by the probability $P_l$ to perform TS;
Step6: Calculate fitness:

a) If a better solution is gotten, then update $P_{id}$;
b) If a better solution of the whole swarm is searched out, then update $P_{gd}$, simultaneously adopt G&T to recreate a new particle instead. Go Step2;

Step7: Show the optimal solution obtained by this algorithm ($P_{gd}$).

### 3.2 TS based on neighbor exchanging of critical operation

Taboo search(TS) algorithm is one of the best algorithms for solving job shop scheduling problem. So far, its running speed is faster, and it may provide a better induct within the whole searching field compared with other algorithms.

In order to obtain better searching results and higher efficiency, neighbors must be highly constrained and can be rapidly assessed. The possibility of moving to high quality solutions should be increased.

The local searching function is TS algorithm. To improve the efficiency of the local searching, we modify the TS algorithm. Firstly, the algorithm reduces the maximum that doesn't evolution. Secondly, a new exchanging strategy of neighbors is proposed based on critical operations so that TS algorithm can rapidly assess neighbors. We firstly indicate the neighbor exchanging based on the critical operation.

The feasible solution of job shop scheduling is usually denoted by the gantt graph. The gantt graph of 6×6 problem is illustrated in Fig. 3. In the figure, x-axis denotes the process time, y-axis denotes the machines, and every rectangular block marked $(i, j)$ denotes the operation $j$ of task $i$, and it is denoted as $O_{ij}$.

The optimized result of job shop scheduling problem is related to the length of the critical paths. The critical path is that the longest path without time intervals between operations in an available schedule. A solution always has many critical paths. For example, in Fig. 3, there are two critical paths. The first one is (4,1) (3,2) (5,1) (5,2) (4,2) (4,3) (5,3) (3,4) (3,5) (6,4) (5,5) (5,6) and another one is (4,1) (3,2) (5,1) (5,2) (4,2) (3,3) (3,4) (3,5) (6,4) (5,5) (5,6).

Furthermore, operations of the critical path can be decomposed into blocks. A block is a set of consecutive operations in a critical path in one machine. For example, operation (5,2), (4,2) and operation (4,3), (5,3), (3,4) in the first critical path forms the block respectively. Operation (5,2), (4,2), (3,3) and (3,4) in the second critical path forms the block respectively. For the two consecutive blocks, the last operation of the anterior block and the first operation of the latter block are always in the same task. For example the operation (3,3) and (3,4) of task 3 are in the same task.

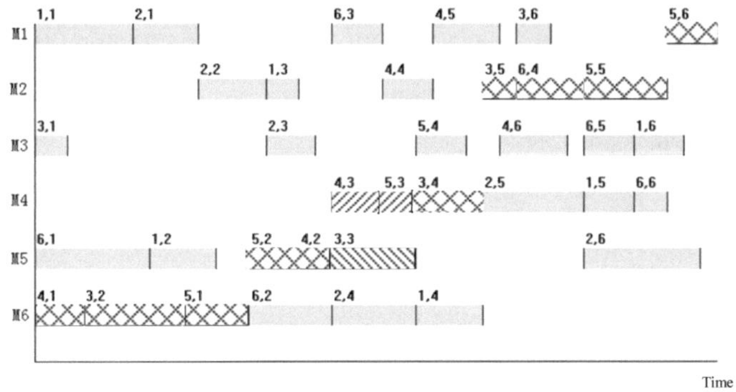


Fig. 3. 6×6 problem solution Gantt figure

Let $J_p(v)$ represent the previous operation of operation $v$ in the same job, and $M_p(v)$ denote the previous operation of operation $v$ processed in the same machine, $S_t(v)$ and $E_t(v)$ denote the start time and the end time of the operation $v$ respectively.

**Definition 6:** Critical operation. In critical path, if operation satisfy the condition that $S_t(v)= E_t(M_p(v))=E_t(J_p(v))$, then $v$ is called a critical operation.

When critical path is not unique, not all the neighbor exchanges can shorten the critical path. For example, Fig. 3 describes a gantt graph which shows the 6×6 problem, the exchange of (4,2) and (3,3) is unable to shorten the critical path. Because St(3,4) = MAX( Et(5,3), Et(3,3) ), operation (3,4) is a critical operation, due to the dependency of critical operation (3,4) on operation (3,3), (5,3), although operation (3,3) is shortened, the neighbor exchange before the critical operation is unable to shorten the critical path.

The method of choosing neighbors based on the critical operations is as follows: when the critical path is sole, exchangeable neighbors in the critical path is considered as a set for neighbor selection; when the critical path is not sole, the exchangeable neighbors between the last critical operation and the last operation is viewed as a set for neighbor selection; TS algorithm selects an exchangeable neighbor (usually the best neighbor) from the above neighbors set to commute. If the set described above is null, then stop the current search with TS.

When TS algorithm search process runs for certain times, the quality of solution can not be improved, then TS algorithm stops.

Because of adopting new exchanging strategy of neighbors based on critical operations, TS algorithm reduces invalid neighbor exchanges, enhances searching efficiency, increases the possibility of escaping from the local optimization, and expands the searching range. Simultaneity, when there is no exchangeable neighbor, it indicates that the cost of improving the solution is too large, or the current solution is already the optimal solution. Then the searching is terminated.

### 3.3 HPSO Convergence

Dr. Van den Bergh has proved that PSO algorithm diverges both in the local region and the global region with the criteria presented by Solis and Wets, under which stochastic search algorithms can be considered as a global search algorithms, or merely locally search algorithms. We analyze the convergence of PSO algorithm with an optimum keeping strategy and TS algorithm by Markov chain theory at a different aspect in this book, and HPSO algorithm based on PSO and TS algorithm is proved to be convergent. First of all, we give an introduction of Markov chain theory as follows.

**Definition 7 (Markov chain)** A stochastic sequence $\{X_n, n\in T\}$ and a discrete temporal series $T=\{0,1,2,\ldots\}$ are given, all state values corresponding to each $X_n$ constitute the set of discrete state $S=\{s_0,s_1,s_2,\ldots\}$. The stochastic sequence $\{X_n, n\in T\}$ is called Markov chain as soon as the conditional probability satisfies the formula (17) as for each integer $n\in T$ and any $s_0,s_1,s_2,\ldots,s_{n+1}\in S$.

$$P\{X_{n+1}=s_{n+1} \mid X_0=s_0, X_1=s_1,\ldots,X_n=s_n\}=P\{X_{n+1}=s_{n+1} \mid X_n=s_n\} \quad (17)$$

**Definition 8 (Transit ionprobability matrix)** The conditional probability $p_{i,j}=P\{X_{n+1}=j \mid X_n=i\}$ is called transition probability of Markov chain $\{X_n, n\in T\}$, where $i,j\in S$. The matrix $\{P_{i,j}:i,j=1,\ldots,k\}$ is called $k\times k$ transition probability matrix.

**Definition 9 (Finite homogeneous Markov chain)** Markov chain is called finite homogeneous Markov chain if conditional probability $p_{i,j}(n)$ of Markov chain $\{X_n, n\in T\}$ has nothing to do with $n$ and its set of state $S=\{s_0,s_1,s_2,\ldots s_k\}$ is finite, where $i,j\in I$. Then $p_{i,j}(n)$ is always regarded as $p_{i,j}$.

**Lemma 1** Markov chain $\{X_n, n\in T\}$ with transition probability matrix $P$ is irreducible if and only if the conditional probability satisfies formula (11) for any $s_i,s_j\in S$, where the set of state is $S=\{s_0,s_1,s_2,\ldots s_k\}$.

$$P\{X_{m+n}=s_j \mid X_m=s_i\}>0 \quad (18)$$

**Lemma 2** Transition probability matrix $P$ is irreducible if $P$ can be turned into the form $\begin{pmatrix} C & 0 \\ R & T \end{pmatrix}$ by the same line and row transformation, where $C$ is a strict positive irreducible stochastic matrix with dimension $m$, $R,T\neq 0$.Then the matrix.

$$P^{\infty}=\lim_{n\to\infty} P^{n}=\begin{pmatrix} C^{\infty} & 0 \\ R^{\infty} & 0 \end{pmatrix} \quad (19)$$

is a stable stochastic matrix, where $R^{\infty}=\lim_{k\to\infty}\sum_{i=0}^{k-1} T^{i} R C^{k-i}$, $P^{\infty}=1'p^{\infty}$, $p^{\infty}=p^{0}P^{\infty}$ has nothing to do with the initial distribution, and $p_i^{\infty}>0(1\leq i\leq m)$, $p_i^{\infty}=0(m\leq i\leq k)$.

In this book, we set the change of the group made up of social collaboration $S$, self adapting $A$ and competition $C$ three basic evolution operations, where social collaboration $S$ means that the particle adjusts its movement by cooperating with the best position $P_g$ of the group; the self adapting $A$ indicates that the particle adjusts its movement at the next moment by cooperation between cognition part $\alpha(P_i - x_i(t))$ and social collaboration part $\beta(P_g - x_i(t))$; All old particles $x_i(t)$ are totally replaced by new particles $x_i(t+1)$ with optimum keeping strategy to update their self best position and group position. Therefore, the course of transformation can be presented respectively by stochastic matrix $P_S$, $P_A$ and $P_C$, and the transition probability matrix of TS algorithm is presented by stochastic matrix $P_T$.

**Theorem 3** The hybrid algorithm HPSO based on PSO and TS algorithm is finite homogeneous Markov chain.

**Proof:** Since the probability of group in next state rests with the current state, which is independent of the past state, HPSO algorithm with the set of finite state $S=\{s_0,s_1,s_2,\ldots,s_k\}$ is Markov chain. Suppose that $P_S$, $P_A$, $P_C$ and $P_T$ are independent of time intervals, then the searching course of HPSO can be noted by a transition probability matrix with one step $P=P_T[P_C(P_SP_A)]$, which is independent of time intervals as well. Therefore, the whole search course of HPSO is finite homogeneous Markov chain.

The design of the neighborhood is the key factor to impact on the quality and efficiency of algorithm as for the neighborhood search algorithm TS. Therefore, we first give two assumptions about the neighborhood structure as follows to ensure the convergence of TS algorithm.

**Assumption 1**: The neighborhood structure is supposed to be symmetrical. That is, $\forall s_i,s_j\in S, s_i\in N(s_j)\Leftrightarrow s_j\in N(s_i), i,j=0,\ldots,k$;

**Assumption 2**: On the point view of the graph theory, the graph $G_N$ is supposed to be strongly connected. Namely, there must be a path from $s_i$ to $s_j$ for any $s_i,s_j\in S$, where $i,j=0,\ldots,k$.

**Theorem 4** HPSO algorithm with the optimum keeping strategy is global asymptotic convergence when time is endless, namely the algorithm will converge to the optimal group.

**Proof:** Compared with the standard PSO velocity update equation, the equation has abandoned the previous velocity $\omega v_i(t)$ of particle $i$, which will make at least one particle of the particle swarm stop evolution of each generation due to its best history position. The optimal strategy algorithm is adopted in this hybrid algorithm. For convenience, the optimal individual reserved from each generation is saved in the left side of the population, but it does not participate in the evolutionary process. The state which contains the same optimal solution is arranged in order as same as which in the original state space, and the one which contains the different optimal solution is arranged in order according to the fitness value. Then new social collaboration transition probability matrix, self adapting transition probability matrix and competition transition probability matrix can be presented respectively as $P_S^*=diag(P_S,P_S,\ldots,P_S)$ ,$P_A^*=diag(P_A,P_A,\ldots,P_A)$ , and $P_C^*=diag(P_C,P_C,\ldots, P_C)$. After the competition, we'll compare the optimal solution of the current population with the optimal solution reserved from the former generation, such an operation is presented by $U=(u_{ij})$. Set $Z^t=max\{f(pop_i^{t+1}), i=1,2,\ldots,N\}$ be the optimal fitness, then the transition probability from $Pop^t=[Z^{t-1},pop_1^t,pop_2^t,\ldots,pop_N^t]$ to $Pop^{t+1}=[Z^t,pop_1^{t+1},pop_2^{t+1},\ldots,pop_N^{t+1}]$ is presented as follows:

$$P(Pop^{t+1}=s_j\mid Pop^t=s_i)=\begin{cases}1 & \text{if } \max\{f(pop_i^t), i=1,2,\cdots,N\}=f(Z^t) \\ & \text{and } (pop_1^t,pop_2^t,\cdots,pop_N^t)=(pop_1^{t+1},pop_2^{t+1},\cdots,pop_N^{t+1}) \\ 0 & \text{otherwise}\end{cases} \quad (20)$$

Thus, there is unique element 1 in every line of the $U$, the others are 0. Meanwhile, $U$ is lower triangular matrix considering that the individual or is replaced by better or remains unchanged. Therefore, $U$ is noted as follows:

$$U=\begin{pmatrix} U_{11} & & & \\ U_{21} & U_{22} & & \\ \vdots & \vdots & \ddots & \\ U_{k1} & U_{k2} & \cdots & U_{kk} \end{pmatrix} \tag{21}$$

Where $U_{ij}$ is $k\times k$ matrix, and $U_{11}$ is unit matrix. That is to say that the transition probability matrix with one step of PSO algorithm is lower triangular matrix.

$$P^* = P_C^* (P_S^* P_A^*)U$$

$$=diag(P_C(P_S P_A), P_C(P_S P_A),\ldots, P_C(P_S P_A))U$$

$$=\begin{pmatrix} P_C(P_S P_A)U_{11} & & & \\ P_C(P_S P_A)U_{21} & P_C(P_S P_A)U_{22} & & \\ \vdots & \vdots & \ddots & \\ P_C(P_S P_A)U_{k1} & P_C(P_S P_A)U_{k2} & \cdots & P_C(P_S P_A)U_{kk} \end{pmatrix} \tag{22}$$

$$=\begin{pmatrix} C & 0 \\ R & T \end{pmatrix}$$

Obviously, $P^*$ is an irreducible stochastic matrix.

In theory, it has been proved that if the search space $S$ of TS is limited, and neighborhood structure satisfies the above assumption 1 and assumption 2, TS algorithm will converge to optimal solutions inevitably. Then the transition probability matrix with one step of TS algorithm is irreducible stochastic matrix as well. Apparently, the transition probability matrix of HPSO algorithm $P=P_T P^*$ is irreducible stochastic matrix. This shows that the probability of individual staying in the non-global optimal solution tends to 0, therefore HPSO algorithm with the optimum keeping strategy will converge to the optimal group when time is endless. Namely, $\lim_{t\to\infty} P(Z_t \in S_{opt})=1$, where $S_{opt}$ is the optimal solution set.

### 3.4 Experiments and Analysis

According to the above analysis, the global asymptotic convergence of HPSO algorithm can be guaranteed theoretically. However, the proof is based on perfect operation situations such as sufficiently large taboo list, infinite time and so on. Considering the reality of computer limitations and the limited time, we just take the convergence theory as the guidance in the specific computational experiments, some relaxations are made in accordance with the actual conditions on aspects of taboo length, search steps. Therefore, the solutions of some problems we obtained can just go nearly to rather than reach the optimal solution.

In this experiment, we apply the HPSO algorithm to 13 typical benchmark job-shop scheduling problems including FT10, LA02, LA21, LA24, LA25, LA27, LA29, LA36, LA37, LA38, LA39 and LA40. The experimental results are shown in Table 1.

In PSO algorithm, the swarm size is set to |O|*200%, where |O| is the number of operations, and maximum of iterative generations is set to |O|*200%; In HPSO algorithm, the swarm size is set to |O|*70% and maximum of iterative generations is set to |O|*70%. We choose the unfeasible solution of PSO algorithm by probability $P_l$ ($P_l$=20%) as the initial solution of TS algorithm, after 50 search steps, algorithm will end if it couldn't find a better solution in TS algorithm. The swarm size of PGA is set to |O|*70% and maximum of iterative generations is set to |O|*70%, where the crossover probability is 0.85, the mutation probability is 0.05.

| Problem | Optimum Makespan | PGA | | | PSO | | | HPSO | | |
|---|---|---|---|---|---|---|---|---|---|---|
| | | *Optimum* | *Average value* | *time/S* | *Optimum* | *Average value* | *time/S* | *Optimum* | *Average value* | *time/S* |
| FT10(10×10) | **930** | 943 | 963.0 | 60.09 | 977 | 996.9 | 24.88 | **930** | 945.2 | 37.49 |
| LA02(10×5) | **655** | **655** | 682.4 | 14.43 | 702 | 734.2 | 3.71 | **655** | 668.2 | 5.05 |
| LA19(10×10) | **842** | **842** | 842.0 | 54.38 | 874 | 884.0 | 10.22 | **842** | 842.6 | 26.21 |
| LA21(15×10) | **1046** | 1058 | 1068.0 | 171.18 | 1254 | 1281.6 | 28.50 | 1078 | 1099.0 | 200.77 |
| LA24(15×10) | **935** | 945 | 949.0 | 165.86 | 1130 | 1149.3 | 27.81 | 947 | 959.4 | 218.16 |
| LA25(15×10) | **977** | 1020 | 1026.5 | 177.18 | 1174 | 1197.0 | 30.43 | 999 | 1018.5 | 217.70 |
| LA27(20×10) | **1235** | 1442 | 1464.9 | 577.59 | 1502 | 1530.1 | 457.35 | 1257 | 1267.4 | 558.50 |
| LA29(20×10) | **1153** | 1305 | 1330.7 | 569.55 | 1439 | 1488.3 | 500.75 | 1198 | 1214.6 | 512.1 |
| LA36(15×15) | **1268** | 1318 | 1326.3 | 687.78 | 1338 | 1356.8 | 758.5 | **1268** | 1283.3 | 599.0 |
| LA37(15×15) | **1397** | 1436 | 1441.1 | 790.57 | 1503 | 1519.2 | 714.01 | 1415 | 1425.8 | 817.2 |
| LA38(15×15) | **1196** | 1242 | 1251.0 | 731.47 | 1262 | 1294.3 | 708.33 | 1208 | 1217.5 | 723.3 |
| LA39(15×15) | **1233** | 1244 | 1247.3 | 720.53 | 1306 | 1320.3 | 709.20 | 1244 | 1246.4 | 614.0 |
| LA40(15×15) | **1222** | 1243 | 1286.4 | 855.50 | 1284 | 1299.8 | 738.17 | 1224 | 1233.1 | 766.26 |

**Note:The bold letters are optimum.**

Table 1. The Average Value of Ten Times Experiments And Optimal Values

The algorithm for JSSP mentioned above can be easily implemented on computer. We program the algorithm in C and run it on CPU AMD2800+with 1G.Table 1 shows that we can find the optimums of problem FT10, LA02, LA19 and LA36 when we apply HPSO algorithm to solve the 13 benchmark problems. We can obtain that there are 10 average value of ten times experiments of HPSO algorithm better than PGA algorithm, and the deviation between the average value of ten times experiments of HPSO algorithm and the optimum is lower than which between PSO algorithm and the optimum by 11.69% . Thus the overall search capability of the algorithm is improved, which make the algorithm get closer to the optimum solution.

We analyze the convergence of PSO algorithm with optimum keeping strategy and TS algorithm by Markov chain theory as for the Job Shop problem, and present a hybrid algorithm called HPSO algorithm with global asymptotic convergence based on the above convergence theory. This algorithm has made full use of the large scale random search capability and the social cooperation of PSO algorithm, at the same time, the local parallel TS algorithm is embedded to improve the local search capability. We apply the above convergence theory to computational experiment and find the optimum of problem FT10, LA02 and LA19 in a short period. When compared with PGA algorithm and PSO algorithm, there are 10 average value in ten times experiments of HPSO algorithm better than PGA algorithm, and the deviation between the average value of ten times experiments of HPSO algorithm and the optimum is lower than that of between PSO algorithm and the optimum by 11.69%. Thus the overall searching capability of the algorithm is improved, which has demonstrated the effectiveness of solving Job Shop Scheduling problem by HPSO algorithm.

### 3.5 Summary

The theorem of No Free lunch (NFL) shows that there isn't a single solution that adapts to all problems effectively. Therefore, a hybrid algorithm of particle swarm optimization and tabu search algorithm (TS) for solving JSSP is proposed motivated by the strong global search capability of PSO algorithm and the good local search capability of TS algorithm. Meanwhile, the convergence of HPSO is proved, and experimental simulation results are given.

## 4. Strategies for Deadlock Elimination for JSSP using PSO

Deadlock is a state that the requests of scheduling transactions which contest resources one another can not be satisfied. Namely, the stagnancy is among transactions waiting for one another appears. When using a PSO algorithm to solve constrained optimization problems, deadlock is one of the key problems need to be solved. In PSO algorithm, we optimize the various operations of jobs based on PSO code. Because the different operations of the same job are studied as the separate object. Different objects can be in different machine queues. So this may be a single legitimate machine queue (two jobs does not occupy the same machine at the same time). Meanwhile, deadlock is latent among the queues on different machines. Efficiency and feasibility are decided by various conditions when using PSO to solve JSSP. Deadlock is a kind of the most important link. The strategies for eliminating deadlock are proposed in the book.

### 4.1 Deadlock Problem of JSSP

We find that the machines as resources are preempted by the jobs in the production process, and the rings that jobs are waiting for one another are created. The state is called as deadlock. Each job is waiting other resources occupied by another job. The utilization rate of the system will decline. If the particle deadlock can not be solved in time, the whole production system will collapse, and automate production will be unable to continue.

The computer scientist first put forward deadlock when dealing with the allocation of resources in the operating system. Coffman has given four necessary conditions for deadlocks as follows:

(1) Exclusion. Resources only can be allocated to a particular task or an idle task. Resources can not be occupied simultaneously by two tasks.

(2) Non-preemption. Resources is non-preemptive. When the corresponding process task is completed, the task will release occupied resources.
(3) Occupation and waiting. The task has been occupying some resources. Meanwhile, the task requests additional resources that have been occupied by other task.
(4) Circular and waiting. There exists a set of requested resources {P1,P2,…,Pn}. Where P1 is waiting for resources occupied by P2, P2 is waiting for resources occupied by P3, …, Pn is waiting for resources occupied by P1.
In the shop scheduling, there exists that the requests of scheduling transactions which contest resources one another can not be satisfied, so a state of stagnancy among transactions waiting for one another appears, resulting in deadlock and emergence of infeasible solutions. When using a hybrid PSO algorithm to solve JSSP, deadlock is one of the key problems need to be solved. To obtain valid hybrid PSO algorithm for JSSP, we study the reasons producing deadlocks in PSO algorithm, and present three countermeasures for deadlock elimination: encoding elimination, detection and reconstruction, and direct reconstruction.
While solving JSSP using a hybrid PSO, we firstly transfer JSSP into encoding denotation based on operations as the result of scheduling. Different operations of identical jobs may be processed on different machines, which can result in deadlock among job queues on different machines, being infeasible solution. In this solution space corresponding to encoding based on operation, it not only contains feasible solutions, but also contains infeasible solutions (namely, solutions with deadlock). For example with a 6×6 shop scheduling problem, Fig. 4 denotes one possible scheduling gantt chart, and the vertical coordinate denotes serial number of processing machine.

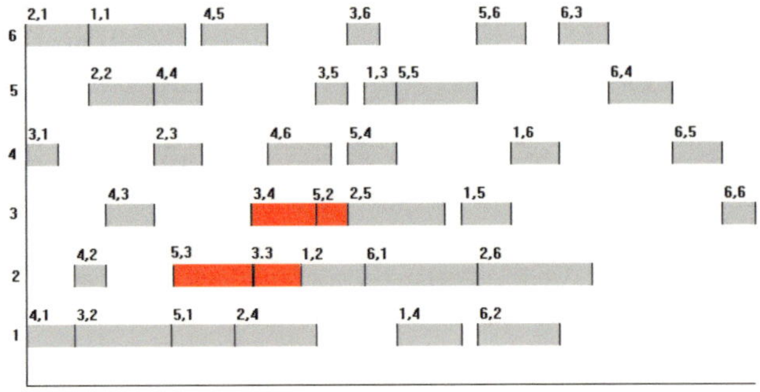


Fig. 4. A possible scheduling gantt of a typical (6×6 scheduling problem)

We can see from Fig. 4 that job queue on both machine 2 and machine 3 have deadlock, namely, operation (5,3)→(3,3) and (3,4)→(5,2) just a typical waiting deadlock. Because before operation (5,3) is processed, it must wait until operation (5,2) has be finished, while operation (3,4) before operation (5,2) must wait until operation (3,3) has be finished, but operation (3,3) is after (5,3), namely, this scheduling has deadlock. Occurrence of deadlock makes particles generated by G&T in hybrid PSO could not go on evolution, because the algorithm simulates course of processing according to scheduling scheme under the above dual restrictions, while valuing individuals, fitness values always are computed according to job's last makespan in processing system, the makespan of the last operation is just circulation ending time of the whole batch of jobs. Deadlock makes the process stagnated at the position of deadlock and could not go ahead, so we could not gain this operation's maximal makespan in a common sense, which forms infeasible scheduling solution in solution space. So, while effectively solving JSSP using PSO or hybrid PSO, the deadlock matter is an obstacle which we must solve.

## 4.2 Deadlock Elimination and Reconstruction

### 4.2.1 Encoding Elimination

In deadlock elimination strategy, we design an encoding denotation based on operations as the result of scheduling. Then we utilize the encoding and decoding to eliminate deadlock (that is, the infeasible solution).

During the optimizing process, $m$ segments of each particle are changed. If we produce solutions by encoding $m$ segments of a particle directly, it is possible to produce some infeasible solutions, which is also called dead lock, and it will lead to bad optimization efficiency. Aiming at an $n \times m$ job shop scheduling problem in which the chromosome is made up of $n \times m$ genes, when the operation position of every machine is changed, the position of operations in the chromosome corresponding to the position of machines is changed, so a feasible solution is obtained, and infeasible solution is avoided, either the characteristic of PSO algorithm is reserved.

The encoding process based on operations is: for the problem that $n$ jobs need to be processed on $m$ machine. The chromosome is an $n \times m$ array that denotes all operations. For a matter of convenience, let m=3, n=3, and chromosome starts with such a segment of genes:[*2, 1, 3, 2, 1, 2, 1, 3, 3*]. We assign the same symbol for the same job's operation, where *1* denotes job $J_1$, *2* denotes job $J_2$, and *3* denotes job $J_3$. Because each job has three processes, each job appears three times in chromosome. The operation of job is equivalent to the order in chromosome.

The decoding process based on the encoding operation is: firstly, the chromosome is transformed to an orderly list that denotes an order of the production process. Secondly, according to the processing order of each operation, scheduling scheme is given. The scheme includes the start time and end time of every job.

For above 6×6 problem, in particle swarm chromosome encoding, there are six dimensions, each of which has a position value. Here, the position value is composed of six segments of chromosomes divided by machine devices, and chromosome's encoding method is an encoding composed of a whole segment chromosome (36 bits genes on it), where each gene which represents a job index is a decimal number. According to its size order, distribute operations over again for the whole segment of chromosome's genes. So, particle's position shift is not restricted to operation shift on a single machine, but within the whole segment of chromosome.

| job | Machine sequence | | |
|---|---|---|---|
| | operation1 | Operation2 | operation3 |
| J1 | 1 | 2 | 3 |
| J2 | 1 | 2 | 3 |
| J3 | 2 | 1 | 3 |

Table 2. 3×3 Job Shop scheduling problem

See table 2, suppose that the chromosome of a 3×3 Job Shop scheduling problem is [2 1 1 3 1 2 3 3 2] , then the process sequence of each machine is as follows: $M_1$ [2 1 3] , $M_2$ [1 3 2] , $M_3$ [1 2 3] . If optimizing it with PSO, the sequence of $M_1$ is changed into [3 2 1] , then the solution become infeasible, because the operations on each machine are all not the first operation. However, by recoding and decoding the chromosome again, new feasible solution can be converted. The new chromosome encoding is [3 2 1 3 1 2 1 3 2] , and the new sequence of each machine by decoding is: $M_1$[2 1 3], $M_2$[3 1 2], $M_3$[2 1 3].

Based on such encoding elimination method for deadlock, it makes this algorithm simplified. Each time dealing with deadlock, we only need to take out the first operation in the waiting schedule, and distribute it to corresponding machine according to the rule of earliest finish time; we take out all bits in chromosome and dispose them, and then we obtain a feasible scheduling scheme, it is not necessary to detect deadlock and tackle fitness value of deadlock scheduling. Although the algorithm is simple, it has good searching capability because the decoding process can create an active scheduling.

### 4.2.2 Detection and Reconstruction

In process of PSO's colony evolution, it can easily bethink of decline strategy, namely, abandon infeasible solution brought by deadlock while only feasible solution is preserved, in which way we could not have to consider the infeasible solutions. This method is available to problems which have weak restrictions, since for weak restrictions feasible solutions have larger proportion in searching solution space; however, this method could still find some good solution from searching space. But, as to JSSP model which is a sort of problem with strong restriction, according to the encoding method in chart 1, feasible solutions have a little ratio in searching space, the complexity of searching for feasible solution is not inferior to the original problem (as to 10×10 problem's first-generation population, the author generates scheduling by particle evolution, in which, the proportion of feasible scheduling is less than 3%, see table 3). It is obvious that, with above encoding method, as to JSSP, only considering feasible solution is not enough.

| | 1 | 2 | 3 | 4 | 5 | 6 | 7 | 8 | 9 | 10 |
|---|---|---|---|---|---|---|---|---|---|---|
| 0 | 0.98 | 0.98 | 0.81 | 0.86 | 0.75 | 0.79 | 0.78 | 0.79 | 0.75 | 0.68 |
| 10 | 0.65 | 0.66 | 0.65 | 0.63 | 0.58 | 0.59 | 0.55 | 0.55 | 0.51 | 0.47 |
| 20 | 0.46 | 0.44 | 0.47 | 0.48 | 0.48 | 0.55 | 0.46 | 0.44 | 0.43 | 0.48 |
| 30 | 0.46 | 0.49 | 0.42 | 0.49 | 0.46 | 0.41 | 0.38 | 0.41 | 0.43 | 0.37 |
| 40 | 0.70 | 0.69 | 0.71 | 0.73 | 0.71 | 0.71 | 0.66 | 0.68 | 0.60 | 0.63 |
| 50 | 0.69 | 0.74 | 0.69 | 0.65 | 0.69 | 0.67 | 0.74 | 0.76 | 0.70 | 0.75 |
| 60 | 0.71 | 0.76 | 0.65 | 0.74 | 0.71 | 0.66 | 0.66 | 0.65 | 0.63 | 0.65 |

Table 3. The proportions of deadlock particles to the total particles in seventy generations of FT10

Note that, one characteristic of this sort of deadlock scheduling, namely, is high quantity and can hardly be found one by one, but its existence can be easily detected. Here, we provide design for deadlock elimination and reconstruction, that is, in routine G&T algorithm "randomly selects an operation in clash set", is modified as "according to the

operations' sequence in deadlock's scheduling generated by PSO algorithm, select operations in clash set $G''$. Consequently, it makes the scheduling being deadlock restored, making the ratio of feasible solution ascend gradually in process of evolution and selection.
As to JSSP, the detecting times of deadlock is denoted as "T", the worst case is that we examine circularly each operation queue once, while we only need to examine m×n times, if there is no deadlock in scheduling, T is m×n times. In process of detection, none but deadlock can make the algorithm process stagnate, that is, the algorithm is blocked by deadlock, in which case we can eliminate deadlock by rebuilding clash set using G&T. In the clash set, according to the order of job index of operation on machine in scheduling having deadlock, select operations, thereby, the scheduling having deadlock gets restoration.

### 4.2.3 Direct Reconstruction

From above analysis for 10×10 problem's deadlock, we can see, in the encoding based on operation, deadlock is in a great deal. In practice, among JSSP's strong restriction problems, deadlock scheduling is in a great proportion in the whole solution space, the effect of the way detecting deadlock is not necessary very good. Because detection need a lot of cost, the worst case of deadlock detection is the same as JSSP, it also a combinatorial blast problem, and deadlock's concrete information is skimble-skamble for JSSP. So, we design a direct reconstruction which directly rebuilds clash set using G&T. In the clash set, according to the order of job index of operation on machine in scheduling having deadlock, we select operations, namely, no necessary detect deadlock for evolving particle and directly reconstruct new solution, which makes the ratio of better solution ascend gradually, finally making the swarm go ahead towards optimization. This method avoids deadlock detection, comparatively, time performance may be simplified.
Seen from the above, the direct reconstruction is different from deadlock elimination reconstruction in that we do not judge whether there is a deadlock, but directly rebuild clash set using G&T. In the clash set, according to the order of job index of operation on machine in scheduling having deadlock, we select operations, thereby, the scheduling having deadlock gets restoration.

### 4.3 Experiments and Analysis

The experiment in this section aims at JSSP, solving 13 typical benchmarks hard problems, such as FT10, LA02, LA19, LA21, LA24, LA25, LA27, LA29, LA36, LA37, LA38, LA39, LA40.
In hybrid PSO (HPSO) using deadlock elimination strategy, the population size of PSO is set as |O|*70%, where |O| is the total number of operation; every evolution generation number is |O|*70%; we select $|O|*P_l$ particles from particle swarm and perform TS search, if though TS's search process has passed half of total operation number we still can not obtain better solution, we end TS process.
The experiment in this section is implemented with C code, the experiment environment is: CPU with Pentium-4 2.4G, and memory with 512M. HPSO1, HPSO2, and HPSO3 are hybrid PSO respectively using encoding elimination, detection and reconstruction, and direct reconstruction. The average fitness values of ten times searching solution of HPSO1, HPSO2, and HPSO3 respectively are 2.66%, 3.13% and 2.62%.
Seen from the experiment results, the effect of hybrid PSO using encoding elimination is worse than that using direct reconstruction, since that, with encoding elimination, the

scheduling schemes generated by particle whose position has changed have many repeated scheduling schemes, namely, there are many particles which have the same position in the particle swarm, and most of them are non-active scheduling, which reduces the diversity of particles, and conflicts with PSO algorithm whose original intention aims at enhancing the diversity of particles, although the running time of the algorithm is short, the searching performance is bad comparatively.

The difference between deadlock elimination reconstruction and directly reconstruction is that, the former detects and judges whether the solution has deadlock, if the solution has deadlock, reconstruct it; the latter doesn't detect solution, but directly reconstructs solution. In deadlock elimination and reconstruction, we total up all infeasible particles in every generation, namely, the proportion of particle being deadlock to the total particle number in the swarm. And we have totaled for the typical problems (FT10, LA02), from which we can see the difference of the two methods.

| | 1 | 2 | 3 | 4 | 5 | 6 | 7 | 8 | 9 | 10 |
|---|---|---|---|---|---|---|---|---|---|---|
| 0 | 0.91 | 0.83 | 0.89 | 0.88 | 0.77 | 0.75 | 0.63 | 0.60 | 0.56 | 0.52 |
| 10 | 0.48 | 0.40 | 0.40 | 0.27 | 0.26 | 0.32 | 0.18 | 0.28 | 0.30 | 0.28 |
| 20 | 0.28 | 0.19 | 0.17 | 0.17 | 0.15 | 0.15 | 0.20 | 0.13 | 0.16 | 0.16 |
| 30 | 0.19 | 0.22 | 0.05 | 0.07 | 0.03 | -- | -- | -- | -- | -- |

Table 4. The proportions of deadlock particles to the total particles in thirty five generations of LA02

We can see from the experiment results, as to JSSP in which the number of job and machine is more or less equal, deadlock elimination reconstruction and direct reconstruction are all square in their search time and search results. As to JSSP in which the number of job is more than the number of machine, it is better employing hybrid PSO using direct reconstruction. Because in JSSP in which the number of job is more than the number of machine, the operation number of each job is oppositely small, thereby, the number of infeasible solution after particle evolution is oppositely small, which can be seen from the statistic result in table 4, since the number of the feasible solution is oppositely large, the detection algorithm runs and does not know the solution is feasible solution until the algorithm runs out, so the algorithm's running time is oppositely long, and after particle's evolution, the number of particles whose fitness value are smaller than the fitness value of the primary particle is oppositely large, which is a disadvantage for obtaining global optimal solution, but as to JSSP in which the number of job and machine is more or less equal, the number of the infeasible solution after particle evolution is oppositely large, for which we can conclude from the statistic result in table 3 that detection algorithm can drop midway, which can reduce the computation time, at the same time, among particles obtained by improved G&T repair algorithm, the number of particles whose fitness value after evolution is smaller than the fitness value before evolution is oppositely small, which makes for the search for optimal solution. The experiment results also indicate that, the hybrid PSO using direct reconstruction has better effect and advantage for solving JSSP.

In seventy generations of FT10 and thirty five generations of LA02, the proportions of deadlock particles to the total particles are shown in table 3 and table 4, in which the proportion is the ratio of deadlock scheduling to the total scheduling in the particle swarm in every generation. The population size of PSO is set as $|O|*70\%$, where $|O|$ is the total number of operation; every evolution generation number are |O|*70%, the total number of operation is m×n, namely, (10×10=100) the generation number is 70.

| Benchmarks problem | Optimal Makespan | HPSO1 | | | HPSO2 | | | HPSO3 | | |
|---|---|---|---|---|---|---|---|---|---|---|
| | | Optimal Value | Average Value | Time (sec) | Optimal Value | Average Value | Time (sec) | Optimal Value | Average Value | Time (sec) |
| FT10(10×10) | **930** | 934 | 939.2 | 44.83 | **930** | 936.7 | 36.83 | **930** | 937.5 | 37.69 |
| LA02(10×5) | **655** | **655** | 667.6 | 1.41 | **655** | 688.4 | 7.275 | **655** | 656.0 | 4.22 |
| LA19(10×10) | **842** | 850 | 883.4 | 19.97 | **842** | 849.9 | 14.604 | **842** | 845.5 | 13.87 |
| LA21(15×10) | **1046** | 1055 | 1067.2 | 211.66 | 1078 | 1088.8 | 255.09 | 1050 | 1080.9 | 244.25 |
| LA24(15×10) | **935** | 954 | 959.3 | 228.62 | 950 | 958.2 | 258.69 | 944 | 950.4 | 250.91 |
| LA25(15×10) | **977** | 986 | 991.6 | 214.48 | 989 | 992.6 | 249.28 | **977** | 983.4 | 250.24 |
| LA27(20×10) | **1235** | 1265 | 1277 | 204.66 | 1269 | 1302.9 | 288.13 | 1260 | 1288.9 | 258.35 |
| LA29(20×10) | **1153** | 1203 | 1210.8 | 257.08 | 1253 | 1273.1 | 280.78 | 1200 | 1227.8 | 289.42 |
| LA36(15×15) | **1268** | 1292 | 1300 | 288.23 | 1274 | 1283.1 | 301.90 | 1280 | 1296.4 | 290.85 |
| LA37(15×15) | **1397** | 1433 | 1448.3 | 310.52 | 1415 | 1439.1 | 359.83 | 1415 | 1435.8 | 360.72 |
| LA38(15×15) | **1196** | 1209 | 1220 | 340.71 | 1212 | 1221.0 | 343.31 | 1204 | 1255.8 | 350.81 |
| LA39(15×15) | **1233** | 1248 | 1264.2 | 350.81 | **1233** | 1260.0 | 340.30 | **1233** | 1260.5 | 364.73 |
| LA40(15×15) | **1222** | 1230 | 1234.6 | 364.21 | 1229 | 1236.2 | 360.77 | 1229 | 1238.6 | 385.62 |

**Note:** boldface is the optimization.

Table 5. The optimal value and average values in ten times' experiments

### 4.4 Summary

To obtain valid hybrid PSO algorithm for JSSP, the reasons producing deadlocks in hybrid PSO algorithm is studied, and three strategies for eliminating deadlocks are proposed. When solving highly constrained combinatorial optimization problems, deadlock is one of the key problems need to be solved. The experiment results show that HPSO algorithm is a kind of feasible and effective method for sloving JSSP. With contrast experiments on 13 hard benchmark problems, both the results of deadlock detection and optimization objective results show that direct reconstruction is more effective, and better than the other two methods in searching quality. The hybrid PSO using direct reconstruction for deadlock problem has more advantages comparatively. The deadlock elimination algorithm, namely, the hybrid PSO for JSSP given in this book, has improved the solution quality of the hybrid PSO for JSSP, and which has provided a feasible and effective method for solving deadlock problem in PSO.

## 5. Other Hybrid PSO Algorithms for JSSP

As the diversity of optimization, the research experience and preferences of researchers often determine the selection of algorithms. The application of algorithms is various, and the new optimal alogorihm is difficlut proposed based on the natural mechanism. Therefore, the hybrid algorithm is an important and effective way to improve algorithm. In the section, we introduce other hybrid PSO algorithm for solving JSSP, including hybrid algorithm of PSO and simulated annealing (SA) algorithm, hybrid algorithm of PSO and genetic algorithm (GA).

### 5.1 Hybrid Algorithm of PSO and SA

#### 5.1.1 Hybrid Algorithm Design

Simulated annealing algorithm (SA) is a kind of stochastic search algorithm based on Monte Carlo iteration search strategy. SA algorithm has the jumping ability and strong universality, and it is easy to be realized. However, the running time of SA is long and its efficiency is low. PSO algorithm has strong universality. However, PSO has the disadvantages of prematurity and the tendency of falling into the local optimization. Therefore, a hybrid algorithm based on PSO and SA algorithm (HPSOSA) is proposed for improving the overall quality of the optimization algorithm.

According to the characteristics of random and large-scale search of PSO, we adopt PSO to construct a group of the initial solutions with good quality and dispersion. At the same time, each particle pursues the process of parallel search of SA in the population $P_l$. SA algorithm not only is a supplement of PSO and beneficial for the local improvement, but also has the probabilistic jumping ability of escaping the local optimization.

#### 5.1.2 Enhanced Simulated Annealing

In theory, SA algorithm can search the global optimal solution with probability 1 only if the parameters of algorithm satisfy the convergence conditions. However, it is impossible that some convergence conditions be met strictly according to SA algorithm convergence theory. The selection of SA algorithm parameters is still a problem. In the book, SA algorithm is used to local search of the hybrid algorithm, and we improve the process and sampling of SA algorithm.

(1) The operational pattern at a single comparison of traditional SA algorithm requests the fully high initial temperature and slow temperature drop. SA algorithm optimizes the part solution of the population each generation of PSO algorithm, and each random searching selects a value from a range as the initial temperature.

(2) The sampling process of traditional SA algorithm requests that sampling time at each temperature is long enough, and the temperature tends to *0* eventually. When the temperature remains unchanged in continuous $n$ steps at the current state, we think that the Metropolis sampling is stable. Then, SA algorithm will terminate calculation in the temperature. If the optimal solution remains unchanged in continuous $n$ steps at cooling process, we think that the algorithm is convergent.

The parameters of SA algorithm:

(1) Initial temperature $t_0$

The higher the initial temperature is, the larger rate of quality solution will be obtained. However, the cost of calculation will increase too. Therefore, we should certainly tradeoffs quality and efficiency when choosing the initial temperature. Before the local search in ESA, we make sure of the biggest difference between two targets ($|\Delta max|$) of PSO. Then according to the difference, the initial temperature is set by function $t_0=-\Delta max/\ln p_r$ ($p_r$ is initial speedups at convergence). If $p_r$ is close to $1$, and the initial random status can express the whole status space, the algorithm will accept all status almost in same probability, will not accept the restriction of the smallest solution completely.

(2) cooling rate$\omega$ ($0<\omega<1$)

The more $\omega$ is close to $1$, it shows that the slower the cooling rate decreases, and vice versa. Moreover, the algorithm has different search depths in different cooling rates. Therefore, SA algorithm uses the strategy of the variable cooling rate for improving the randomness of search by the random changing $\omega$ values in the search process.

(3) Iterative times $L$

The iterative times of each temperature is fixed value. When the temperature is high, the algorithm will accept all status almost in same probability, and the iterative times can reduce. When the temperature is gradually decreasing, the algorithm will reject most of status almost in same probability. If the iterative times reduce, the objective function will converge to a local optimization prematurely. In the enhanced algorithm, SA has different iterative times in different temperature. When the temperature is decreased, the iterative times of the same temperature will increase.

## 5.2 HPSOSA Algorithm

Considering normality of the objective function, the optimal solution is situated in an active schedule. In HPSOSA algorithm, Giffler-Thompson (G&T) algorithm is adopted to construct the initial solution of PSO, which has the ability of random and large-scale search. According to the convergence of PSO, each particle pursues the process of SA parallel searching in the degressive population $P_l$. Then, the solutions which aren't selected and new solutions by SA conduct the search solutions of PSO in the current generation. Adopt G&T algorithm to regenerate the particle, when $X_{id}(t)=P_{id}=P_{gd}$ in generation $t$.

In HPSOSA algorithm, we not only utilize parallel SA algorithm improving the search area, but also use PSO algorithm ensuring the convergence. The hybrid algorithm give attention to the accuracy and efficiency of optimization.

HPSOSA algorithm is described as:

```
Begin
PSn ← DetermineSizeOfParticleSwarm()
Pgd ← NULL; Pid ← NULL; Pl ← 0.2;
ConstructionInitializeParticleSwarm ( PSn )
while termination conditions not satisfied do
      CalculationParticleValidity( Vid )
      CalculationParticleNewPosition( Xid )
      if ( rand( 0, 1 ) < Pl ) then
                    ESASearch( )
      else
            ApplyLocalSearch( )
          end if
      CalculateParticleFitnessValue( )
      Update( Pid )
      Update( Pgd )
          if ( Pid== Pgd ) then
            GenerateNewParticle( i )              //G&T algorithm
          end if
end while
Output  Pgd
End
```

### 5.3 Hybrid Algorithm of PSO and GA

#### 5.3.1 Hybrid Algorithm Design

Genetic algorithm (GA) is a kind of algorithm for solving JSSP, and the algorithm has the ability of large-scale and global-convergence search. However, after the evolution to a certain evolutional generations, the objective function value (satisfaction) almost does not change, falling into the local optimization. If there is no external interference, it is difficult to leave the local optimum. Meanwhile, if external intervention can break the balance, the probability of searching the optimal solution will increase.

PSO algorithm is a kind of search algorithm based on iteration search strategy, and PSO algorithm has the characteristics of global optimization. In the last few years, along with application study is further, PSO has expanded its application to solve JSSP. It is found that PSO algorithm is good in the early evolution for solving JSSP. However, particle populations will quickly lose diversity, and PSO algorithm will cause premature convergence or slow the global convergence.

We should choose the algorithms with different characteristics, make them integrate mutually, give full play to their advantages, and generate the better efficiency of optimization. It is undoubted that it is an effective way to solve JSSP. Therefore, the hybrid parallel GA and PSO (HPSOGA) algorithm is established for solving JSSP. Considering the difference of the search mechanism and characteristics of PSO and GA algorithm, the parallel asynchronous hybrid method is adopted as the hybrid method. There are the global search in two populations individually, using the principle of different search algorithms.

Simultaneity, migration Operator is adopted to achieve the intercommunication between PSO and GA algorithm. When the optimization algorithm of one population falls into the local optimization, another algorithm disturbs the first population. Through exchanging individuals in the evolutionary process, the search area and accuracy are improved.
The HPSOGA algorithm is established for solving JSSP, its parallel hybrid model is illustrated in Fig. 5.

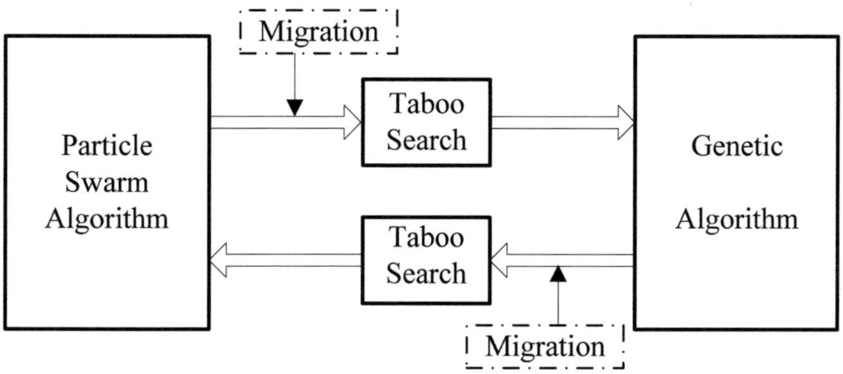


Fig. 5. HPSOGA parallel hybrid model with TS

### 5.3.2 Migration Operator Design

The key cycle of the hybrid algorithm is the designing of communication medium, which can make different algorithms exchange their information with each other so as to cause populations to share high quality seed. Thereby, introduce migration operation, which not only improves the diversity of GA population and enlarges the search scope of solution space, but also enhances the convergence of PSO population and deals with particles evolution halt problem. As two different algorithms are employed parallel hybrid to search optimization solution, it is too unilateral that only a single migration strategy is adopted. According to different specificity of both algorithms, two migration strategies related with two different migration occasions are designed, corresponding a migration strategy in each migration occasion is employed.
(1) Considering the fact that a particle stops evolution in the process of PSO evolution, if the particle which stops evolution is detected out, a solution randomly from GA population is selected, and it is diverted to PSO and replaced with the particle which stops evolution; if there are several particles which stop evolving in PSO population, the other GA solutions which have different fitness value are selected and diverted to PSO, and the corresponding particles which stops evolution are replaced; the particle which stops evolution is migrated to GA population and replaced with the individual whose fitness value is lowest in GA population.
(2) According to the fact that GA is prone to convergence, in the process of GA population evolution, the convergence factor *cf* is detected every certain generation in this algorithm. When a convergence factor is smaller than the preset value, select several good solutions whose fitness value are different from PSO population, and divert them to GA population so as to disturb GA evolution; meanwhile, select several good solutions whose fitness value

are different from GA population, and then diverts them to PSO population. So it can be achieved that two populations exchange their good individuals. The computation function of convergence facto *cf* is formulated as formula (16), in which $f_i$ denotes the fitness of chromosome *i*.

$$cf = \frac{\sum |f_i - \bar{f}|}{n} \tag{23}$$

During the initial phase of migration operation, smaller the migration rate and larger migration intervals are set. The main reason is to maintain the diversity of population, not destroy the inherent evolution mode of the algorithm. Then, with the increase of evolving generation, migration rates is increased and the migration interval is decreased gradually, which benefits improving of good individual spread in the whole population, convergence speed, accelerating solving speed, and achieving new balance. Thus it assures that it can find best solution of JSSP. This is a dynamic migration strategy.

For each migration strategy, if local search method is employed to conduct deeply search for migrated good individuals, on the one hand, the approximately best or the global best solution of the problem can be found as soon as possible, on the other hand, the better guide for migrating objective population can also be provided. The tabu search (TS) is employed as the local search operator.

A list structure with *1* in length is employed in TS algorithm. In the process of TS search, if neighbor which improves solution is found, the neighbor is saved in the list with probability $\gamma$ $(0<\gamma<1)$. When the list is full, replace the neighbor in the list with the new neighbor needed to be saved; when better solution is still not found within the maximal generation, pop the neighbor saved in the list, and go on searching. The function of recording neighbor which is potential to improve solution is performed by the list. Meanwhile the neighbor is put into the list once, $\gamma$ will be decreased once by a certain rate $\lambda$ $(\lambda<1)$. Because the more deeply the search is conducted, the less the chance of improving solution is. So it make the probability of input list descend, which can lower the chance of the same neighbor is input to the list repeatedly, avoid useless search, and dynamic memory function of the list is achieved.

### 5.3.3 HPSOGA Algorithm

HPSOGA algorithm is described as:

$GAn \leftarrow$ DetermineSizeOfGAPopulation()

$PSn \leftarrow$ DetermineSizeOfParticleSwarm()

$GAS_{bs} \leftarrow$ NULL ; $P_c \leftarrow 0.85$; $P_m \leftarrow 0.1$;

$cf = 0$ ; $iter = 0$; $mig = 0$; //mig is migration parameter

$P_{gd} \leftarrow$ NULL; $P_{id} \leftarrow$ NULL; $S_{bs} \leftarrow$NULL;

$ps = 0$ ; //ps is the number of stopping evolvement

```
Begin
ConstructionInitializePopulation ( GAn )
ConstructionInitializeParticleSwarm ( PSn )
while termination conditions not satisfied do
    RouletteWheelSelectionOperation( )              //selection operation
    CrossoverOperation( Pc )
    MutationOperation( Pm )
    CalculatePopulationFitnessValue( )
    if (iter % 5 == 0) then
        cf ← ComputeConvergenceFactor( )
        if ( cf < 0.05 ) then
            mig++
        end if
    end if
    MigrationFromParticleSwarm( mig , TS )          // migration operation
    Update( GASbs )
    CalculationParticleValidity( Vid )
    CalculationParticleNewPosition( Xid )
    CalculateParticleFitnessValue( )
    Update( Pid )
    Update( Pgd )
    ps ← NonEvolvementParticleNumber ()
    if ( ps > 0 ) then
        MigrationFromGAPopulation( ps , TS )        // migration operation
    end if
    iter++
end while
Output Sbs ← max{ AI(Pgd), AI(GASbs) }
End
```

### 5.4 Summary

In the section, we introduce another two different hybrid PSO algorithms for solving JSSP. When constructing the hybrid PSO algorithm, the key is how to use the different optimized mechanism of algorithms, making up for deficiencies each other. With the deepened study of the intelligent optimization problem, effective hybrid algorithms will continue to emerge for solving JSSP.

## 6. Summary

In the chapter, we describe PSO algorithm, propose HPSO algorithm for solving JSSP, analyze the convergence of HPSO, and provide three strategies for deadlock elimination. Then, we present other hybrid PSO algorithm. The chapter can give readers comprehensive research results on hybrid particle swarm optimization algorithms for JSSP, which will promote research and application of JSSP.

## 7. References


Byung Joo Park, Hyung Rim Choi, Hyun Soo Kim.(2003). A hybrid genetic algorithm for the job shop scheduling problem. *Computers & industrial engineering,* Vol.45 (4),597-613

Cagnina L, Esquivel S, Gallard R.(2004). Particle Swarm Optimization for Sequencing Problems: A Case Study, *Proceedings of 2004 Congress on Evolutionary Computation, Oregon, Portland,* pp. 536-540

Chang Fang, Liaw.(2000). A hybrid genetic algorithm for the open shop scheduling problem. *European Journal of Operational Research,* Vol.124(1), 393-407

Cheng R, Gen M.(1996). Atutorial survey of job-shop scheduling problems using genetic algorithms-I. Representation. *Computers and Industrial Engineering,* Vol. 30(4), 983-997

Christian B, Dirk C M.(1999). Production scheduling and rescheduling with genetic algorithms. *Evolutionary Computation,* Vol.7(1), 1-17

Eberhart R, Kenned J.(1995). A New Optimizer Using Particles Swarm Theory, *Proc Sixth International Symposium on Micro Machine and Human Science,* pp. 39-43. Nagoya, Japan: IEEE service Center.Piscataway

Eugeniusz Nowicki, Czeslaw Smutnicki.(1996). A Fast Taboo Search Algorithm for the Job Shop Problem. *Management Science,* Vol.42 (6), 797- 813

Giffler, B. and Thompson, G. L.(1960). Algorithms for Solving Production Scheduling Problems. *Operations Research,* Vol.8(4), 487-503

Isaacson D L , Madsen R W.(1976). *Markov chain theory and applications,* John Wiley&Sons Inc

Jain A S, Meeran S.(1999). Deterministic job-shop scheduling: past, present and future. *European Journal of Operational Research,* Vol.113(2), 390~434

Jerald J, Asolcan P, Prabaharan G, et al.(2004). Scheduling Optimization of Flexible Manufacturing Systems Using Particle Swarm Optimization Algorithm. *Advanced Manufacturing Technology*

K ennedy J, Eberhart R.C.(2001). *Swarm Intelligence,* San Francisco: San Francisco Morgan Kaufman Publishers

Kirkpatrick, S., Gelatt, C. D. (Jr) and Vecchi, M. P.(1983). Optimization by Simulated Annealing, *Science,* Vol.220(5), 671-680

Kolonko, M.(1998). Some New Results on Simulated Annealing Applied to the Job Shop Scheduling Problem, *the European Journal of Operational Research*

Leticia Cagnina, Susana Esquivel, Raul Gallard.(2004). Particle Swarm Optimization for Sequencing Problems: A Case Study, *Proceedings of 2004 Congress on Evolutionary Computation, Oregon, Portland,* pp. 536-540

Lian Z G, Jiao B, Gu X S.(2006). A similar particle swarm optimization algorithm for job shop scheduling to minimize makespan. *Appl. Math. Comput.* , Vol.183, 1008-1017

Liu H B, Abraham A, Choi O, et al.(2006). Variable neighborhood particle swarm optimization for multi-objective flexible job shop scheduling problems. *LNCS,* Vol.4247, 197-204

Ph. Preux, E.-G. Talbi.(1999). Towards hybrid evolutionary algorithms. *International transactions in operational research,* Vol.6 (6), 557-570

Radcliffe N .J, Surrv P D.(1995). *Fundamental limitations on search algorithms,* U K: Cniversity of Edinburnh

Sha D Y, Hsu C Y.(2006). A hybrid particle swarm optimization for job shop scheduling problem. *Comput. Ind. Eng.* Vol.51, 791-808

SongXiao-Yu, Cao Yang, Meng Qiu-Hong.(2008). Study on particle swarm algorithm for Job Shop scheduling problems. *Xi Tong Gong Cheng Yu Dian Zi Ji Shu/Systems Engineering and Electronics*, Vol.30(12), 2398-2401

Van Laarhoven, P. J. M.,Aarts, E. H. L., Lenstra, J. K.(1992). Job Shop Scheduling by Simulated Annealing, *Operations Research*, Vol.40,113-135

Wolpert D H, Macready W G.(1995). *No Free Lunch theotrems on search algorithms,* U K: Univercity of Edinburnh

Xia weijun, Wu zhiming, Zhangwei, et al.(2004). A New Hybrid Optimization Algorithm for the Job Shop Scheduling Problem. *Proceedings of the 2004 American Control Conference Boston, Massachusette*, pp. 5552-5557

Xia W J, Wu Z M.(2005). An effective hybrid optimization approach for multi-objective flexible job shop scheduling problems. *Comput. Ind. Eng.* Vol.48, 409-425

Xia W J, Wu Z M.(2006). A hybrid particle swarm optimization approach for the job shop scheduling problem. *Int. J. Adv. Manuf. Technol.* Vol.29, 360-366

Xu Gang, Wu Zhi ming.(2002). Deadlock-free scheduling method using Petrinet model analysis and GA. *Proc of the IEEE Int Conf on Control Application*, pp. 1153-1159

Yamada, T. and Nakano, R.(1996). Job-Shop Scheduling by Simulated Annealing Combined with Deterministic Local Search, *Meta-heuristics: Theory and Applications, Kluwer Academic Publishers*, Boston, MA, USA, Chapter 15, pp. 237-248

Yamada, T. and Nakano, R.(1996). Scheduling by Genetic Local Search with Multi-Step Crossover, *PPSN'IV Fourth International Conference on Parallel Problem Solving from Nature*, pp. 960-969, Berlin, Germany, Sept 22-26

Yamada, T. and Nakano, R.(1996). A Fusion of Crossover and Local Search, *ICIT'96 IEEE International Conference on Industrial Technology*, pp. 426-430, Shanghai, China, Dec 2-6

Yannakakis, M.(1990). The Analysis of Local Search Problems and their Heuristics, *Proceedings of the 7th Annual Symposium on Theoretical Aspects of Computer Science*, pp. 298-311.

Yannakakis, M.(1997). Computational Complexity of Local Search, in Aarts, E. H. L. and Lenstra, *J. K (eds) Local Search in Combinatorial Optimization, Wiley, Chichester, Chapter* 2, pp. 19-55